International Symposium on Chemical Reaction Engineering, 5th, Houston, Tex., 1978

Chemical Reaction Engineering Reviews—Houston

Dan Luss, EDITOR

University of Houston

Vern W. Weekman, Jr., EDITOR

Mobil Research and Development Corporation

The Fifth International Symposium on Chemical Reaction Engineering co-sponsored by the American Chemical Society, the American Institute of Chemical Engineers, the Canadian Society for Chemical Engineering, and the European Federation of Chemical Engineering, held at the Hyatt Regency Hotel, Houston, Texas, March 13–15, 1978.

ACS SYMPOSIUM SERIES 72

AMERICAN CHEMICAL SOCIETY

WASHINGTON, D. C. 1978

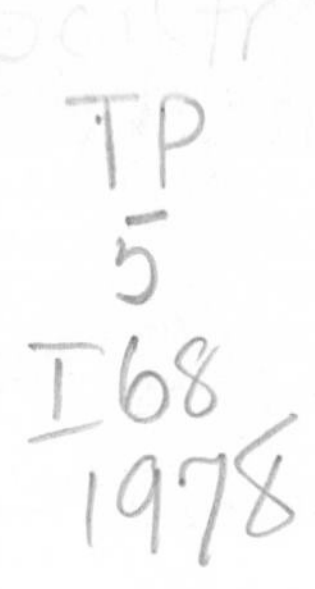

Library of Congress CIP Data

International Symposium on Chemical Reaction Engineering, 5th, Houston, Tex., 1978.
Chemical reaction engineering reviews—Houston.

(ACS symposium series; 72)
Includes bibliographies and index.

1. Chemical engineering—Congresses. 2. Chemical reactions—Congresses.
I. Luss, Dan, 1938- . II. Weekman, Vern W. III. American Chemical Society. IV. Title. V. Series: American Chemical Society. ACS symposium series; 72.

TP5.I68 1978 660.2'9'9 78-8477
ISBN 0-8412-0432-2 ASCMC8 72 1–331 1978

PRINTED IN THE UNITED STATES OF AMERICA

ACS Symposium Series

FOREWORD

The ACS SYMPOSIUM SERIES was founded in 1974 to provide a medium for publishing symposia quickly in book form. The format of the SERIES parallels that of the continuing ADVANCES IN CHEMISTRY SERIES except that in order to save time the papers are not typeset but are reproduced as they are submitted by the authors in camera-ready form. As a further means of saving time, the papers are not edited or reviewed except by the symposium chairman, who becomes editor of the book. Papers published in the ACS SYMPOSIUM SERIES are original contributions not published elsewhere in whole or major part and include reports of research as well as reviews since symposia may embrace both types of presentation.

CONTENTS

Organizing Committee
for the
Fifth International Symposium on
Chemical Reaction Engineering

Dan Luss, EDITOR

Vern W. Weekman, Jr., EDITOR

Members: Chandler H. Barkelew (Shell Development Co.)
K. B. Bischoff (University of Delaware)
John B. Butt (Northwestern University)
James M. Douglas (University of Massachusetts)
Hugh M. Hulburt (Northwestern University)
Donald N. Miller (Dupont Co.)

PREFACE

The Fifth International Symposium on Chemical Reaction Engineering has, as in past symposia, provided an excellent forum for reviewing recent accomplishments in theory and applications. This international symposium series grew out of the earlier European Symposia on Chemical Reaction Engineering which began in 1957. In 1966, as part of the American Chemical Society Industrial and Engineering Chemistry Division's Summer Symposium series, a meeting was devoted to chemical reaction engineering and kinetics. This meeting highlighted the great interest and activity in this field in the United States and led the organizers to join with the American Institute of Chemical Engineers and the European Federation of Chemical Engineers in organizing International Symposia on Chemical Reaction Engineering. The first symposium was held in Washington in 1970 and was followed by symposia in Amsterdam (1972), Evanston (1974), and Heidelberg (1976).

One of the most important features of all the symposia has been the invited plenary review lectures. This latest symposium is no exception, with nine review authors being carefully chosen so as to cover a broad range of subjects with current interest in chemical reaction engineering. In keeping with the international flavor of the meeting, four of the nine authors of the plenary review lectures were from Europe.

The meeting format allowed three plenary review lectures each morning and three parallel, original paper sessions in the afternoon. The 48 original research papers have been published under the title "Chemical Reaction Engineering—Houston" as volume number 65 in the American Chemical Society Symposium Series (1978).

We acknowledge financial support from the National Science Foundation, the American Chemical Society—Petroleum Research Fund, the Exxon Research and Development Co., Mobil Oil Corp., and Shell Oil Co.

DAN LUSS
University of Houston
Houston, Texas

VERN W. WEEKMAN, JR.
Mobil R & D Corp.
Princeton, New Jersey

March, 1978

1

Chemical Reactor Modeling—The Desirable and the Achievable

R. SHINNAR

City University of New York City College, 140th St. and Convent Ave., New York, NY 10031

I. INTRODUCTION

A significant number of reviews on reactor modeling were recently published (1,2,3). However, my intent is not to review current literature. What I want to do is present a rather personal overview of reactor modeling as a tool in the hand of the designer and practitioner, and the role the academic can play in helping the practitioner to improve his skills.

A few years ago, a panel of industrial practitioners convened by NSF (4) came out with a statement, that most academic work in reaction modeling has become both irrelevant and unnecessary since the practitioner now has most of the tools he needs. While I agree that chemical reactor modeling has made tremendous progress in the last thirty years, I would hardly agree that our job is really finished. There are a great number of challenging and interesting problems that await solution. But there may be justification to take stock as to what we can and cannot do with our present tools while we try to identify the problems that merit more attention.

II. GOALS OF MODELING

The concept of reactor modeling is used in a rather broad sense by different people for different purposes. But there is one underlying thread to industrial use of models. It is a method to translate existing information and data to useful predictions for new conditions. Such predictions might involve:

1. scaleup from pilot plant to a large reactor
2. behavior of different feedstock and new catalysts
3. effect of different reaction conditions on product distribution
4. prediction of dynamic trends for purposes of control
5. Optimization of steady state operating conditions
6. better understanding of the system that may lead to process and design improvements

0-8412-0432-2/78/47-072-001$09.00/0

These goals are in no way exclusive, but we should not overlook the fact that our goal is sound predictions. We want to be able to predict reactor conditions and outputs for inputs and design conditions for which we have no previous data. Otherwise a table of previous results would be just as informative for a practical application. To apply such predictions to design we also need to know the confidence limits of our predictions or the risk we take in basing our design on them.

The first question the reader might ask is: why do I bring this up? If we can develop a proper mathematical description of a chemical reactor we should be able to use it for any purpose we want. This brings us to the basic problem of chemical reactor modeling. Most chemical reactors of interest are too complex to be given an exact mathematical formulation. We assume that there exists a true mathematical relation,

$$Z = M^*_{KR}(X) \tag{1}$$

that relates the output variable of the reactor Z to all the input variables. We include here in X the properties of the feed X_F, the fixed design variable of the reactor itself X_D, and the adjustable parameters of the reactor X_R. We also assume that M^*_{KR} can be broken up into a relation describing the reaction rates as a funtion of local concentration M^*_K and a relation M^*_R describing all the transport processes occurring in the reactor.

But while M^*_{KR} (X) exists, it is in most cases inaccessible, and we must settle for an approximate description $\bar{M}_{KR}$. We often do not know the complete dimension of the vector X. What distinguishes chemical reaction modeling from many other fields of engineering such as distillation, servo mechanisms and electronic equipment, is the fact that the difference between M^*_{KR} and $\bar{M}_{KR}$ is much larger. This applies both to M_K and M_R. Even simple chemical reactions involve complex mechanisms and after 50 years we still do not completely understand the formation of ammonia, one of the first industrially used catalytic reactions. But we have the data and tools to describe that reaction, for the purpose of modeling an ammonia reactor (5,6,7). While we have in many cases a much better theoretical formulation for M_R that does not mean that M^*_R is accessible. The solution of the Navier Stokes equation for a stirred tank is too complex to be useful. Again we have to settle for some simplified models that will, in most cases, work remarkably well. The term modeling implies just that. We describe a complex system by a more simple one, which, hopefully, contains the essential features of M^*_{KR}.

The fact that $\bar{M}_{KR}$ can be considerably different from M^*_{KR} and still give useful predictions, implies that finding a useful approximation $\bar{M}_{KR}$ is not a unique process. In fact, successful modeling, just as many other engineering design activities, is part science and part art. We always present it in our papers as a straightforward process whereas, in reality, it is an iterative process involving considerable judgment.

Engineering is different from science in the sense that we design equipment with incomplete information. Modeling is the art of predicting reactor performance based on incomplete knowledge of M^*_{KR}. It is an art in the sense that it requires experience and judgment to formulate $\bar{M}_{KR}$ and evluate its reliability. And just as the art content of engineering can be reduced (though not eliminated) by proper research and a scientific approach, so can the art content of modeling. But if we want to progress further we require better understanding of the modeling and the design process itself.

III. CLASSIFICATION OF MODELS

Some of the confusion in many discussions on the value of academic modeling research is caused by the fact that we apply the term chemical reaction modeling to procedures with rather different goals. Let me try to classify them.

1. Predictive Models

Our overall goal in applications is always a prediction, and final models are judged by their predictive capabilities. Here I distinguish between two cases:

a) Interpolation models
b) Design models

In the first case we have a large number of data and we want to organize them in proper form. The basic feature of an interpolation model is that the vector X for which a prediction is desired is surrounded by values of X for which Z is measured. In the design case this is not true. We want not only to correctly predict Z far outside the X range where measurements are available, but also to minimize the risk of making a wrong prediction.

2. Learning Models

Modeling is an iterative process. We do not set up a final model immediately; we develop it in stages. I refer to any intermediate step as a learning model. A learning model is helpful in setting up a proper final model in several ways.

a) It provides us with proper structural relations between the variables. This is important in any statistical correlation, and in identifying a model based on observation.

b) It helps us understand the modeling process itself by illuminating the relations between simpler models and more complex systems.

c) It provides guidance for efficient experimentation.

a) Structure of Relations

The main difficulty in deriving a correct predictive model for many chemical reactors is the complexity of the system. There are two types of complexity in modeling. One is mainly computational. An electronic filter constructed from carefully designed and checked subsystems is a good sample. Here, we can construct deterministic models that reliably predict the behavior of the system. In the other type, such as in models for social and economic problems, the main difficulty is proper identification, and the best we can often hope for is some simple statistical correlations. Many systems are somewhere in between these extremes. In Table I I tried to provide a ranking of different complex systems.

Deterministic Models			Statistical Correlations
Electronic Filters Servomechanisms	Distillation Columns Piping net works	Chemical Reactors	Social Sciences econometrics

Table I - Classification of Models

Well designed distillation columns are close to electronic filters and while complex their performance can be well predicted from thermodynamic data. The complexity of chemical reactors varies from case to case. The performance of a well designed methanol or ammonia reactor can be predicted as well as that of a distillation column, whereas in more complex systems the risks of predictions is greater. In some sense identification of chemical reaction models has some features of a purely statistical correlation. Even in a simple well defined system such as isomerization of xylene (25) where we can measure all reaction rates and their activation energy accurately, predictions are only reliable within the range of temperatures for which we have data. At higher temperatures other reactions may become important. There is however a very basic difference between statistical identification in econometrics and reaction modeling. In the latter we know much more about the structure of the system, and furthermore, we have a much greater ability to conduct carefully controlled experiments. With enough effort we may move any chemical reactor model fromthe right to the left of the scale. It is in providing an inside to the structure of the system that purely deterministic learning models play a very significant role in reaction modeling. To illustrate that I bring a rather simple example. Consider James Wei's attempt to draw an elephant using Fourier transform.

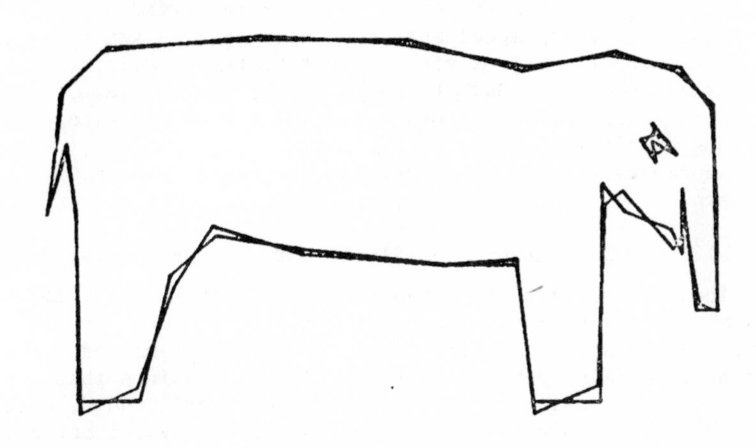

An intelligent eight year old kid will recognize in this drawing the main features of an elephant, provided he has been to the zoo or at least seen several pictures of elephants. Our human mind has a great capacity of pattern recognition, but it needs a storage of relevant information. The experienced engineer acquires this in the course of his work. Well defined intelligent deterministic models of chemical reactors played a significant role in advancing the state of the art and providing us with a structural framework for deriving predictive models. They provide guideline for the experimentation required to set up models both for chemical reactions ($\bar{M}_K$) as well as for the reactor ($\bar{M}_R$ and $\bar{M}_{KR}$).

If we want to understand the effect of temperature on reaction rate it is generally more efficient to plot ln r versus 1/T, where r is the rate to be correlated and T is the temperature, even if the overall mechanism is complex and the relation is far from being a straight line. Obtaining suitable structures for semiempirical correlations is one of the most important goals of theoretical modeling. One very valuable result of mechanistic models for chemical reactions was to suggest a good structure for correlating reaction rates (8). Proper modeling will also give us an idea of the range over which a relation will hold (9). We are especially interested if relations have sudden steep changes or discontinuities. Thus, it is helpful to know if a reactor may have multiple steady states, and what the approximate form of the relations governing these states are. This is important even if we do not have sufficiently detailed knowledge to exactly predict them.

To be useful as a guideline in experimentation a good deterministic learning model should be more complex and richer than the simpler model we often have to settle for in actual industrial modeling. But it should not be too complex otherwise it becomes useless for pattern recognition. Construction of learning models with just the right kind of complexity for a given purpose is an art, and here we all owe a debt to N. Amundsen.

b) The Relations Between Simple and Complex Models

Academic modeling is often concerned with the way simplified models approximate more complex systems. Here, we are trying to understand the nature of the complex systems and the way simplified models approximate them. The constraints that apply to successful predictive models do not necessarily hold, since our goals are simply to understand the overall system better. By understanding the more complex model, we also learn about the nature of the simplification process.

In predictive modeling of reactors we are always faced with the choice of a proper compromise. We require a model complex enough to contain the essential features of the process and simple enough to allow us to measure the parameters in a reliable way.

We are helped in this simplifying procedure by setting up a much more detailed model of the process, guessing a reasonable range of values for its parameters. This will show us which parameters are important and which have a small effect, and which timescales are important (10). If the timescale of any process within the more complex framework is small we can either neglect this effect or make the assumption of a pseudosteady state. And if one parameter shows little effect over the range of values that it will reasonable have, we can skip the effort to measure it accurately. These are essential steps in every modeling process. But this more detailed model can also be useful in testing our identification modeling and design procedure. Having formulated it we can pretend we don't know it and that it is inaccessible to us in the same way the real process is. We can then perform on it, using a simulation, the same identification modeling and design procedure that we would apply to the real case. If our method works for a set of such models, we gain added confidence that it will work for the real unknown M^*_{KR}. I found this very helpful in evaluating design of process controllers (11, 12) and I will cite briefly two other examples for illustration.

In the design of stirred tank reactors the model of an ideal stirred tank is a useful approximation. To understand when it breaks down we can set up more complex models of the mixing process (13) and of the kinetics and estimate the effect of mixing time. The fact that our complex model does not represent a real reactor is not that important, since all we want is an estimate of the importance of the mixing time. On the other hand, we can not use this complex model to predict the effect of imperfect mixing in a quantitative way.

Another example is a dynamic model for an exothermic reaction in a catalytic packed bed reactor such as an ammonia reactor or a hydrotreater (14, 15). In the steady state we can neglect the complex transport processes between the gas and the particle and inside the particle and lump them into an overall local reaction rate.

In a dynamic model, we can no longer neglect the heat and mass transfer processes between the gas and the catalyst. A real detailed description of the unsteady state process inside a particle is rather complex. We have a number of dynamic models for packed bed reactors (14, 15). What we can learn from them, is that the dominant timescale is the heating time of the catalyst phase. The response of the catalyst phase to changes in temperature is slow as compared to the residence time of the gas. This simplifies our modeling. We can set up a model which treats the catalyst as an inert heatsink (14) and assumes that the reaction occurs with the rate that would prevail at steady state at the local average temperature and concentration of the system. This gives approximately the same response as a more complex model which takes into account the effect of both heat and mass transfer processes inside the catalyst particle (15) (see Fig. 1 from ref. 16). This is reasuring as otherwise we would have trouble in estimating the proper data for our model. Now let us remember: None of the models, even the most complex ones, describes accurately a specific reactor.

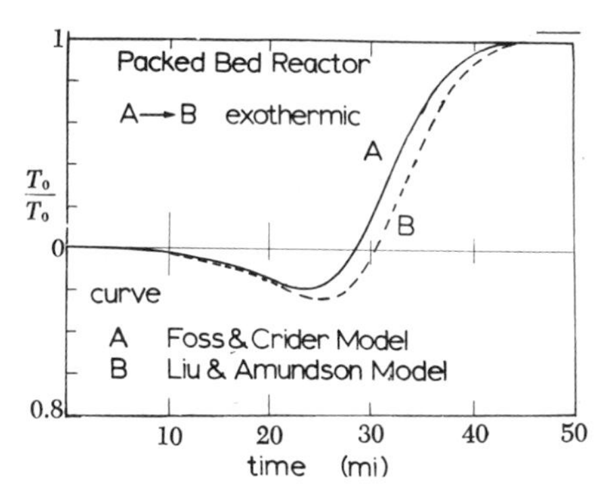

Figure 1. Response of an adiabitic exothermic packed bed reactor to a step-input in feed temperature

We often make the mistake of assuming that more complex models describe reality more accurately, which is sometimes but not always true. But as I noted previously, such models still serve a very important purpose.

c) Guidance for Efficient Experimentation

Models are an important tool in guiding the experimentation required to set up a predictive model. We have some limits as to what and how many experiments we can carry out. The most efficient procedure is iterative in which initial results are used to determine the range of conditions for which further results are required and what space of X_F and X_C should be investigated to give us confidence for our predictions. By trying to model the final reactor we will find that certain parameters have very little effect, whereas others are critical. In that sense proper modeling is an iterative procedure. We already mentioned the problem of finding a proper structure for kinetic relations, and the problem of estimating the dominant parameters and timescales.

Consider, for example, the vast literature on transport processes inside catalyst particles (10, 17, 18). If we want to predict the actual performance of a specific reactor we will find it very hard to get all the parameters that some of the extensive models require. But they will provide us with several important bits of information. In catalytic reaction we cannot directly measure intrinsic kinetics and they guide us in designing experiments in which diffusion effects are minimized (13, 14). They will also tell us that for certain types of reactions we can safely predict the performance of a packed bed reactor on the basis of a micro reactor, while for others we cannot. If the reaction is highly exothermic and there is a chance of multiple steady states, we might require experiments at flowrates similar to the large reactor. We often cannot accurately predict the occurrence of multiple steady states in a catalyst particle, nor do we want to. We want to avoid them, and to assure that we need reliable experiments. Theory provides a framework for them. It is also useful in guiding the development of better catalysts.

Learning models are also used as diagnostic tools. When a large reactor does not perform as expected, we try to find out why. Experiments in large reactors are expensive and we can make them more effective by investigating potential reasons for the deviation by modeling. We are interested here in identifying which of the potential causes would have the proper timescale and magnitude.

One of the most important uses of learning models is in suggesting new experiments. We do not perform experiments at random, and any suggestion as to what might lead to an improvement is very valuable. Mechanistic models for catalysts or chemical reactions are not very reliable to actually predict reaction rates or even the effect of variables on reaction rate.

But in development we are content with useful leads, and here proper modeling and understanding of the fine scale processes inside a catalyst particle (17, 18), the mechanism and potential intermediates, can be very helpful.

Sometimes such leads prove to be successful despite the fact that the underlying theory is proven to be wrong. As a theoretician, I find it sobering to remember that some of the most important inventions came from wrong models or theories. My favorite example is the case of unstable combustion in a solid propellant which was a serious problem in World War II. The problem was solved by a totally unrelated experiment. A wrong computer program predicted that a small percentage of aluminum powder would increase the specific impulse of such propellants. The specific impulse actually decreases for small percentages of aluminum but the investigators noted that the pressure trace was uncommonly smooth. Addition of aluminum powder solved the problem of instability and in large percentages even increased the specific impulse. There are a number of theories explaining this, one of them mine, but I am not convinced anyone is really correct.

We therefore have to carefully distinguish between learning models and design models. Many practitioners judge academic work by their standards and do not realize the value of learning models. Perhaps it is also our fault as we seldom take the time to translate the meaning of our results in a way that makes them easily accessible to the practitioner.

We will devote the rest of the paper to the methods and problems of deriving predictive models and will start with models based on simple statistical correlations.

III. CORRELATION MODELS

In some cases purely correlational models are quite effective. Understanding the nature and limitations of correlation models also helps us to better understand the limitations of reaction modeling itself.

If we have an industrial reactor operating we can get a purely empirical model that will correlate outlet concentrations as a function of pressure, inlet temperature, inlet concentration and other operating variables. For some reactors such a model properly constructed can be rather good, as long as the new X is within a certain space of X. If the function $Z=M_{KR}(X)$ is well defined and smooth for a subspace of X, then we can find successive approximations, either linear

$$Z_i = a_{ij}X_j \tag{2}$$

or quadratic

$$Z_i = \sum_j a_{ij}X_j + \sum_{jk} b_{ijk}X_jX_k \tag{3}$$

which give a good approximation of the real model M^*_{KR} and I know some rather successful industrial models of this type.

Models of this type are sometimes useful in optimization when the real relations are complex such as in polymerization reactors. There it is often hard to express desired properties of the final product in terms of reactor variables (19).

In the case of an adiabatic ammonia reactor M^*_{KR} is not a well defined smooth function as it has a discontinuity of blow out condition. But even in the range where M^*_{KR} is well defined, approximations of M^*_{KR} of the type given by 4a and 4b have strong limitations as they refer only to fixed designs.

It is always preferable to use models which have a stronger content of our knowledge about the physics and chemistry of the process. We normally do that by breaking up M_{KR} into a kinetic model M_K that relates local reaction rates to local concentrations, temperature, pressure and catalyst conditions, and a reactor model M_R that quantitatively describes the transport processes and heat and mass balances in the reactor. While in many cases we have the tools to derive an approximation for M_R from first principles, we can very seldom do that for M_K. Aside from a few cases of gaseous reactions, we cannot predict M_K. For the reactor modeling we have to accept the fact that reaction rate expressions are empirical correlations. What the study of reaction mechanisms has contributed to us is some a priori information as to what form M_K might have.

In some sense many of our successful reaction models, especially for complex cases are really refined correlation models. That does not mean that we should obtain them by a straightforward statistical approach. A proper M_K should contain not only the stochiometric and thermodynamic constraints, but should be based on a maximum of physical information. In this way we obtain reliable information for a wide space of X with a much smaller number of experimental measurements, than a straightforward correlation would require. What I mean by a refined correlation model is that if we have to extrapolate far outside the range of measured X, we will get estimates which are valuable as guidelines for experimentation, but not reliable enough for design. We cannot a priori predict what side reaction would occur at higher temperatures or pressures.

By looking at models in the way a statistician looks at building models and confirms them, one can gain some insights and it might be time to start to combine the best features of both approaches. Let me cite some examples:

a) Spurious Correlations

One important problem in all model building is in identifying and verifying the important parameters. Choosing the set of X which predicts Z is already a model and it is probably the most important decision we make. If we just empirically try different sets of X, we might get clues as to which parameters are important; but we also can get wrong results that have a totally different meaning than implied by our relation and are therefore of little predictive value. For example, in a recent paper (20), the author tried to statistically correlate the results of coal gasification rates from various sources with greatly different coals. This correlation is plotted in Fig. 2. According to (20) the main parameter controlling the gasification rate is the molar flow rate of total oxygen (O_2+1/2H_2O). Pressure temperature and space velocity are claimed to have only minor effects. What makes this correlation work is the fact that investigators only publish data in the range where there is significant gasification. They will adjust oxygen to steam ratio, temperature, pressure and space velocity to hit that range. The correlation in Fig. 2 simply indicates that once they find such conditions they are correlated by stoichiometry.

I cite this example because such spurious correlations are one of the main problems in any modeling procedure. They are normally less obvious and appear in a more subtle way. We often conduct experiments in a very constrained space of X, and sometimes along a single trajectory of X. This is especially true in pilot plant operation and can be misleading. To get confidence in a model we need data over a wider space of X than we intend to operate.

b) Alternate Models

In statistical inference we are very well aware of the

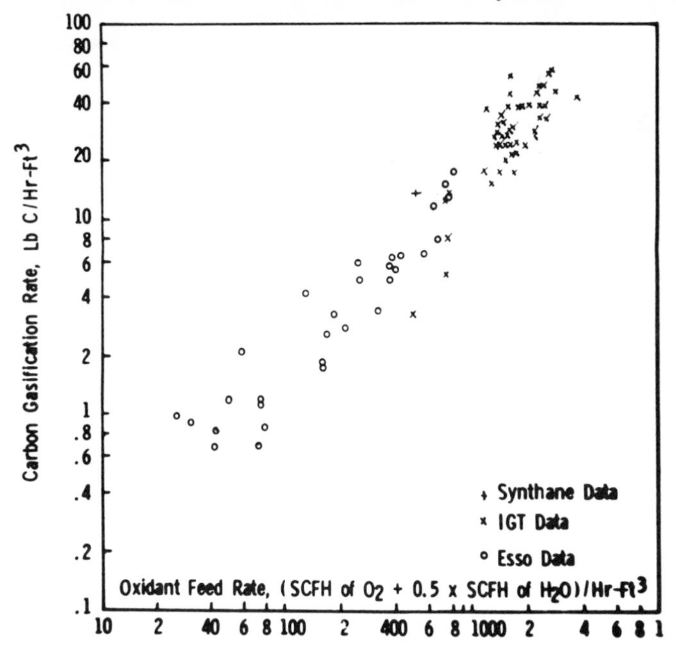

Figure 2. Relationship of carbon gasification rate and oxidant availability

fact that an experiment is in agreement with a given hypothesis is often not enough. If the results confirm some a priori predictions this is valuable information. We also must ask: What is a reasonable alternative hypothesis and did we distinguish between the two? Model identification is there, strongly related to statistical inference. I will elaborate on this in the next section.

IV. MODEL FIT AND IDENTIFICATION

If we have a set of data $(Z,X)_M$ we can construct a model M_{KR} such that the square deviation between measurement and model

$$\langle E^2 \rangle = \langle Z_M - M_{KR}{}^{(*)}(X)^2 \rangle \tag{9}$$

is not significantly different from experimental error. We would say then that the data are in agreement with the model. However, we can also ask how many alternative models are in agreement with the same set of data. If we ask the question in an unconstrained way it has no plausible answer. We can always find a large number of models which will have a smaller error than the experimental error.

So we normally ask the question in two ways. Either we try to distinguish between two alternative models, both based on some knowledge of what we suspect the process to be, or we fix the model but leave one or more of the parameters free and ask what is the range of parameters that is consistent with our results considering the experimental error.

To get reasonably reliable predictions it is not only important that the error be small but also that the model makes sense.

We have to be careful as to what we mean in saying a model agrees with the data. Since our approximate model M_{KR} is different from the true physical relation M_{KR}^*, we can always create an experiment that will show M_{KR} to be incorrect. What we mean by a correct model is that:

a) it contains the correct physics and chemistry of the process, and therefore all important variables are contained in the X, specified;

b) it predicts trends correctly, such as that the signs of $\partial Z_i/\partial X_j$ are correct, over the range of X for which the model applies;

c) the errors are reasonably small, not just for $\langle E^2 \rangle$ but for all individual measurements Z.

For interpolation models it is important that the space of X is well covered. For design and scaleup models this is impossible, and if we extrapolate we require some additional constraints on M_{KR}.

a) We want to know if there are any terms that contribute little to Z in the measured space, but that will dominate in the space of desired extrapolation. We can check this by modeling, and if we cannot avoid this problem we will have to admit that we have no reliable prediction. We therefore try to avoid terms that cancel each other out in the measured space of X, or higher order terms with small coefficients.

We do not necessarily want to minimize E^2. The model with the lowest value of E^2 is not the best predictor and the statistical inference is full of problems of this sort (22). However, little has been done to apply this systematically to "approximate" modeling.

However, some general principles apply to our case. Fitting data using an assumed structure of M_K involves the estimation of a set of parameters C_j, that are contained in M_K. We measure the error in the Z space, but we also need an estimate of the error in the C space, as we need confidence limits on C.

A good identification means that the confidence limits on C_j are narrow. We can only achieve this if the dimension of C_j is reasonably small as compared to the number of independent observations and in kinetic experiments the number of really independent observations is not easy to estimate and is considerably lower than the number of rate measurements. Reliable simultaneous estimates of a large number of C_j are difficult, even if we have a large number of data. It is therefore important to reduce the dimension of C_j estimated simultaneously. We should try to set up experiments which measure a single C_i, or decouple the system in a way that allows us to get independent estimates for a subspace of C_i. For example, we try to measure M_K decoupled from M_R by using isothermal reactors with clearly defined M_R (no mass or heat transfer restrictions, etc.) For certain monomolecular or pseudo-monomolecular systems there are formalized ways to decouple the rate matrix into smaller subsets (23, 24, 25). In other cases this cannot be done rigorously and we use heuristic methods.

For a complex M_K^* or M_R^* we always face a compromise using the full set of X and a more realistic model will lead to difficulties in estimating C_i. We gain reliability by reducing the dimension of X and simplifying the model. The advantage is that compared to straightforward correlation we can justify the simplification by proper understanding of the system, and we can test our simplifications in an independent way. Again, this points out the iterative nature of modeling.

b) We want to have maximum confidence in our model in the sense that the underlying physical assumptions are correct. A good way of testing this is by looking carefully at where the model fails or what D. Prater (21) calls "operating a model in a failure mode."

When we design a plant we do not want to learn from failures (though we regretfully do) but in the model building process we want to set up the model and the experiment in such a way that we clearly notice where it fails. This is one reason why I prefer simple models.

We want to test the model in the learning stage where wrong predictions do not involve a large penalty. Model building should therefore be successive in a manner that we derive both the form of the model and the coefficients C_j from a limited set of Z_m, X_m. We should then test it on another set, preferably outside the space of x_m used to derive it.

One example of this approach is given in reference 26 for setting up a kinetic model for the water gas shift reaction. I will illustrate it by an example from my own work.

Operating a Model in the Failure Mode

The Mixed Crystallizer. One problem that concerned our research for many years was the understanding of the cyclic behavior of some crystallization and polymerization reactors (27, 28). Under certain conditions the particle size undergoes strong cyclic fluctuations, which severely upset operation and control. To understand this behavior we set up a simple model proposed by Hulburt and Katz for a stirred tank crystallizer, which is simply a particle balance of the crystallizer as well as a mass balance of the solute. For our purposes it is not necessary to go into the details of the model, but rather to deal directly with its implications. The only important kinetic parameters of this model are the linear crystal growth rate G and the nucleation rate B. The simplest assumption we can make about B and G are to assume that they are functions of supersaturation only.

If we look at the stability of this system we find that it is determined by a ratio $b_c/g_c = \partial \ln B/\partial \ln G$

where

$$g_c = \frac{1}{G}\frac{\partial G}{\partial c} = \frac{\partial \ln G}{\partial c}$$

and

$$b_c = \frac{1}{B}\frac{\partial B}{\partial c} = \frac{\partial \ln B}{\partial c} \tag{5}$$

A mixed crystallizer is unstable if b_c/g_c is larger than 21. For a given residence time we can measure B and G by measuring the particle size distribution in the product. For our simplified model it is exponential and in this case we can compute G and B from the average particle size $\langle r \rangle$. (29). We can then compute b_c/g_c from $\partial\langle r\rangle/\partial\Theta$ by the relation

$$\frac{\partial\langle r\rangle}{\partial\theta} = \frac{b_c/g_c - 1}{b_c/g_c + 3} \qquad (6)$$

For A value of $B_g/g_c = 21$ corresponds to a value of $\partial\langle r\rangle/\partial\theta = 0.3$. If $\partial\langle r\rangle/\partial\theta = 0$ then $b_g/g_c = 1$ and the system is very stable.

In the beginning we were happy with several aspects of our results. The form and time scale of the particle size fluctuations agreed well with what we observed. We also had a case for which $\partial\langle r\rangle/$ was in the correct range. Furthermore, our model predicted correctly the effect of product classification on stability (30). But when we started a closer investigation of real data, (and here I am limited by proprietary considerations) we found that there are systems exhibiting cyclic instabilities for which $\partial\langle r\rangle/\partial\theta$ was close to zero. The correlation between $\partial\langle r\rangle/\partial\theta$ and the dynamic behavior sometimes went in the wrong direction. First we examined the effect of our simplifications. The crystallizer is not really mixed, and contains regions of higher supersaturation. This we found increases stability. Also G is a function of particle size. This can decrease stability but the effect is minimal if $\partial\langle r\rangle/\partial\theta$ is close to zero. We also looked at the effect of secondary nucleation in which nucleation is promoted by the area of crystals present and again we found that this stabilizes the system. We looked for alternative models (31) and were actually helped in this by a more complex model which we had set up, for the polymerization of acrylonitrile in an aqueous dispersion (32). Here polymerization occurs both in the aqueous phase as well as in monomer adsorbed by the polymer. New particles are formed by coalescence of polymer molecules formed in the aqueous phase. Small particles are colloidally unstable and coalesce, whereas large particles which are stable seldom coalesce with each other. They can, however, adsorb small unstable clusters. The formation of a new particle depends on the ability of small polymer clusters to grow by coalescence to a critical size before they are adsorbed by an existing stable particle. We can express this by writing

$$B = B[(c-c_s);a] \qquad (7)$$

where a is the area of the crystal magna per unit volume. We now define

$$b_a = \frac{\partial \ln B}{\partial a} \quad \text{and} \quad b_c = \frac{\partial \ln B}{\partial c}$$

and if we repeat the analysis for this model, we obtain

$$\frac{\partial \langle r \rangle}{\partial \theta} = \frac{b_c/g_c - 1}{b_c/g_c + 3 - b_a} \tag{8}$$

If we look at reference (8) we find that we cannot get b_g/g_c from $\partial\langle r\rangle/\partial\theta$. If we measure the supersaturation, then, if $\partial\langle r\rangle/\partial\theta$ is small and b_g/g_c is large, we know that our simple model is wrong. But if b_g/g_c is small we will not detect any effect of b_a in steady state experiments. We checked this by more extensive modeling over wider ranges of θ.

A stability analysis shows (31) that positive values of b_a stabilize whereas negative value destabilize and a system with b_g/g_c of unity or less could be unstable if b_a is less than minus 20. There is no way we would notice that by measuring $\langle r\rangle$ versus θ.

We also found that Glassner (33) had independently shown by studying nucleation on a microscale that there are crystallization systems in which nucleation occurs by a mechanism in which nucleation occurs by coalescence of clusters and is inhibited by capture of clusters at existing stable crystals.

Once we accept that B and G are functions of the properties of the crystal magma there are a number of models that will have similar behavior. For example, the case where nucleation is promoted by collision of large particles is quite similar to the case of negative area dependence. There is no way that we can prove any specific model from the overall behavior of the crystallizer. But if we carefully organize our experiments we can disprove certain mechanisms and thereby increase our understanding of the system. There are several lessons I learned from this example.

a) Keep the model as simple as possible

In the age of the computer one is always tempted to put up large complex models. The requirement for simplicity that time put on us has passed. While we can often gain a lot by studying the full nonlinear simulation of the system, it has some significant drawbacks at the learning stage. Our mind does not work in 10 dimensions and it is hard to present the results of a complex model so that we clearly note where it fails. By keeping the model simple and slowly increasing the complexity as additional data require, we learn much more about the physics, then if we start to fit a large model. This is important in another way. In statistical inferences as well as in any experimentation there is a big difference between the fact that a model fits the data on which it is based, and where it predicts data in a new space of X and Z.

B) Check carefully what information is really contained in your experiment

Often we cannot increase parameters independently, though, if possible, we should make an effort to do so. We set up a model and measure its parameters by an experiment in a way that strongly depends on the model itself.

While the results may fit our model, they may not contain the information we seek, regardless of accuracy. We therefore have to be very careful to study what we are really measuring; what information is directly contained in our measurement, as distinguished from the information we extract by setting up a model.

In our example, all we measured was $\partial \langle r \rangle / \partial \theta$ at steady state and despite the large amount of literature based on this approach it does not give sufficient reliable information on the kinetics of crystallization. A mixed crystalliser is not a reliable tool to get kinetic data for crystallization. This has recently been demonstrated by other investigators (34).

c) Increase the space of X and Z and avoid constrained trajectories in component space

One of our problems with the crystallizer was that we tried to extract information from a series of experiments in which only space velocity was varied. This is standard procedure in pilot plant design.

This severely constrains the space of X. If we had changed feed conditions and measured supersaturation, we would have gotten additional information. A batch experiment would have been even more illuminating. For the polymerization reaction we got the important clues from batch experiments.

Frequently, we cannot cover the space of X, especially in design models. Proper modeling can guide the experimenter as to the space of X required in reliable predictions.

Consider for example a continuous stirred tank dispersion polymerization reactor operating with a low viscosity emulsion. We cannot maintain exactly the same fluid dynamics in a large scale reactor as in a pilot plant. No modeling can overcome that. But we know the direction of these changes. Dispersion of feed will be slower. Mixing and temperatures will be less uniform, etc. We can investigate in the pilot plant the sensitivity to temperature. We can experimentally measure the sensitivity to mixing by building a pilot plant of the type described by Shinnar (19) (see Fig. 3). We can also investigate the potential effect of these parameters by modeling. In the end what we want from this are limited estimates on the time scales of nonuniformities and mixing processes for which the process becomes sensitive to mixing. If we can design the large reactor so that the mixing times associated with the reactor are small enough so that the process is insensitive to their exact magnitude, then we have a safe scaleup. If not, modeling will not be very reliable, whatever we do. We note that while we do not measure Z for the desired X, we somehow increased our confidence by increasing the

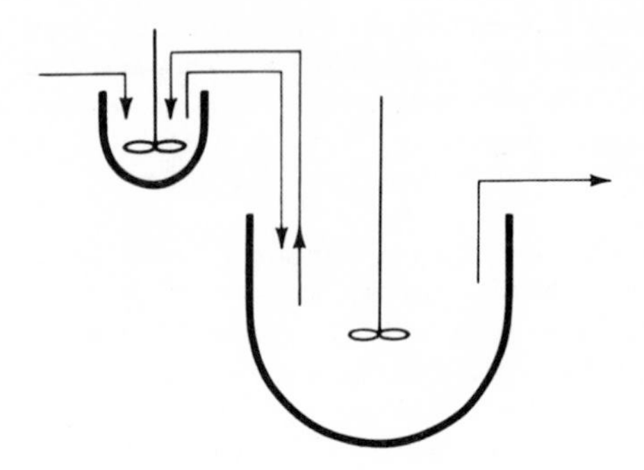

Figure 3. Simulation of finite mixing in a pilot plant

space of X so that we included a maximum of the important properties of the final X in our experimental design.

We do not achieve this confidence by setting up an accurate model for mixing in the stirred tank. We do achieve it trying to confine our design to regions in which the unknown features of turbulent mixing do not have an undue influence, and by performing proper experiments which were outside the normal operation of the pilot plant.

The most common mistake that I meet in pilot plant operation and modeling is a lack of understanding of what it means to get a proper wide space of X. By Murphy's law, pilot plants are invariably shut down when a commercial plant is designed. Looking over the data I find that the operators concentrated on getting good reproducible data in the pilot plant searching for optimum conditions which are often in a narrow space of X. No thought was given to the fact that this is not the sole or main function of a pilot plant. Very seldom can we reproduce the exact conditions in a large plant, and we need a much larger space of X in the pilot plant to give reliable predictions for scaleup. The termscaleup is a misnomer. We scale down in a pilot plant from the large reactor. Therefore modeling and design of the commercial plant should start before we design the pilot plant and definitely before we close it. It is an iterative process. We have to estimate in advance whether the set of X obtained in the pilot plant will give us sufficient information for the range of X encountered in the final design. Proper guidelines for efficient pilot plant operation are still a challenging research problem for the academic.

Model Identification and the Concept of Model Space

The concepts presented in the previous sections can be formalized in a more rigorous way. If we have a physical system M^*_{KR}, and a set of observations $(X,Z)_m$ we can always find a set of approximate models M^i_{KR} such that

$$Z_M - M^j_{KR}\ (X_M) < E \qquad (9)$$

for any member of the set. We noted before that E contains both the experimental and the model error. The nature of reactor modeling is such that the set of reasonable M_{KR} is large and may be infinite. We call the set of M_{KR} that is consistent with any given set of experiments and an acceptable error E the permissible model space. It is vaguely defined in the sense that the number of models that we permit does not only depend on $(Z,X)_m$ but also on our physical knowledge about the system. In that sense it is a fuzzy set. But we can often formulate a clearly defined model space in which the set of vectors

$$|Z^j_i| = |\bar{M}^j_{KR}\ (X)| \qquad (10)$$

bounds the space in which the real $Z_i = M^*_{KR}$ (X) will be, for the desired prediction, $\underline{Z}_p = M^*_{KR}(X_p)$. While we cannot identify M^*_{KR} we can identify a set $\bar{M}_{KR}$ that has the property, that

$$Z^j_{i\ min} < Z^*_{,} < Z^j_{i\ max} \qquad (10a)$$

for each Z; Here Z_i min and Z_i max are the minimum and maximum values of Z; in the whole set defined by 10.

We also require that $Z_{i\ max} - Z_i{}^{min}$ should be less than the permissible error E. To do this we need to know first of all the permissible value of E, which varies from case to case along with the penalty for a large value of E.

We can often construct a set of $|\bar{M}|$ that fulfills equ. (10a) by defining a fixed structure for $\bar{M}$ and varying the constants (11, 12). Or we can use a set of models with different structures that bound the properties of M^*_{KR}. For example, Zwietering's method of bounding the conversion of second order reactions used the residence time distribution as an example of the latter type.

Adequate model identification means that we have narrowed the properties of the set $|\bar{M}_{KR}|$ so that for all desired regions of predictions the error of Z in the space defined (by 10a) is sufficiently small for our purposes.

An interpolation model in that sense means a model space in which the desired prediction $(Z,X)_p$ falls in a space which is well covered with measured vector pairs $(Z_m X_m)$. Here, narrowing the space of $|\bar{M}_{KR}|$ is not critical, as any member of the set will give a reasonable prediction provided

a) M^*_{KR} is smooth and well defined

b) X is complete in the sense that the contribution to Z of parameters not included in X is negligible

c) we choose a form of $\bar{M}_{KR}$, which is also smooth in the space under consideration.*

While we can define what we mean by a space being well covered by X, doing it efficiently requires judgment and here more theoretical work could be very useful. The problem of completeness is more difficult. An industrial feed may contain impurities with strong catalytic action that we do not encounter in our pilot plant. In recycle operation we can at least foresee this possibility, but the only way we can protect ourselves is by extensive and expensive experimentation. Earlier, we discussed the problem that the set X has to be wide and that constrained trajectories are preferably avoided.

In design models the problem is more difficult. We have to evaluate the likely range of conditions X_p that could be encountered in the full scale plant. Here estimation of the space $|\bar{M}_{KR}(X_p)|$ requires more judgment. We cannot cover the space of X in the same way, as the large plant operates in a different flow regime, and we therefore need a larger degree of confidence in our model. We have to ask what features of X will affect the prediction and how we can narrow the set $|\bar{M}_{KR}|$ in such a way that it is likely to bound M^*_{KR} (X) within a sufficiently small space.

In that sense all predictive modeling is a bounding procedure. We have to approach it from two sides. First we have to ask what properties of $|\bar{M}_{KR}|$ are going to be the most important in determining Z in the desired region of predictions.

For example, in control problems the main features of M^*_{KR} we need are the main time constants of the response, the sign of $\partial Z_j/\partial X_i$ and the magnitude of the final response. The most critical is the derivative. If the other two are accurate within 30%, or even a factor of two, we can compensate for this by pro-

---

* Some polynomial fitting functions lack that property and minimize (E2) in a way that leads to strong fluctuation in between the measured points.

per design of the feedback control.

We have to ask what properties of the space $|M_{KR}|$ we defined in our experiment. Consider, for example, a tracer experiment leading to a residence time distribution F(t). F(t) is a linear property of the set $|\bar{M}_R|$. As M_R^* and $\bar{M}_R$ are nonlinear there is no way we can reconstruct a model $\bar{M}_R$ from F(t), regardless of how accurately we perform the experiment. But we can use residence time distributions efficiently in the failure mode. Let us look at the tracer experiments in Fig. 4 taken from Murphrees' paper on hydrotreaters (35). Figure 4 gives the results replotted in terms of the intensity function (6).

Note that the pilot plant is very similar to a packed bed where both traces from the industrial reactor show a maximum in the intensity function which indicates a bypass phenomenon in the flow. This could be due to a badly designed liquid distributor. But it could equally be due to instabilities caused by the flow inside the bed, which is in different flow regime than the pilot plant. This type of tracer experiment does not allow us to distinguish between these alternatives and we would need additional experiments or clues to do so.

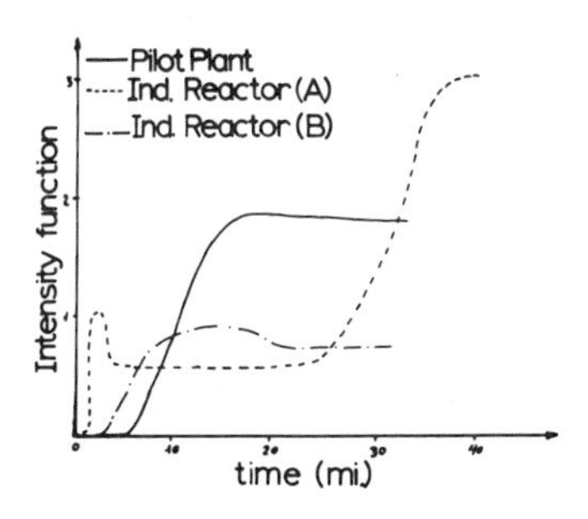

Industrial and Engineering Chemistry, Product Research and Development

Figure 4. Tracer experiments in a trickle bed reactor (35)

If we want to get a better approximation of M_R^*, we should carry out a set of kinetic experiments in the pilot plant and the industrial reactor, using different feedstock temperatures and feed rates. This would improve our knowledge of $\bar{M}_R$ and provide valuable information for future scalings. If all we want is to predict the behavior of new feedstocks, this would probably be enough. Modeling of existing reactors does not require a detailed knowledge of M_R^*. But if we want to understand how to improve the reactor for future designs, we need much more information, and we would require detailed studies of the flow processes inside such a large reactor as a function of process conditions and design. This could be done in a cold flow model.

While M_R^* and M_K^* have some similarity in that they are inaccessible, there are some basic differences. We can deliberately design our reactor so that the space of $\bar{M}_R$ is narrow, or, in other words, the flow models are well defined. We can achieve this even in turbulent flows or in fluid beds by designing the reactor very close to a well defined asymptotic case such as a stirred tank or a plug-flow reactor. We can achieve this despite the complex non-linear form of the transport equations. If the deviations from the clearly defined asymptotic case are small, we can describe them by an expression around the asymptotic case. Happily, most processes are not sensitive to small deviations from the asymptotic ideal flow.

We cannot do this for M_K. On the other hand, M_K itself is not a function of scaleup, so if we understand what properties of M_K we need to know for safe scaleup, we can cover the space of M_K well enough in the pilot plant, to give reliable predictions.

We will discuss both problems in the following sections.

VII. FLOW MODELS FOR REACTORS. THE EFFECT OF DESIGN ON PREDICTABILITY

One important tool of the designer, using modeling intelligently, is to design the reactor so that its performance is easier to predict. Good modeling is a two-way street or an iterative process. It should effect our design decisions.

There is an analogy to this in control and electronic equipment design. For reliable complex electronic control equipment, the designer will choose components with reproducible linear behavior over a wide range. If needed, he will pay a premium for this, to ensure well predictable performance in complex systems. We cannot make our reactors linear, and we are limited in what we can do about M_K. We should be happy to have a selective catalyst at all. But we can spend money to make M_R more predictable.

The concept is already quite familiar in the design of kinetic experiments. Here we want to have conditions where M_R is well-defined and the reaction rates can be measured directly without being disguised by the effect of transport processes. In practice, for most reactions this translates into designing a micro-reactor which is either close to isothermal plug-flow (51), or to a stirred tank (13). In both problems the effort that we spend to approach the ideal case should depend on the sensitivity of the reaction to mixing. In most instances we can approach the ideal case sufficiently to get reliable rate measurements.

We check this by a bounding procedure. For small deviations from plug-flow we can express the deviation from plug-flow by the variance of the residence time distribution. We can estimate this variance, or measure it, either by a tracer experiment or by using a reaction sensitive to mixing (52). We can then use any simple model, such as a stirred tank, to estimate the effect of a small deviation from plug-flow, and design the reactor with a variance small enough to have a negligible result on the kinetics. We achieve this by measuring the length of the micro-reactor and decreasing the particle size of the catalyst. Since this is expensive, we have to decide if it is necessary. We approach the problem of assuring isothermal conditions in a similar way. It is, therefore, an iterative procedure because only after we have the first results can we evaluate whether our micro-reactor was good enough.

We can apply the same approach to design of industrial reactors. While in a pilot plant the goal is to measure M_K accurately, here our goal is to reduce the risk of scaleup.

We previously discussed the case of the polymerization reactor. In a similar vein, in alkylation the sensitivity to mixing is reduced by pre-blending the feed with recycle.

I would like to add here another, less known, example - the design of a catalytic fluid bed reactor. At low flow velocities such reactors are hard to scaleup. Large bubbles are formed and the exact behavior of the bed is hard to predict (38, 39). This is well illustrated by the available data. In Fig. 5 conversion is plotted for a first order reaction as a function of bubble size for a specific case (40). It is very sensitive to bubble size. In that sense bubble models are learning models and not predictive models, inasmuch as we cannot predict bubble size that accurately. At higher velocities bubbles become less distinct and gas solid contact improves considerably (41, 42). The risk of scaleup decreases.

The most important lesson of bubble models, not stressed sufficiently, is that one should avoid this regime for catalytic reactor design. Present catalytic reactors operate, therefore, at much higher velocities than those used in most academic investigations. We can also approach plug-flow by properly baffling the bed (see Fig. 6). This will introduce a staging effect, break up the bubbles and further reduce the hydraulic diameter. This also reduces the distance betwen X_p and X_m. Such a bed will be very close to plug-flow, though the catalyst efficiency will be lower than predicted from a packed bed reactor (43). I know of one case where a 99.5% conversion is achieved at a space velocity

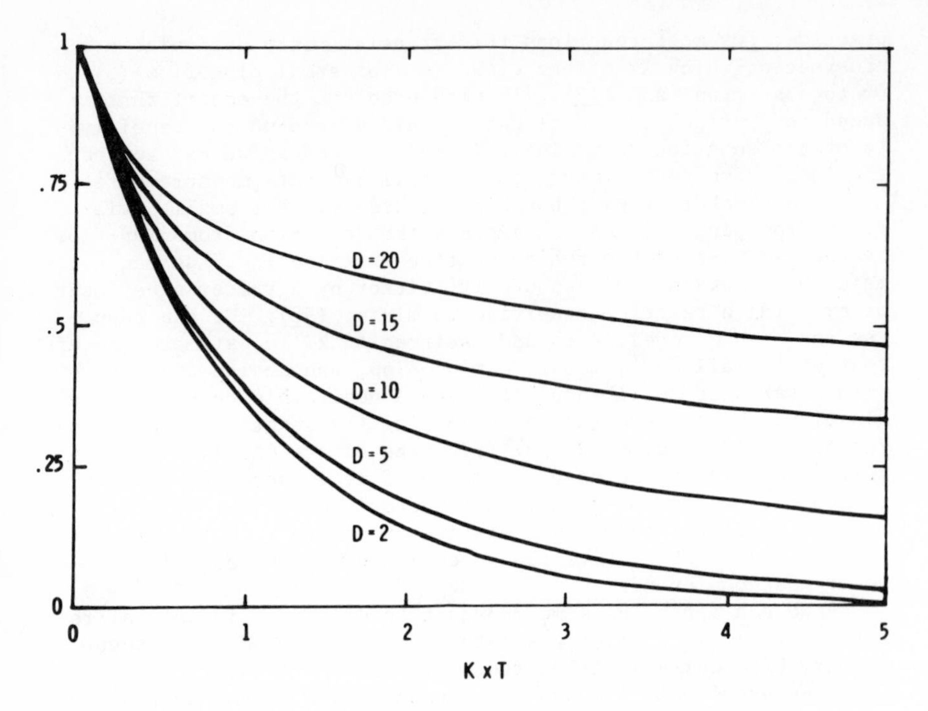

Figure 5. Conversion of a first-order reaction in a bubbling fluid bed as a function of bubble size

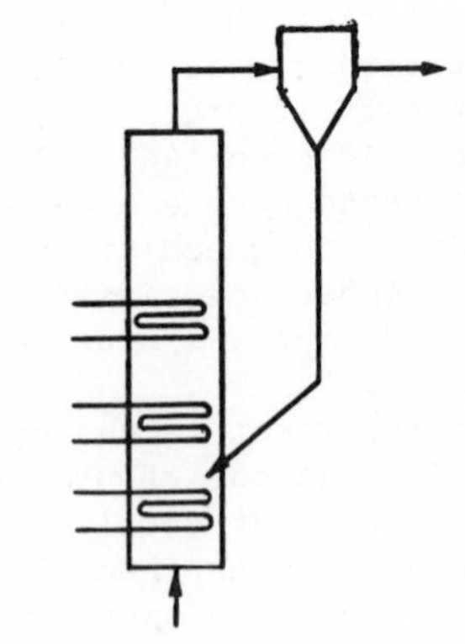

Figure 6. Fluid bed baffled by sets of horizontal tubular heat exchangers

which is only lower, by a factor of two, than in a packed bed. Here we achieve reliable scaleup at an increased reactor cost. Regretably, operating at a high velocity requires a tall pilot plant, which is expensive. It also reduces the range of operability of the bed.

There are many other cases in which design changes will reduce the risk of scaleup and improve our capability of prediction. The question the designer has to ask is whether the added cost is worth the reduction in risk. Simulation and modeling can help estimate the risk, but they cannot eliminate it. Therefore, a designer has to understand clearly what the basis and confidence limits of such predictions are.

The ability to approach an ideal case is not only valuable in terms of reducing the risk of scaleup. For many cases the ideal plug-flow or stirred tank reactor is also an optimal reactor (44), and we often improve the yield by approaching plug-flow more closely.

In no way do I want to create the impression that we cannot or should not handle more complex flows. What I want to point out is not only that this is more complicated, but that it also involves larger uncertainties or wider bounds on M_R.

I concentrated here on the effect of design on predictability. Modeling should enter design considerations in many other ways, since it indicates the compromises the designer faces. One problem that has until now been given too little attention is the way design affects dynamic behavior and control of the reactor. Good examples of this are recent studies dealing with design of catalytic after-burners (45, 46) and hydro-crackers (16).

VIII. KINETIC MODELS

Finding a proper M_R is a more complex problem and what I want to do here is to illustrate some of the general principles outlined before, and mention some of the problems that require solution.

Let me start with a simple case: the conversion of nitrogen and hydrogen to ammonia. On the one hand, it is a complex reaction and we do not know the mechanism. On the other hand, with reasonable effort we can get an all-purpose M_K, a rate relation accurate enough so that we do not have to ask about its use. Such expressions are not necessarily unique. Ferrari (45) lists 24 rate expressions of the form

$$r = \frac{\Pi a_i^{\alpha_i} - 1/K^{*}\, \Pi a_i^{\beta_i}}{\Sigma C_j \varphi_j}$$

where

$$\varphi_j = \Pi a_j^{\nu_{ij}} \tag{11a}$$

$$C_j = A_j e^{-\frac{E_j}{RT}} \tag{11b}$$

which fit a wide range of data with an $\langle E^2 \rangle$ of less than 10%. All of them are based on different mechanistic models for the individual steps on the catalyst surface. It is obvious that such overall data cannot differentiate between complex mechanisms. The best fit was obtained by the rate equation

$$r = \frac{a_{N_2}a_{H_2}^2 - a_{NH_3}^2/a_{H_2}^3 K^2}{C_1 a_{H_2}^2 + C_2 a_{NH_3}^2/a_{H_2}^2 + C_3 a_{H_2} a_{NH_3}} \tag{12}$$

which has an error of less than 1.0%. As Ferrari only gives $\langle E^2 \rangle$, it is difficult to judge between the models as $\langle E^2 \rangle$ is an insufficient criterion. Since he had, at his disposal, a large amount of data, the models are all probably good. In my view, all of them are too complex and contain more parameters than this case justifies. We could get a model reliable enough for any purpose of reactor modeling by a simpler model.

For estimating M_K we want structural relations that contain our knowledge of the physics and which have the property that they map efficiently back between concentration space and parameter space without increasing the error. We know too little about that, as well as how we incorporate and test our approximate knowledge of M_K which is often consistently into the estimation procedure.

Estimation procedures for $\bar{M}_K$ show one special feature. Compared to estimation problems in other areas of engineering we know much less what the accurate physical relation is. Often we cannot use straightforward deterministic modeling, as our knowledge of M_K^* is insufficient. We also have experimental errors.[a] On the other hand, we know much more about M_K^* than we do in the areas for which purely statistical model-building procedures have been developed. What we really need is a new approach that incorporates into statistical estimating procedures the considerable knowledge of the thermodynamic constraints (55

---

a) Ref. (47) in this volume shows that even if we knew the exact structure, recovering the coefficients requires a great effort.

and M. Feinberg in ref. 2) and the physical features of M_K that we know. The application of the concept of model space and bounding may provide a starting point.

We should also realize that in the ammonia case the most important rate information is contained in the following implicit assumptions:

1) The only product promoted by the catalyst is ammonia. The dimension of Z is always the most important information contained in M_K and requires careful checking for new catalysts.

2) The rate is a function of measurable concentrations. We know, or believe, that intermediate species play a role. The fact that we can express a series of steps, depending on the formation of unstable intermediates as a function of concentration of measurable species, is a strong assumption based on experience. It does not always hold and requires experimental verification.

3) We know that other feed componen ents,such as CH_4, are inert. The fact that CO and H_2S affect the rate by poisoning the catalyst is taken care of in equation (12) by assuming that we eliminated them from the feed.

Let me now go to a more complex case. As the first example I chose the isomerization of Xylene over a silica glumina catalyst studed in reference (25). Here, Xylene isomerizes with a monomolecular reaction:

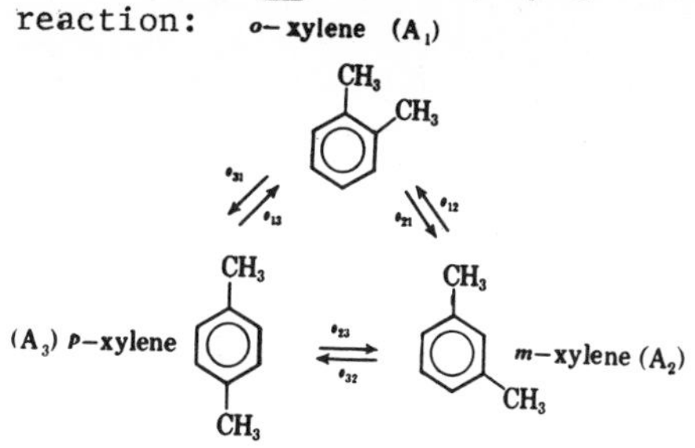

o-xylene (A_1)

(A_3) p-xylene $\underset{k_{32}}{\overset{k_{23}}{\rightleftarrows}}$ m-xylene (A_2) (13)

and similarly undergoes a disproportionation reaction

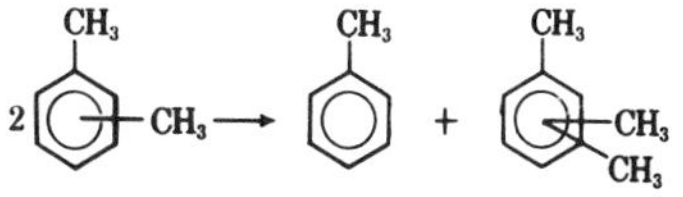

$$2\,C_6H_4(CH_3)_2 \longrightarrow C_6H_5CH_3 + C_6H_3(CH_3)_3 \qquad (14)$$

The second reaction is pseudo-monomolecular; the reverse reaction is bi-molecular and in their cases are assumed to be negligible. As long as we are far from equilibrium, we can model the backward reaction as pseudo-monomolecular with a small mistake. The paper is a fine example of how to simplify and decouple M_K in order to simplify the estimation of the rate matrix. Using methods derived in (23, 24) it also reveals some other problems. We can write the whole reaction as a single pseudo-monomolecular rate matrix:

$$\frac{da_j}{dt} = K_{ij}a_i$$

There are several assumptions involved in this. The reaction rate expression probably contains a nominator. For the isomerization case it is justified to take the nominator in the front of the matrix and include in it terms related to catalyst activity. However, this is less clear for the disproportionation. Doing so is equivalent to assuming that all reactions occur at the same type of catalytic site. This was true for their case, but is an undesirable assumption in general. In fact, we hope it is not true. If it is correct, the main parameter effecting selectivity is temperature. On the other hand, should the disproportionation reaction be favored by a different type of site, we could hope to improve selection by changing reaction concentration or finding a selective poison for this site.

For scaleup this is less relevant and I could probably get by with taking out the nominator in front, provided my data space is wide enough.

There is another problem here. For many objectives there is little reason to map into parameter space. Here my rate matrix is really a selectivity matrix (the time element has been taken out in front), I can use the measured trajectories in concentration space directly for most design purposes. In fact, they might be preferable for scaleup, since what I really care about is the curvature of the trajectory and the sensitivity to temperature.

For reactor design the fact that a complex reaction has a pseudo first order nature is in itself important information, perhaps more than the accurate matrix. I would define a pseudo first order reaction in the following way (44, 48):

a. Mixing phenomena can be approximated by averaging space time histories of different aggregates with the same history. Consider, for example, the following second order reaction schemes:

1) $A + A \rightarrow C$

and

2) $A + A \rightarrow B \rightarrow C$

in a pre-mixed feed.

If I treat them as pseudo first order reaction, I can estimate the conversion from the residence time distribution reasonably well (at least for conversions below 80%) and by averaging, will introduce no error in the product distribution.

For both cases a plug flow reactor is optimum (44), and for a prescribed temperature distribution, mixing is detrimental; reducing conversion and reducing the curvature of a curved trajectory in component space.

Consider, on the other hand, the reaction scheme

$$A \rightarrow B \rightarrow C \rightarrow D \qquad 3)$$
$$A + D \rightarrow E$$

Here, mixing will not just straighten a curved trajectory in component space, it will change its direction and could introduce a product D that might not be noticed in a plug flow reactor. Complex reactions of type 3 are especially common in polymerization. Alkylation is another example when the trajectory is totally changed by mixing.

In complex reactions it is, therefore, important to get some information as to how important strong interactions are. For scaleup this is all we need. Again, our method of setting up a simple model from one set of experiments and trying it in a different reactor (for example, a backmixed reactor with high conversion) will provide this information.

For safe scaleup we also need to know the effect of temperature, and in good pilot plant practice temperature should be changed over a far wider range than operation of the pilot plant requires. We cannot control temperature as easily in a large plant as transfer and mixing processes are slower. Again, we will be guided here by modeling the final reactor at the early stages of setting up M_K. We should always match the identification procedure of M_K to the design procedure, so that M_{KR} (X) has a sufficient narrow bound.

As a last example I want to discuss the catalytic cracking of oil in a riser cracker. Weekman (49, 50) proposed a simple scheme:

$$\text{Gas-oil} \xrightarrow{r_1} \text{gasoline} \xrightarrow{r_2} + \begin{matrix}\text{gas} \\ \text{coke}\end{matrix} \qquad \text{(Gas-oil} \xrightarrow{r_3} \text{gas + coke)} \qquad (13)$$

r_1 is second order in mole fraction of gas-oil, and r_2 and r_3 are first order reactions. This simple scheme contains all the essential features of the cracking reactions. It gives a curved trajectory in component space typical of consecutive reactions. It also shows that at low conversions formation of coke and gaseous products is a parallel reaction. We can learn a lot about FCC design from such a simple model. We can get a better representation of composition and feed-stock effects by more refined lumping schemes, such as the one given by Jacobs (52). (See Fig. 7).

Here, the estimation of the first order matrix is more difficult, as most of the reactions are nearly irreversible and the lumps are improper lumps Further, any modeling will involve much stronger simplification assumptions as compared to the xylene use. The method of Wei and Prater therefore does not apply (23). We have very useful results for the pitfalls of lumping (53, 59), but few guidelines for efficient improper lumping, or lumping in systems with strong interactions. We also lack knowledge as to how to put thermo dynamic into improper lumps.

Lumping is not a unique process, nor are all interesting lumping problems pseudo-first order. What we want to know is: Are there any strong interactions between lumps, especially between product and feed? Such interactions could significantly change the reaction path in component space. We know that some olefines will polymerize, but the mistake in representing this as a psueudo-first order lump is small.

We could check this by obtaining results both in a plug-flow as well as in a continuous stirred tank or recycle reactor, but this is not easy. The case of the cracker also illustrates the big difference between interpolation modeling and design modeling.

A large riser reactor is hardly a one-dimensional flow. M_R is not well defined. At the bottom hot catalyst (1300°F) mixes with oil (400-700°F). The mixing time is of the order of magnitude of 0.5-1 second and the total residence time is on the order of 4-10 seconds. The effect of the non-mixed zone is greater than appears from the ratio of mixing to residence time. The catalyst activity decays along the riser by a factor of 100 and the contribution of the non-isothermal mixing zone should therefore be considerable. There is no way we can exactly simulate this in a pilot plant. We would require data over a wide range of temperatures to indicate if this will significantly effect our results. The first riser cracker design involved a large risk. The fact that we have large units today allows the modeler with access to them to check how well pilot plant experiments predict real units. Again, the fact that they do it for one catalyst does not mean this will always apply. That is what makes the life of a reaction engineer interesting.

Several facts reduce the modeling of the cracker, at least partially. It is well to remember them if we want to extend these methods to other cases:

1. The reactors exist and correlations are checkable in large reactors.

2. FCC reactors are extremely flexible and the operator has at his disposal a large number of manipulated variables. Even if

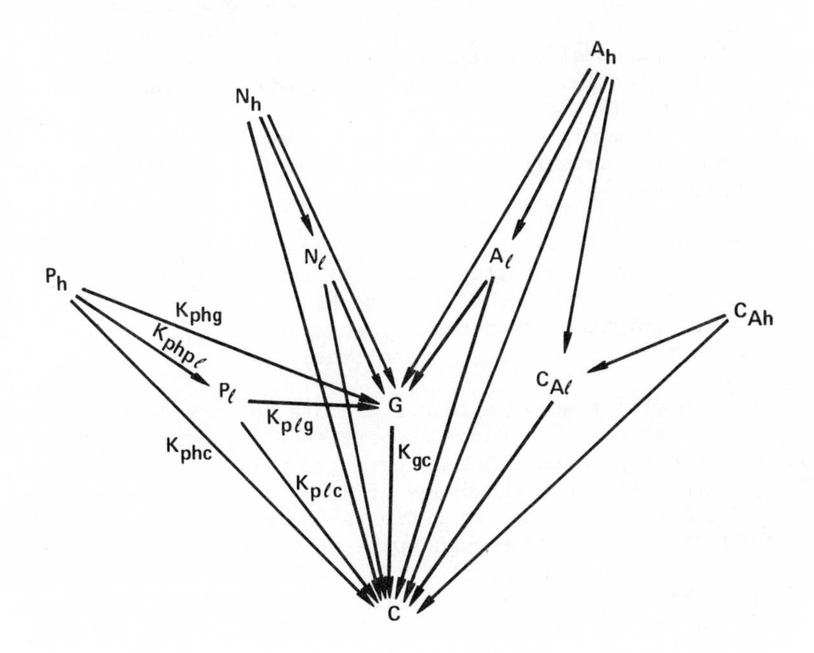

American Institute of Chemical Engineers Journal

Figure 7. Detailed lumping model for catalytic cracking reaction (52).

P_l = Wt. % paraffinic molecules, (mass spec analysis), 430°–650°F; N_l = Wt. % naphthenic molecules, (mass spec analysis), 430°–650°F; C_{Al} = Wt. % carbon atoms among aromatic rings, (n-d-M method), 430°–650°F; A_l = Wt. % aromatic substituent groups (430°–650°F); P_h = Wt. % paraffinic molecules, (mass spec analysis), 650°F$^+$; N_h = Wt. % naphthenic molecules, (mass spec analysis), 650°F$^+$; C_{Ah} = Wt. % carbon atoms among aromatic rings, n-d-M method, 650°F$^+$; A_h = Wt. % aromatic substituent groups (650°F$^+$); G = G lump (C_5–430°F); C = C lump (C_1 to C_4 + COKE); $C_{Al} + P_l + N_l + A_l$ = LFO (430°–650°F); $C_{Ah} + P_h + N_h + A_h$ = HFO (650°F$^+$). Adapted nomenclature for rate constants is detailed in the above figure for the paraffinic molecules. Similar rules apply for the other reaction steps.

we put in a wrong catalyst, we can exchange it without shutting down. Let us remember that this is not always true.

In obtaining M_K for a complex case the interaction between M_K and M_R and the exact definition of the goal are always going to be important considerations, if we require a proper balance between the potential value of increased knowledge and the cost involved in getting it.

There is a big difference among complex lumped kinetic models that provide sufficient information for one of the following purposes:

a) safe scaleup
b) efficient characterization of new catalysts
c) characterization of feedstocks
d) optimization of design procedures
e) effect of catalyst deactivation

With a finite effort we want to concentrate on improving the accuracy of our estimate for these properties of M_R which are most relevant to our problem. Often these are very specific and far simpler than a complete knowledge of M_K^*.

For scaleup we need to estimate the probable space of X that will give us a wide enough space of M_K, that allows us a safe prediction for the expected range of M_R. Characterization of feedstocks or complex mixtures requires some good chemical knowledge as to what makes them similar enough so that we can hope to get a simplified M_K as well as when such simplifications break down. The most difficult and challenging is probably the case of efficient search procedures for new catalysts in complex reactions. Here it is especially important that we keep our identification model simple enough so that the researcher can be guided by the information obtained, and complex enough that we don't throw out the baby with the water.

SUMMARY

In the previous sections I have elaborated my views on what I consider the important problems in modeling today. I would summarize my approach as follows:

a) Reactor modeling is somewhat different from many other modeling problems in engineering due to the complexity of the systems. Reaction models are rather imperfect approximations of the real system. They are also different from models used in economics or statistical model-building, since in reactors we know much more about the physics of our system. We have as yet to find a proper synthesis between the two approaches. An attempt to find such a synthesis is outlined in this paper.

b) Proper modeling must match the effort involved with the goal. We need a better understanding in what way simplified approximate models can describe a more complex system. We have to understand how errors and uncertainties affect our final predictions, and what the value of better information is.

c) Applied modeling involves a match between identification, modeling and the required accuracy of the prediction. It is, by nature, an iterative process. The most challenging problems are in more efficient identification procedures of complex systems, which integrate our physical knowledge of the system with modern identification procedures.

Most of my ideas were outlined in this paper, and I would like to avoid repeating them. I would end with a few general observations.

My experience with industrial modeling practice has led me to conclude tht it sometimes suffers from organizational problems as well as from a too narrow concept of its application. By putting modelers in a special group, one often limits tjem to a specific type of modeling; large models for existing reactors. While this is valuable, it is not the only, nor even the most, powerful use of modeling. Proper modeling is a powerful tool in the hands of a good designer, and should be coupled with process development and pilot planting. In those stages the chance of a significant contribution is greater than when a reactor is standing there. Proper cooperation with designers and developers requires that our results should not come "deus ex machina" from a computer, but should be clearly intelligible so that they reinforce and extend their intuition and experience.

That does not mean we cannot use complex computer program. We just have to take the trouble to construct them in a way that we can clearly explain their results and the uncertainties involved in their predictions. We also have to admit our limitations.

Looking back on my own work in the academic area, I admit that I did not always put sufficient effort into making clear, even to myself, what my results really meant and how they could be used by the practitioner. I am not alone in that. Recently we tried to deal with the design process itself in a number of papers. It is time that some of us paid less attention to learning models, and devoted our efforts to improving the art of predictive modeling and design. We ought to look in depth at concrete examples which include the incomplete knowledge we have in real problems. We could try to increase the scientific content of this art and improve its methodology. I tried to outline some of the problems, and I admit that any such suggestions or overviews are a risky undertaking in themselves.

Nomenclature

Z output variables

X input variables

M^*_{KR} correct mathematical description of reactor

$\bar{M}_{KR}$ approximation of M_{KR}

$E+$ square average error of prediction

M_R reactor flow model

M_K kinetic model

C_j parameters of model

ACKNOWLEDGEMENTS

The author is grateful to Dr. C.D. Prater, Dr. Paul Weisz, Dr.V. Weekman, and Dr. F. Krambeck from Mobil Research Company, as well as Prof. H. Hulbert, for many helpful discussions and advice in preparing the paper and shaping the opinions expressed in it.

REFERENCES

1. Weekman, V.W. Jr. (1975) Advances in Chemistry, No. 148, p. 98.
2. Denn, M., Vol. 15 Control of Dynamic Systems, Advances in Theory and Application, Academic Press, 1978.
3. Lapidus, L. and Amundsen, N.R., editors; Chemical Reactor Theory, a Review, Prentice Hall, 1977.
4. Keane, Th. R., Manogue, W.H., Russel, Frazer, T.W. and Weekman, V.W. Jr.; Feedback on Reaction Engineering Techniques, Report to the National Science Foundation (1971).
5. Ferrari, C.B. et al., (1976), Chem. Eng. Science, 29, 1621.
6. Shah, M.J. (1967) I&EC, Vol. 59, p. 81.
7. Baddour, R.F., et al., Chem. Eng. Sci. 20, 281 and 297.
8. Hougen, O.A. and Watson, K. (1934) Ind. & Eng. Chem. 35, 529.
9. Aris, R., Re, k and Π, A conversation on some aspects of mathematical modeling, B. Levitch conference, Oxford, July 1977.
10. Weisz, P. (1973) Science 179, p. 433.
11. Kestenbaum, A., Thau, F. and Shinnar, R. (1975). I&EC Process Dev. & Des.
12. Palmor, Z., Shinnar, R., Design of Controllers for Sampled Data Systems, AIChE meeting, Nov. 1977.

13. Fishman, G.F. "Concepts in Discrete Event Digital Stimulation" J. Wiley, 1973.
14. Lin, S., Amundson, N.R. (1962) Ind. Eng. Chem. Fund. 1, 200, 2, 183.
15. Crider, J.E. and Foss, A.S. (1966). AIChE Journal 14, 77.
16. Silverstein, J., Shinnar, R. Paper given at AIChE meeting, November 1977.
17. Aris, R., The Mathematical Theory of Diffusion and Reaction in Permeable Catalyst, Clarendon Press, Oxford (1975).
18. Jackson, R., Transport in porous catalysts, Elsevier, 1977.
19. Shinnar, R. and Katz, S. 1972, Advances in Chemistry No. 109, 56.
20. Crawford (1976), Fluid Bed Char Gasifier, ERDA Report FE2213-5.
21. Prater, C.D., Private communication.
22. Ephron, B. and Morris, C. Stein's paradox in statistics, Scientific American, May 1977.
23. Wei, J. and Prater, D. (1962) Advances in Catalysts 13, 203.
24. Prater, C.D. and Silvestri, J.A. and Wei, J. (1967), Chem. Eng. Sci. 22, 1587.
25. Silvestri, J.A. and Prater, C.D. (1969), Journal of Physical Chemistry, 68, 3268.
26. Hulburt, H.M. and Srini-Uasan, C.D. (1961) AIChE Journal 7, 143.
27. Sherwin, M.B., Shinnar, R. and Katz, S. (1967), AIChE 13, 1191.
28. Thomas, W.M., Matheson, W.C. (1961), Petr Ref. No. 5, 211.
29. Randolph, A.D., Larson, M.A. (1971). Theory of Particulate Processes, Academic Press.
30. Sherwin, M.B., Katz, S., Shinnar, R. (1969). Chem. Eng. Symposium Series No. 95, 65, 75.
31. Liss, B. and Shinnar, R., AIChE (1976) Symposium series, Vol. 72, No. 153, p. 28.
32. Liss, B. Ph.D. Thesis (1975) City College of New York.
33. Glasser, D. (1973), Mat. Res. Bull. 8, 413.
34. Bryson, A.W. et al. (1976) Hydrometallurgy 57, 1.
35. Murphree, E.V., Voorhies, A., Mayer, F.X. (1968), I&EC Proc. Des. Dev. 1968 3, 381.
36. Naor, G., Shinnar, R. (1963) Ind. Eng. Chem. Fund. 2, 278.
37. Silverstein, J. and Shinnar, R. (1975) I&EC Proc. Des. & Dev. 14, 127.
38. Kumi, I. and Levenspiel, O., Fluidization Eng., Wiley, 1969.
39. Davidson, J.F. and Harrison, D., Fluidized Particles, Cambridge University Press, 1973.
40. Krambeck, F. Personal communication.
41. Lanneau, K.P. (1960) Trans. Inst. Chem. Eng. 38, 125.
42. Argyriou, D.I., List, H.L. and Shinnar, R. (1971) AIChE J. 17, 122C.
43. Krambeck, F., Katz, S., Shinnar, R. (1969) Chem. Eng. Sci. 24, 1497.
44. Weekman, V.W. Jr. (1968), I&EC Process Des. Dev. 1, 90.

45. Wei, J. Catalysis and Reactors (1969) CEP monograph series, vol. 65.
46. Kuo, J.W. et al., XIV Automotive Technical Congress, FISITA, London June 1972.
47. Cropley, J.B. This symposium.
48. Shinnar, R. Tracer experiments in reactor design, paper given at B. Levitch Conference, Oxford, 1977.
49. Shinnar, R., Glasser, D. and Katz, S. (1973) Chem. Eng. Sci. 28, 617.
50. Voltz, S.E., Nace, D.M., Jacob, S. and Weekman, V.W., Jr. (1972) *¢EC Process Des. Dev., 11, 261, 10, 538, 10, 530.
51. Glasser, D., Katz, S., Shinnar, R., (1973) I&EC Fundamentals 12, 165.
52. Jacob, S.M., Gross, B., Volk, S., Weekman, V.W. (1976) AIChE Journal 22, p. 701.
53. Wei, J. and Kuo, J.C.W. (1969), Ind. Eng. Chem. Fund, 8, 114.
54. Golikeri, S.V., Luss, D. (1974) Chem. Eng. Sci. 29, 845, 26.
55. Krambeck, F.J. (1970) Ann. for Rat, Mech. and Analysis 38, p. 317.

RECEIVED March 30, 1978

2

Liquid-Supported Catalysts

C. N. KENNEY

Department of Chemical Engineering, Cambridge University, Pembroke Street, Cambridge, England CB2 3RA

This review is a survey of the applications and properties of supported liquid phase catalysts (SLP). By a supported liquid phase catalyst is meant the distribution of a catalytically active liquid on an inert porous support and the behaviour of such systems raises many interesting questions on catalyst chemistry, mass transfer in catalysts and reactor design. It is noteworthy though that such systems have been employed in the chemical industry for many decades - indeed for over a century in the Deacon process for obtaining chlorine from hydrogen chloride - and of almost equally respectable antiquity are the vanadium based catalyst systems used for sulfuric acid manufacture: but the recognition of SLP catalysts as possessing features of their own is much more recent.

Their relation to catalysts as a whole is indicated in Figure (1). In this paper principal attention will be given to the types of SLP catalyst which have been identified together with a discussion of some of the mass transfer and liquid distribution problems which are clearly important in these systems. A related type of catalyst which will not be discussed here is the supported homogeneous catalyst where the active agent, usually a transition metal complex, is actually bound chemically to the support, often a porous polymer matrix. It is fairly clear that with these latter systems the interaction between the catalyst and the support have important effects on modifying the catalyst properties which fortunately do not usually arise in the two, and sometimes three phase reaction systems which occur with SLP systems. The survey will be highly selective since the related topic of molten salt catalysts has been recently reviewed by Kenney (1) and Villadsen and Livbjerg are publishing a survey of (SLP) catalysts in general (2).

1. High Temperature Reactions

1.1 Reactions involving chlorine. Amongst the earlier kinetic studies of potential (SLP) catalysts are those of Gorin in 1948 (3) - these studied the rate of chlorination of methane in

0-8412-0432-2/78/47-072-037$05.00/0

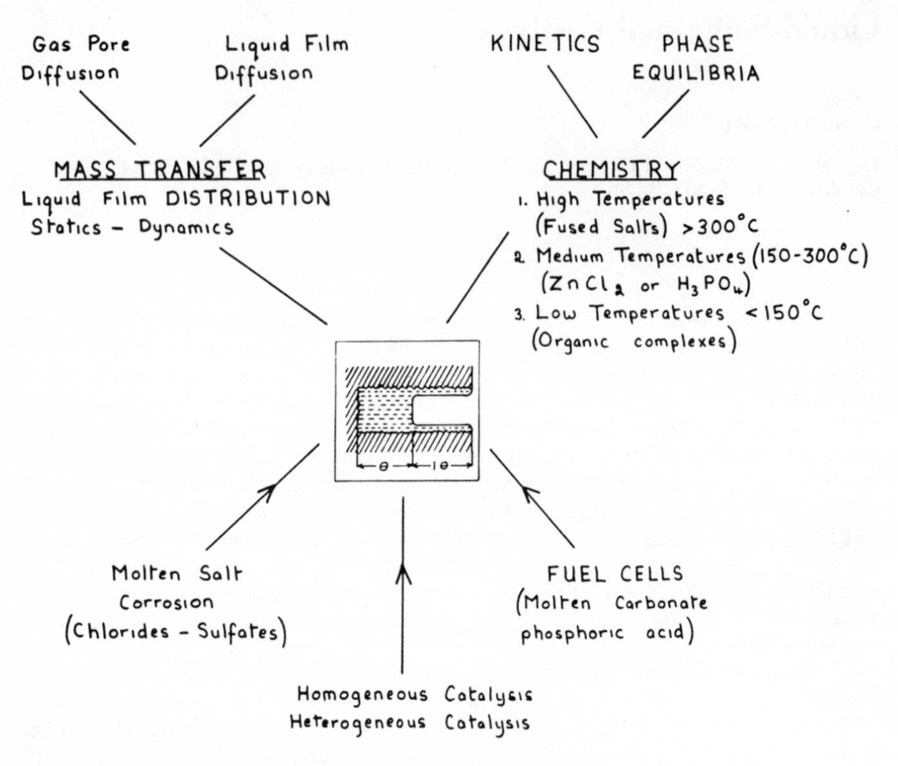

Figure 1. SLP catalyst systems

KCℓ/CuCℓ melts (4). The reaction was found to be zero order with respect to methane. A complicating factor which can obscure mechanistic studies of such systems is the formation of methyl chloride by a parallel gas phase reaction, and conditions can be found which give more highly halogenated species. A wide range of related investigations have been reported.

Imperial Chemical Industries states (5) that the chlorination of paraffins, olefins, benzene, and benzene homologs can be effected by contacting them with a LiCℓ/$PdCℓ_2$ melt with alkali or alkaline earth additives to keep the melting point below 275°C. Here the chlorine for the chlorination originates from the $PdCℓ_2$ which is reduced to metallic palladium, the $PdCℓ_2$ being regenerated with fresh HCℓ. The chlorination of butadiene has also been described in a Japanese process (6) using copper and potassium chlorides. A kinetic study has been made by Altshuler *et al.* (7) using copper chlorides, and Sumitomo (8) reports the formation of carbon tetrachloride and tetrachloroethylene.

Most industrial interest, however, has been attached to the use of melts containing $CuCℓ_2$ for the production of vinyl chloride and related compounds from ethane and ethylene. British Petroleum (9)(10) claims the formation of vinyl chloride from ethylene directly in KCℓ/$CuCℓ_2$/CuCℓ melts.

The formation of 1,2-dichloroethane from ethane and ethylene is described in patents issued to National Distillers (11) and I.C.I. (12), respectively. Englin *et al.* (13) report the formation of vinyl chloride from chloroalkanes using catalysts containing melts of cuprous and cupric chloride. The details of the mechanism and kinetics of many of these reactions are still unresolved. It appears, though, that copper chloride can function effectively as a catalyst for chlorination and dehydrochlorination as well as being able to participate in oxychlorination reactions.

The Lummus Corporation has recently summarized the numerous patents on their Transcat process for vinyl chloride production (14) which again uses CuCℓ/$CuCℓ_2$/KCℓ melts. The organic feed is ethane with or without ethylene, and the inorganic feed is chlorine and/or HCℓ with air and/or oxygen. The process presumably involves several reactions simultaneously. At least three distinct steps are postulated: (1) chlorination of organic feed, (2) catalyst regeneration, and (3) dehydrochlorination and various modifications involving a range of reactor systems have been given.

1.2 The Deacon reaction. The conscious reaction of a molten phase in a catalyst is very apparent in the post war development of the Deacon process. Work on this system has highlighted the subtle interactions that can occur in the promoted catalyst containing $CuCℓ_2$/CuCℓ/KCℓ together with rare earth halides such as lanthanum or neodymium trichloride. Deacon's original catalyst of copper chlorides supported on pumice was not particularly active and most of its disadvantages were due to low activity requiring a high temperature of operation (450°C to 500°C). This resulted though in (1) low equilibrium conversion, (2) rapid deterioration of the catalyst due to volatilisation of copper chloride and (3) severe corrosion.

The interest in recovering chlorine from the hydrogen chloride formed in the making of organo-chlorine compounds has led to the development of improved catalysts in recent years. The active components are usually oxychlorides or chlorides of transition metals, particularly copper. The promotional effect of rare earth chlorides has been known for some time (15) and the Shell catalyst used in a fluid bed process contains rare earth and alkali metal chlorides in addition to copper chloride. The (SLP) nature of the catalyst is clearly spelt out in the patents (16). A large surface area of the support may increase the activity but only insofar as the reactant can easily enter the pellet. Excessive catalyst loading can lead to pore blocking and an anticipated maximum activity for a specific loading is noted.

Fontana (4) recognised the importance of melt composition and Ruthven and Kenney (17)(18)(19) carried out kinetic and equilibrium studies on the melt system. The reaction rate is found to be largely independent of $HC\ell$ partial pressure but is second order in $[Cu^+]$ (or some chloride complex of cuprous ion). There is also strong product inhibition. These findings lead to a mechanism of the form

$$4\ CuC\ell_2 \rightleftharpoons 4\ CuC\ell + 2\ C\ell_2 \quad \text{fast equilibrium}$$

$$2\ CuC\ell + O_2 \rightarrow (CuC\ell)_2O_2 \quad \text{slow irreversible}$$

$$(CuC\ell)_2O_2 + 2\ CuC\ell \rightarrow 2(CuC\ell)_2O \quad \text{fast irreversible}$$

$$2(CuC\ell_2)O + 4\ HC\ell \rightarrow 4\ CuC\ell_2 + 2H_2O \quad \text{fast irreversible}$$

$$4\ HC\ell + O_2 \qquad 2\ C\ell_2 + 2H_2O$$

It was shown (17)(18) that kinetic data may be interpreted by assuming that the equilibrium

$$\underset{(\ell)}{CuC\ell_2} = \underset{(\ell)}{CuC\ell} + \underset{(g)}{\tfrac{1}{2}C\ell_2}$$

is maintained at all times between catalyst melt and gas and that the overall rate of hydrogen chloride oxidation is determined by the rate at which the melt absorbs oxygen. This leads to a rate equation of the form

$$R,\ \text{rate of hydrogen chloride oxidation}\ (\text{gmol/cm}^2\text{sec}) = \frac{4kj^2k^2a^2P_{O_2}}{jK + \sqrt{P_{C\ell_2}}^{\,2}}$$

where k is rate constant for oxygen absorption, a is total mole fraction of copper in the melt, K is the equilibrium constant: $jK = \frac{[Cu^+]}{[Cu^{++}]} \cdot \sqrt{P_{Cl_2}}$, j is the activity coefficient ratio for Cu^{II}/Cu^{I}.

Rate constants calculated from the overall kinetic data agree well with values obtained from a study of the kinetics of oxygen absorption.

This equation can be linearised and correlates the data satisfactorily as shown in Fig.2.

$$\sqrt{\frac{P_{O_2}}{R}} = \frac{1}{2a\sqrt{k}} + \frac{1}{2ajK\sqrt{k}} \cdot P_{Cl_2}$$

The activity of the catalyst is promoted by the addition of lanthanum chloride and it was shown that this effect is due to a strong catalysis of the oxygen absorption rate. The complex effect of temperature on the rate of the overall reaction may also be accounted for satisfactorily in terms of this mechanism. For melts containing equimolar amounts of potassium and copper chlorides it was found that the apparent activation energy is approximately 28 kcal/mol of hydrogen chloride oxidised.

The activity coefficient of cuprous ions is obviously important in determining the kinetics of this reaction and fortunately the thermodynamics of molten chlorides have received considerable attention. From vapour pressure studies it was shown (19) that the binary liquid system $CuCl_2$-$CuCl$ is approximately ideal and that the addition of zinc or lanthanum chloride does not cause pronounced deviations from ideality. The addition of potassium chloride to the system $CuCl_2$-$CuCl$, however, leads to decrease in the entropy change for the reaction given by the above equation and it appears that this effect is due to an increase in the partial molar entropy of cupric chloride in the ternary system. This is a surprising effect since quite large entropy differences (6,0e.u) are involved and this implies a significant change in melt structure.

Sachtler and Helle (20) drew attention to the fact that the widely used alkali promoters should be expected to lower the free energy of chlorine release by

$$2\ CuCl_2 \rightarrow 2\ CuCl + Cl_2$$

but that Ruthven and Kenney's data led to the opposite conclusion and that Cl_2 over such a melt was reduced by KCl addition through a reduction in the reaction entropy.

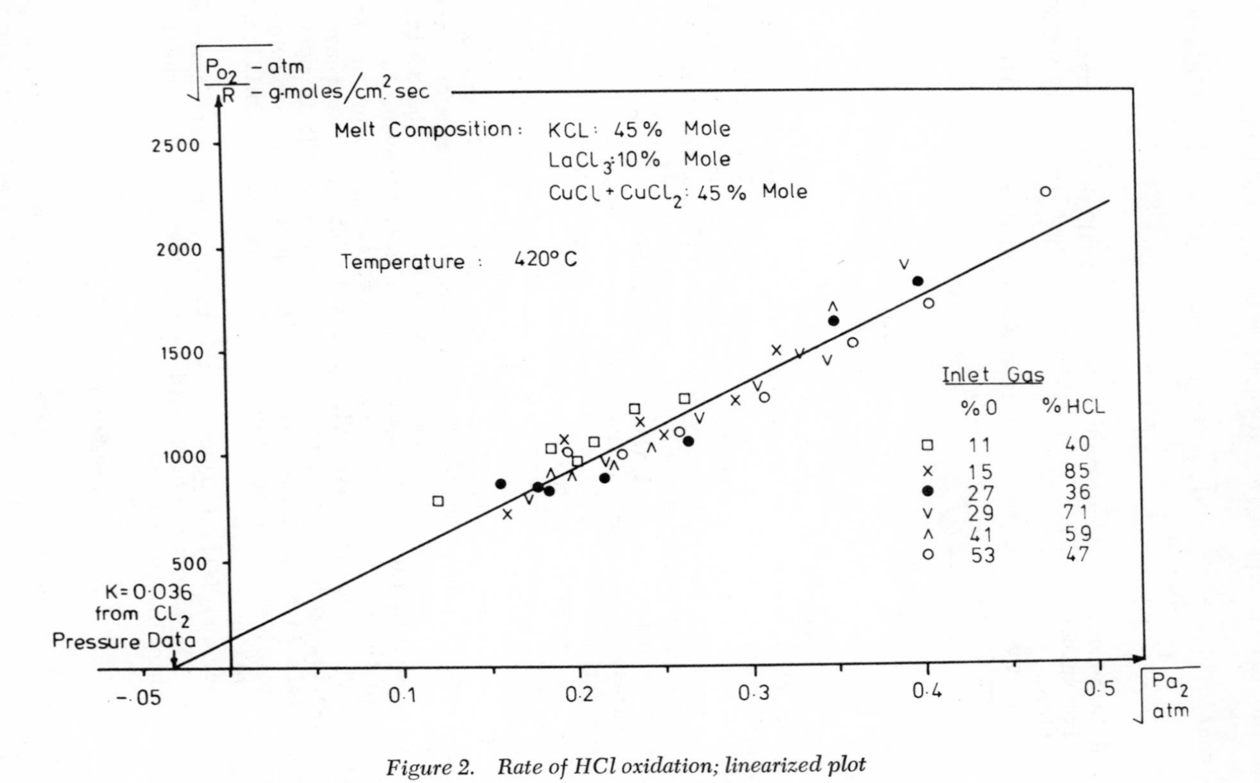

Figure 2. Rate of HCl oxidation; linearized plot

$$\Delta G = \Delta H - T\Delta S = -RT \ln \frac{P_{Cl_2} \cdot (CuCl_2)^2}{(CuCl)}$$

They resolved this paradox by noting that under reaction conditions the catalyst could be a mixture of melt and undissolved solids such as K_2CuCl_3. In $CuCl_2/CuCl$ melts any increase in temperature causes a release of chlorine and a reduction in the $CuCl_2/CuCl$ ratio in the melt. If, however, undissolved solids containing cupric copper have increasing solubility with increased temperature the Cu^{++}/Cu^{+} ratio can increase. Further, when a solid phase containing divalent copper disappears this causes an abrupt change in the slope of the P_{Cl_2}-temperature curve. This is indeed found, with breaks occurring at 329° for K_2CuCl_4 and 362°C for $KCuCl_2$. The change from catalytic promoter to anti-promoter can thus be resolved in terms of classical thermodynamics and depends on the differing solubilities of $CuCl_2$ and its potassium complexes in CuCl containing melts. We shall not dwell on the variety of complexes which can occur in the melt but it is noteworthy that the understanding of this reaction depends on a wide range of studies, phase data, electrochemical and spectroscopic measurements obtained over many years by workers not at all concerned with the Deacon reaction, together with the fact that there is a clear rate determining step, the oxidation of cuprous copper to the cupric species. Few other SLP catalysts have been subjected to such detailed study with such a satisfactory conclusion.

The similarities between catalysts used for oxidation of HCl and oxychlorination of olefins are apparent from the catalyst patents. A very high selectivity (98-99%) for 1,2 dichloroethane is said to be obtained and the catalyst contains $CuCl_2$, alkali metal, rare earth metals and sometimes small amounts of other metals such as silver (21)(22) or gold (23).

A marked difference between oxychlorination of alkenes and the Deacon process is the considerably lower reaction temperature (200-300°C) of the former process. At temperatures ∿230°C where almost no chlorine is produced in the Deacon process activity for oxychlorination of ethylene remsins high. At these lower temperatures a different reaction path (24) takes place.

$$2CuCl_2 + C_2H_4 \rightleftharpoons Cu_2Cl_2 + (CH_2Cl)_2$$

$$CU_2Cl_2 + 2HCl + O_2 \rightarrow 2CuCl_2 + H_2O$$

The overall activation energy is much smaller (15 kcal/mol) than that of the Deacon reaction (28.6 kcal/mol) and free chlorine is not evident at such low reaction temperatures.

Oxychlorination of saturated hydrocarbons (e.g. methane) (4)(25) takes place in a much higher temperature interval 350-450°C than the corresponding process with an olefin. Chlorine reaction occurs by substitution, and the liberation of chlorine appears to be a reaction step since it inhibits the reaction in an analogous way to the Deacon reaction and the activation energies are similar. The rate is proportional to p_{CH_4}, independent of $p_{HC\ell}$ and approximately proportional to p_{O_2}.

A number of other processes involving chlorination via HCℓ on a Deacon type catalyst are listed by Villadsen (2). Phosgene, benzonitrile, chlorinated anilines can all be manufactured on $CuC\ell_2$-KCℓ catalysts which are partly or completely molten at the reaction temperature.

1.3 The oxidation of sulfur dioxide to sulfur trioxide. The catalysis of sulfur dioxide oxidation by V_2O_5 deposited on a porous support is of outstanding industrial importance. Such catalysts are almost invariably vanadium pentoxide promoted with potassium and related oxides which under reactor conditions give a pyrosulfate melt containing vanadium oxides distributed in the pores of an inert material, commonly diatomaceous earth. Recent work on this system is discussed by Villadsen *et al.* (26), but some historical perspective is in order here. In 1940 Frazer and Kirkpatrick (27) and Kiyoura (28) reported that the promoting action of the alkali metals was due to the formation of higher sulfates. These pyrosulfates, e.g. $K_2S_2O_7$, have lower melting points than the corresponding sulfates, can form lower melting eutectics with sulfates and will dissolve appreciable amounts of vanadium oxides.

The catalytic activity of the melt was confirmed conclusively by Topsoe and Nielsen (29) in 1947 when they bubbled sulfur dioxide and oxygen through a column packed with Raschig rings and containing molten potassium pyrosulfate in which 14% V_2O_5 was dissolved. The continued appearance of research papers for many years afterwards, discussing the reaction as if only a solid phase is present, shows that ideas and experimental findings can be as slow to diffuse as matter or heat! Also after three quarters of a century the mechanism of this highly important reaction is still not understood in spite of a formidable amount of industrial know-how on sulfuric acid converters, and a highly efficient process giving over 99% conversion. Interest in the mechanism has been heightened recently because of increasing constraints on the SO_2 allowed in stack gas. Pollution legislation requires this to be less than about 500 ppm, and under typical operating conditions thermodynamic equilibrium requires temperatures in the final bed of below 390°C. Unfortunately, vanadium bed catalysts rapidly lose activity below 420°C, hence the need for an improved catalyst. Industrial vanadium catalysts are activated by passing an SO_2/O_2 mixture through the bed for many hours at 450-500°C. During this activation process the catalyst absorbs large amounts of sulfur oxides - more than 1 mole/mole K_2SO_4 - and V^5 is partly reduced to V^4. The activated catalyst is a mixture of K_2SO_4, SO_3, and largely unknown V^4 and V^5 species, complexed with potassium and sulphate.

Tandy (30) examined systems of alkali metal sulfates in equilibrium with SO_2-SO_3-air mixtures. His experiments covered a temperature range of 380°C to 600°C with V_2O_5 and metal sulfates (Ba, K, Rb and Cs). He stated that in the range between 440 and 600°C a liquid is produced that is a vanadium compound dissolved in alkali pyrosulfate-sulfate melt, with the melting point of the mixture increasing with increasing atomic weight of the alkali metal. Through weight increase measurements, Tandy found, with a metal/vanadia mole ratio of 2.5, that the normal pyrosulfate, $M_2S_2O_7$, and probably vanadyl sulfate, $VOSO_4$, are formed (M = alkali metal). With Rb_2SO_4 and CS_2SO_4 there was evidence of partial formation of higher sulfates, $M_2S_3O_{10}$. The extent of reduction of vanadium pentoxide was less with the alkali metal sulfates of higher atomic weight. Thus the ability to stabliize vanadium in the pentavalent state appears to be greatest with rubidium, and decreases according to the order Rb>Cs>K>Na. Tandy believed that this accounted for the selection of potassium salts as promoters for commercial catalysts.

Pure $K_2S_2O_7$ melts at about 420°C and addition of only 5% V_2O_5 lowers the melting point to almost 300°C (31)(32), but the complexity of the resultant melt is reflected in its readiness to form glasses. All V_2O_5-$K_2S_2O_7$ mixtures with K/V>2 are molten at 380°C, which may provide some explanation of the promotional effect of K and of the activation process by which sulfur oxides are absorbed into the catalyst. The accompanying reduction of V^5 to V^4 is unavoidable and appears to be a key factor in the loss of activity. The degree of reduction is strongly dependent on the activation temperature. At 380°C it is almost complete, while it is less than 0.15 at T = 500°C (33)(34)(35).

Coates (36) and Penfold (37)(38) in quite different studies examined the corrosion of boiler tubes in power plants, which is ascribed to formation of H_2SO_4 by catalysis in a vanadium containing Na_2SO_4, K_2SO_4 deposit in the tubes. They found that sulfur uptake increases with V_2O_5 content in the deposit until a maximum is reached at K/V (or Na/V) = 3-5. After the maximum there is a sharp decrease in sulfur uptake. The degree of reduction follows the sulfur uptake - at 470°C and with a 9/1 Na_2SO_4, 70% of the vanadium is reduced while only 30% is reduced when Na/V = 1.

It is clear the system is much more complex than that in the Deacon process, partly because the 'solvent' ($K_2S_2O_7$) is itself thermally unstable and attempts at a kinetic analysis are centred, at present, around a three step mechanism.

$$(V^5) + SO_2 \rightarrow (V^5 - SO_2) \quad \ldots\ldots\ldots (1)$$

$$(V^5 - SO_2) \rightarrow (V^4) + SO_3 \quad \ldots\ldots\ldots (2)$$

$$(V^4) + \tfrac{1}{2}O_2 \rightarrow (V^5) \quad \ldots\ldots\ldots (3)$$

There is a general agreement that (1) is fast and SO_2 is fairly soluble in the melt. If (1) and (2) are combined and (3) is rate determining the widely employed Mars-Maessen rate expression presented in 1964 results (39). Glueck and Kenney (40) in 1966 suggested the decomposition (2) might be rate determining. Boreskov (41) in 1971 argued that oxygen absorption and reaction might be rate controlling. Holroyd and Kenney in 1971 (42) studied the rate of oxygen and SO_2 absorption into such melts and concluded the active film thickness was a few hundred Angstroms. Supporting work published in the previous year by Boreskov (56) examined the way in which the rate varied in thin films (i.e. non-pore clogging) on supported catalysts. Within the last five years Villadsen and co-workers have embarked on an extremely thorough study of this system and a series of definitive papers are appearing, the latest of which will be presented at this meeting (26). An intriguing and challenging facet of this reaction system is that it provides the best documented example of the 'inert gas effects', documented by Hudgins and Silveston (43) for which there is no unequivocal explanation as to whether these begin physical and/or chemical interactions with the melt ('salting out?') or heat and mass transfer interactions in the gas phase.

1.4 Organic compounds. Rony (44) has mentioned the attempted oxidation of ethylene and propylene on supported melts of $CuC\ell$-$KC\ell$ containing $PdC\ell_2$ but few details are given. Also, Lummus (45) have claimed that acetaldehyde can be produced by contacting hydrocarbons with molten metal oxychlorides. In 1959 I.C.I. (46) obtained a patent for the vapor phase oxidation of aromatic hydrocarbons using fused salts. More recently, BASF patented (47) the formation of phthalic anhydride from naphthalene and p-xylene. The salt system employed was $K_2S_2O_7$-V_2O_5, and Satterfield and Loftus (48) carried out a kinetic study that showed the major product was o-methyl benzaldehyde. It should be noted that organic oxidations with this melt are likely to display rather complex kinetics since $K_2S_2O_7$ has an appreciable dissociation pressure of SO_3 above 400°C. Since this is a powerful vapor phase oxidizing agent there are difficulties in separating the contribution of homogeneous gas phase oxidation by SO_3.

Butt and Kenney (49) studied the oxidation of naphthalene over an unsupported molten salt, a eutectic of V_2O_5-K_2SO_4 (mp 432°C). A particularly interesting feature of the system is that the reactions and their kinetics can be studied at temperatures below and above the catalyst melting point. The main products of this oxidation are phthalic anhydride, 1,4-napthaquinone, carbon dioxide, and water. The most likely site of reaction for the melt would appear to be the surface, and the kinetics of the overall reaction for both phases of the catalyst are interpreted with particular reference to a simple two-step mechanism involving (1) the removal of active oxygen from the catalyst by reaction with naphthalene and (2) the replenishment of the active oxygen from the gas phase. There are indications that the latter step is rate limiting. The data from the overall kinetic studies with the solid catalyst in the temperature range 300 to 340°C and with the melt over the temperature interval 440 to 470°C were fitted to and are consistent with the following rate equation:

$$\text{rate of reaction} = \frac{k_1C_Nk_2C_o}{k_2C_o + Bk_1C_N}$$

The activation energy for the reaction over the solid catalyst decreases with an increase in temperature in the range 340 to 380°C, and there is a dramatic fall in the activity and selectivity of the solid catalyst as the temperature is raised from 380 to 400°C. It seems probable that the latter behaviour is a consequence of a structural change in the catalyst prior to melting.

The data obtained from separate studies of the reduction and re-oxidation of the catalyst show that naphthalene can react with catalyst oxygen which can be replenished from the gas phase. However, the results do not prove that naphthalene reacts exclusively with catalyst oxygen in the overall oxidation reaction, and an alternative explanation is of a surface reaction obeying a Langmuir-Hinshelwood mechanism. In spite of the apparent desirability of operating below 400°C in many industrial processes the hot spot temperature in a fixed bed can be well above the melting point of the catalyst. The widely employed addition of a few ppm of SO_2 raises many unanswered questions.

2. Medium and low temperature reactions (<300°C)

Another established SLP catalyst is the widespread use of phosphoric acid, typically supported on kieselguhr. This is widely employed in a range of hydrocarbon rearrangements such as propylene dimerisation and the alkylation of both aliphatic and aromatic compounds. The formation of ethylbenzene, a precursor of styrene, and cumeme prepared from benzene and propylene are examples of major industrial importance. Reactions of this type appear to occur by classic carbonium ion mechanisms and have attracted relatively little academic study but the patent literature is vast and has been reviewed (50). The corrosive Lewis acids such as aluminium trichloride fall into this category and references are available (1)(2). Zinc chloride too probably functions in an analogous manner when used for the elimination of hydrogen chloride from alkyl halides. Kenney and Takahashi (51) showed that the rate of the reaction with unsupported binary melts of zinc chloride and other metal halides correlates well with the polarising strength of the second metal.

The intensive research on transition metal complexes in recent years has led to experiments in ways of improved contacting of these often expensive compounds. Acres, Bond *et al.*(52) described the use of rhodium carbonylation reactions. Related and extended studies of the activity of rhodium complexes have recently been given by Rony and Roth (53). These clearly demonstrate that a wide range of complexes can be effectively used as (SLP) systems. Obviously compounds such as $[(C_2H_5)_4N][SnC\ell_2]$ and the corresponding chlorogermanates described by Parshall (54) with melting points of about 150°C and below could be used in a similar way. One of the few comparative studies of an SLP catalyst has been made by Komiyama and Inoue in an investigation of the Wacker oxidation of ethylene to acetaldehyde (55).

3. Diffusion and catalytic reaction in SLP

Any discussion of the catalysis of a gas reaction by an (SLP) must consider the site of catalytic activity since this may be far from obvious. If strong specific interactions occur, appreciable gas solubilities may be found and the catalysis can be regarded as homogeneous. For many liquids, however, gas solubilities are extremely small, solute concentrations are high, gas-liquid diffusion coefficients could be low, and the reaction can, in principle, take place on the surface of the liquid.

Three broad regimes are thus possible: (1) reaction takes place on the surface of the liquid, (2) reaction takes place homogeneously in the bulk of the liquid catalyst without diffusional restrictions, (3) reaction takes place homogeneously in the liquid catalyst but is limited to a layer near the liquid surface because of diffusional effects.

Reaction between a gas and a liquid normally involves absorption and physical solution of the gas followed by homogeneous reaction between the dissolved species. The problem of gas absorption with chemical reaction has been extensively studied and in such systems the observed rate of gas absorption will be a function of the chemical reaction rate, the diffusion of the dissolved gas, and possibly fluid dynamics of the system (the rate of surface renewal) if surface tension driven or other circulation effects occur. There is no evidence of these so far in the thin films employed in practical catalysts.

In regimes (1) or (2), if the concentration of catalyst is constant the formal rate equations are the same as that used by Mars and van Krevelen (61) to model partial oxidation reactions.

Here we have

$$A(\text{gas}) \text{ and } B(\text{liquid catalyst}) \xrightarrow{k_1} C(\text{gaseous product}) + D(\text{intermediate}) \quad \ldots\ldots (a)$$

$$D(\text{intermediate}) + E(\text{gas}) \xrightarrow{k_2} B(\text{liquid catalyst}) \quad \ldots\ldots (b)$$

B and D are two forms of the catalyst which in a redox process would be different valency states of a transition metal ion, or alternatively they could represent unstable complexes of the dissolved catalyst with the reactants A and E.

If the reaction is fast enough, relative to the transport processes, then liquid phase diffusional transport is important: this possibility is highly relevant in SLP catalyst behaviour because one of the main incentives is to disperse the liquid and obtain a high liquid-gas surface area.

When the reaction is rapid, most of the reaction takes place in a thin film near the interface, and the bulk concentration of the physically dissolved gas remains low. In that case the rate of gas absorption and reaction will then be proportional to the interfacial surface area and given by an expression of the type provided all other parameters are held constant

$$\text{Rate of reaction} = AC^*(Dk_o)^{1/2}$$

where A = interfacial surface area, cm^2, C^* = saturated solubility of the physically dissolved gas (g-mole/cm^3)atm, D = diffusion coefficient of the dissolved gas, cm^2/sec, k_o = pseudo-first-order rate constant, sec^{-1}.

Estimates for the depth of penetration of the dissolved gas can be obtained from

$$\lambda = \frac{D^{1/2}}{k_o}$$

where λ = penetration distance, cm, D = diffusion coefficient of the dissolved gas, cm^2/sec, and k_o = pseudo-first-order rate constant.

These equations are applicable for a reaction proceeding under pseudo-first-order conditions, i.e. when the concentration of the solute species is constant right up to the gas/liquid interface. It is thus possible to examine the possibility that reaction may occur in a film for the catalyst reoxidation and reduction reactions separately, if the two-stage redox mechanism is appropriate. The penetration theory leads to a series of coupled nonlinear partial differential equations which have to be solved numerically with appropriate boundary conditions. For example, if y is the distance from the melt surface, the equation governing the concentration of species B in time and space is

$$\frac{\partial C_B}{\partial t} = \frac{D\partial^2 C_B}{\partial y^2} - k_1 C_A C_B + k_2(C_{B_o} - C_B)C_E \ .$$

This problem becomes quite involved since there is a similar equation for each diffusing/reacting species, all of which have to be solved simultaneously. Shah and Kenney (56) have considered the solution of these equations. The steady state concentrations of all species are obtainable once the liquid distribution in the particle is specified and can be derived either by solving equations of the above form with the time dependent terms set to zero, or, often just as convenient, computationally determining concentration distributions as $t \to \infty$. Villadsen and Livbjerg (2) have recently adopted the former approach and derived only slightly more complex analogues for diffusion-controlled reactions for limiting regimes such as that in which one gaseous reactant penetrates deeply into the melt but the other virtually reacts on the melt surface, as well as presenting plots of the familiar Thiele modulus - effectiveness factor type for intermediate regimes. Particular interest again attaches to the SO_2 oxidation catalyst and the thickness of the active film. Using the oxygen absorption kinetics in an unsupported melt, Holroyd and Kenney obtained a value of 900°A (42). Boreskov *et al.* (60) approached this problem in a different way by depositing known weights of melt on an inert granular support and determining the reaction rate as a function of the thickness of the melt film. They suggested that the rate-controlling step was diffusion of oxygen into the melt which penetrated 100 to 200 Å.

The calculations of Villadsen do however show that measurements with thick layers of unsupported melts could give erroneous predictions when transposed to the much thinner SLP's, but the inevitable uncertainties in liquid film thickness and distribution in an actual SLP are likely in any experiments to give errors which are comparable in size. Further consequences of concentration changes in dissolved species arising from the partial precipitation of V^{4+} species are discussed in the paper of Villadsen *et al.* (26). The detailed understanding of such processes requires more data on gas solubilities and gas diffusion coefficients. Comtat and Mahenc (57) show how the latter can be obtained in pyrosulfate melts using electrochemical methods.

4. Liquid distribution within the porous support

The behaviour of an (SLP) will depend not only on the amount of liquid dispersed within the porous solid but also on whether the liquid is dispersed as a thin film, a plug or some more complex and realistic distribution. Ideally all pores will be coated with a uniform thin film of liquid. As the liquid film thickness is increased, assuming gas and liquid phase transport resistances remain negligible, the rate/unit volume of catalyst will rise. However, at some stage of liquid loading, flooding will occur, clumps or clusters of liquid can form, the area of liquid exposed to the gas will fall, liquid film thickness is then appreciably greater and liquid diffusion effects could become important. The reaction rate will thus be less than that obtained with a non-transport limited liquid filled catalyst with the same liquid loading or alternatively the rate can actually go through a maximum. These effects were first recognised by Rony (58) in a simple cylindrical pore model and his and other findings have been collated by Villadsen and Livbjerg (2) which confirm these qualitative predictions very clearly (Fig.3). These show the actual rate of reaction per unit volume of liquid catalyst relative to the expected rate if all the catalyst were uniformly accessible to the reactants as a function of α, the fraction of pore volume filled with liquid. The systems are (a) Hydrogenation of propylene with a solution of tris (triphenyl-phosphine) rhodium chloride/silicagel at 24°C and 20 psig, (b) Isomerization of but-1-ene with solution of rhodium chloride/silicagel at 25°C, (c) Hydroformylation of propylene with solution of bis (triphenylphosphine) rhodium carbonyl chloride/silicagel 136°C and 490 psig, (d) SO_2 oxidation on molten V_2O_5-$K_2S_2O_7$/porous SiO_2 glass, pore diameter 586Å, 480°C, (e) Conversion of ethylene and water to acetaldehyde with a hydrochloric acid solution of $PdCl_2$ and $CuCl_2$/alumina at 70°C, (f) as (d) with pore diameter 3060 Å.

A more precise picture of an (SLP) distribution in a porous network raises serious theoretical problems because of the difficulty in defining analytically the pore structure. Thermodynamically the total surface free energy is compounded from contributions from three interfaces

$$G^S = A_{C\ell_2}\gamma_{SL} + A(1-a_L)\gamma_{SG} + A_{LG}\gamma_{LG}$$

where A is the total support surface area, A_{LG} the gas/liquid area, a_L is the fraction of A covered with bulk liquid and γ the surface tension. Limiting behaviour arises if $\gamma_{SG} - \gamma_{SL} < 0$ implying a contact angle >90° and giving non-wetting behaviour. Alternatively $\gamma_{SG} - \gamma_{SL} < 0$ when spontaneous wetting can occur. A wealth of more complex distributions can arise in practice depending on the pore structure. For example, liquid can accumulate at the contact points of non-porous pellets, in the irregularities that form in pore walls or in the formation of islands. These latter clusters were reported by Villadsen and co-workers at an earlier meeting (62) and are more fully discussed in their recent review (2) with special reference to the V_2O_5-K_2O sulphuric acid catalyst. Again, film thicknesses

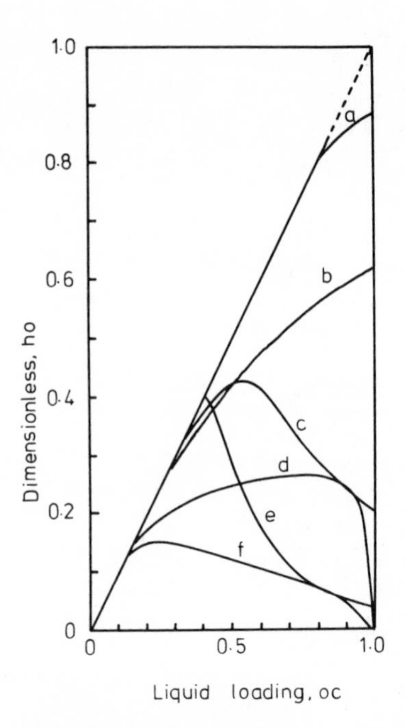

Figure 3. Effectiveness factor, h_o as function of dimensionless liquid loading α_o

of about 1000Å form the theoretically and technically-interesting range. Another model of (SLP) distributions has been given by Abed and Rinker (59).

Theoretical problems abound in this area because for a given pore structure there are a large number, if not an infinity, of thermodynamically stable liquid distrbutions. Hysteresis effects are well known in capillary phenomena. The dynamic behaviour of creeping liquids is a relevant problem too, however inconvenient and intractable. Industrial reports and folk-lore speak of the decrease of the active area of sulphuric acid catalyst during operation, or fused lumps of catalyst pellets being found in phthalic anhydride converters. Villadsen (2) quotes an example of the 'suction' of pyrosulfate-V_2O_5 melt from an impregnated silica support of 3000Å pore diameter into a 160Å pore support. Kenney and Alexander (unpublished) found molten carbonates can climb out of ceramic crucibles, analogous problems being well known to those developing porous electrodes for high temperature fuel cells. Such dynamic effects arising from physical and chemical interactions between an SLP and the support may be responsible for some of the arguments as to whether the support itself plays any role in the catalytic reaction. Anyone embarking on research in this area requires some familiarity with fields as diverse as liquid distribution in gas chromatography supports and the dynamics of moisture movement in clays and soils.

References

1. Kenney, C.N., Cat. Rev. Sci. Eng. 11, 197 (1975)
2. Villadsen, J. and Livbjerg, H. Cat. Rev. Sci. Eng. 1978 (in Press).
3. Gorin, E., Fontana, C.M. and Kidder, G.M., Ind. Eng. Chem., 40, 11 (1948)
4. Fontana, C.M., Ind. Eng. Chem., 44, 363 (1952)
5. Belgian Patent 705, 244 (1968)
6. Japanese Patent 4850/68 (1968)
7. Altshuler, S.V., Flid, R.M., Englin, A.L., Pet. Chem. USSR, 11, 59 (1971)
8. German Patent 2, 246, 904 (1973)
9. British Patent 1, 213, 402 (1970)
10. British Patent 1, 284, 267 (1972)
11. U.S. Patent 3, 720, 723 (1973)
12. German Patent 1, 468, 085 (1972)
13. Englin, A.L. et al., 281, 433 (1970)
14. Riegel, H., Schindler, H., and Sze, M., A.I. Che.E. Meeting New Orleans, March 1973
15. Balcar, F.R., Air Reduction Co., U.S. Patent 2, 271, 056 (1942)
16. Engel, W.F., and Wattimena, F., U.S. Patent 3, 260, 678 (1966)

17. Ruthven, D.M., and Kenney, C.N., Chem. Eng. Sci., 22, 1561 (1967)
18. Ruthven, D.M., and Kenney, C.N., Chem. Eng. Sci., 23, 981 (1968)
19. Ruthven, D.M., and Kenney, C.N., J. Inorg. Nucl. Chem., 30, 931 (1968)
20. Sachtler, W.M.H. and Helle, J.W., Chemisorption and Catalysis, Inst. of Petroleum (London) 1970 p.31
21. Wolf, F., et al. Chemische Technologie 25, 37, 89, 156 (1973)
22. Wolf, F., et al. DDR Patent 90, 127 (1972)
23. Komihami, N., German Patent 1, 618, 030 (1971)
24. Todo, N., Kurita, M. and Hagiwara, H., Kogyo Kagaku Zashi 69, 1463 (1966)
25. Agulin, A.C., Bakshi, M.Y. and Gelbshtein, A.I., Kinetika i Kataliz 17, 581 (1976)
26. Grydgaard, P., Jensen-Holm, H., Livjberg, H., and Villadsen, J., A.C.S. Symp. series (65) 1978, 597
27. Frazer, J.H., and Kirkpatrick, W.J., J. Amer. Chem. Soc., 62, 1659 (1940)
28. Kiyoura, R., Kagaku (Tokyo), 10, 126 (1970)
29. Topsoe, H.F.A., and Nielsen, A., Trans. Dan. Acad. tech. Sci. 1, 18 (1947)
30. Tandy, G.H., J. Appl. Chem. 6, 68 (1956)
31. Hähle, S., and Meisel, A., Z. anorg. allg. Chemie 375, 24 (1970)
32. Holroyd, F.P.B., Ph.D. thesis (Cambridge) (1968)
33. Boreskov, G.K., Davydova, L.P., Mastikhin, V.M., and Polyakova, G.M., Dokl. Akad. Nauk. SSSR 171, 760 (1966)
34. Mastikhin, V.M., Polyakova, G.M., Zyulkovskii, V., and Boreskov, G.K., Kinetika i Kataliz 11, 1219 (1970)
35. Boreskov, G.K., Polyakova, G.M., Ivanov, A.A. and Mastikhin, V.M., Dokl. Akad. Nauk. SSSR 210, 423 (1973)
36. Penfold, D., Journ. Inst. Fuel 43, 151 (1970)
37. Coats, A.W., Journ. Inst. Fuel, 42, 75 (1969)
38. Coats, A.W., Dear, D.J.A., and Penfold, D., Journ. Inst. Fuel 41, 3 (1968)
39. Mars, P., and Maessen, J.G.H., Proc. 3. int. congr. catal. Amsterdam 1964 paper I, 7 p. 266
40. Glueck, A.R. and Kenney, C.N., Chem. Eng. Sci. 23, 1257 (1968)
41. Polyakova, G.M., Boreskov, G.K., Ivanov, A.A., Davydova, L. and Marochkina, G.A., Kinetika i Kataliz 12, 586 (1971)
42. Holroyd, F.P.B., and Kenney, C.N., Chem. Eng. Sci. 26, 1963 and 1971 (1971)
43. Hudgins, R.R., and Silveston, P.L., Cat. Rev. Sci. Eng., 11, 167 (1975)
44. Rony, P.R., Ann. N.Y. Acad. Sci., 172, 238 (1970)
45. German Patent 2, 247, 343 (1973)
46. U.S. Patent 3, 012, 043 (1958)
47. British Patent 1, 082, 326 (1967)
48. C.N. Satterfield and J. Loftus, AIChEJ, 11, 1103 (1963)

49. Butt, P.V., and Kenney, C.N., Proceedings 5th International Congress on Catalysis London 1976
50. McMahon, J.F., Bednars, C., and Solomon, E., Adv. Petr. Chem. Ref. 7, 285 (1963)
51. Kenney, C.N. and Takahashi, R., J. Catal, 22, 16 (1971)
52. Acres, G.J.K., Bond, G.C., Cooper, B.J., and Dawson, J.A., J. Catalysis 6, 139 (1966)
53. Rony, P.R.and Roth, J.F., J. Mol. Catal., 1, 13 (1975)
54. Parshall, G.W., J. Am. Chem. Soc., 94, 8716 (1972)
55. Komiyama, H. and Inove, H., J. Chem. Eng. Japan 8, 310 (1975)
56. Shah, Y.T.,and Kenney, C.N., Chem. Eng. Sci., 28, 325 (1973)
57. Coste, J., Comtat, M., and Mahenc, J., Bull. Soc. Chimique de France No. 3 p. 767 (1971)
58. Rony, P.R., J. Catalysis, 14, 142 (1969)
59. Abed, R., and Rinker, J. Catal. 31, 119 (1973)
60. Boreskov, G.K., Dzisko, V.A., Tarasova, D.V., and Belaganskaya, G.P., Kinetika i Kataliz 11, 144 (1970)
61. Mars, G.P. and van Krevelen, D.M., Chem. Eng. Sci. (special supplement) 3, 41 (1954)
62. Livjberg, H., Sorensen, B., and Villadsen, J., Adv. Chem. Ser. 133 (1974)

RECEIVED March 30, 1978

3

Coal Conversion Reaction Engineering

C. Y. WEN and S. TONE

Department of Chemical Engineering, West Virginia University, Morgantown, WV 26506

Coal is clearly our most abundant fossil resource and will play a key role in supplying energy and chemicals well into the next century.

Coal combustion, gasification and liquefaction processes are presently in various stages of development, ranging from those that are commercially available or are now being tested at pilot plant scale to those that are formulated conceptually and are yet to be tested. Coal gasification plants based on the first generation processes, which are mostly German processes or improved versions of them, are now being built or planned for construction. A number of second generation processes are being tested in large pilot plants or are being prepared for building demonstration plants. In addition, there are several recent discoveries now in research stage that show exciting promise. Thus, we are encouraged by the advance in the technical status of coal conversion processes because present and planned activities promise to place technical operability of these processes within reach.

In recent years considerable advances have also been made in our understanding of the physical and chemical properties of coal and of the coal conversion reactions. These advances are due in large part to an intensive research and development effort aimed at improving coal conversion technologies to meet societal, economic and environmental requirements.

Relatively few attempts have been made, however, to systematically organize the subject and critically evaluate the vast amount of information available in literature based mainly on the chemical reaction engineering point of view. A major difficulty in accomplishing this task is due to the complexity and heterogeneity of coal's structure and its behavior under different environments, which precludes any attempt to draw generalizations. Additional difficulty stems from lack of understanding the intrusion of complex physical phenomena, such as hydrodynamics and mass and heat transfers, on chemical rate processes in coal

0-8412-0432-2/78/47-072-056$13.10/0

conversion reactors, particularly at high temperatures and high pressures.

It is the goal of this paper to attempt a systematic organization of material in order that mechanistic as well as phenomenological models useful for design and scale-up of coal conversion reactors can be developed. The approach taken here is to examine the current status of coal conversion reactions on a single particle and to select a realistic yet sufficiently simple model capable of describing the phenomena. The deficiencies and limitations of each model are presented carefully.

In the selection of a proper rate expression of a single particle reaction, we often prefer a simpler form that represents the experimental data over a more complicated form, even at the expense of somewhat reduced accuracy. Since all rate expressions are empirical or semi-theoretical at best, it seems unnecessary to add complications by applying analysis that originated with the Langmuir adsorption isotherm or other type of isotherms (Freundlich, Templins, etc.). Such complication becomes more apparent when the rate expression must be incorporated with varying heat and diffusion effects of the system. In fact, when the rate expression contains more than four parameters that must be determined experimentally for different operating conditions and for different types of coal, it becomes not only impractical to use such an expression but also extremely difficult to apply it in reactor design and simulation. This is quite obvious in view of the fact that in addition to the hydrodynamics, heat transfer and diffusion effects, a number of simultaneous reactions occurring in the reactor must be taken into consideration.

The single coal particle models selected are then combined with reactor flow models and heat and mass transfer characteristics of a multiparticle system. These analyses are applied for reactor design stressing the current state of knowledge and uncertainties in the supporting data. Here rather than attempting to cover the numerous individual coal conversion reactors developed or being developed, they are classified according to their unique fluid-solids contacting modes (i.e., moving bed, fluidized bed, entrained bed, slurry bed reactors, etc.) in order to emphasize the similarity as well as dissimilarity of the coal conversion reactors. Fig. 1 presents an overall flow diagram of coal conversion reaction engineering, which illustrates interrelations and sequences of the subject matters desirable in organizing this field into a systematic and coherent branch of chemical reaction engineering.

COAL CONVERSION MODEL FOR SINGLE PARTICLES

In developing a coal conversion reaction model for a single particle system, it is very important to recognize the complexity and heterogeneity of the structure of coal. Coal is a complex,

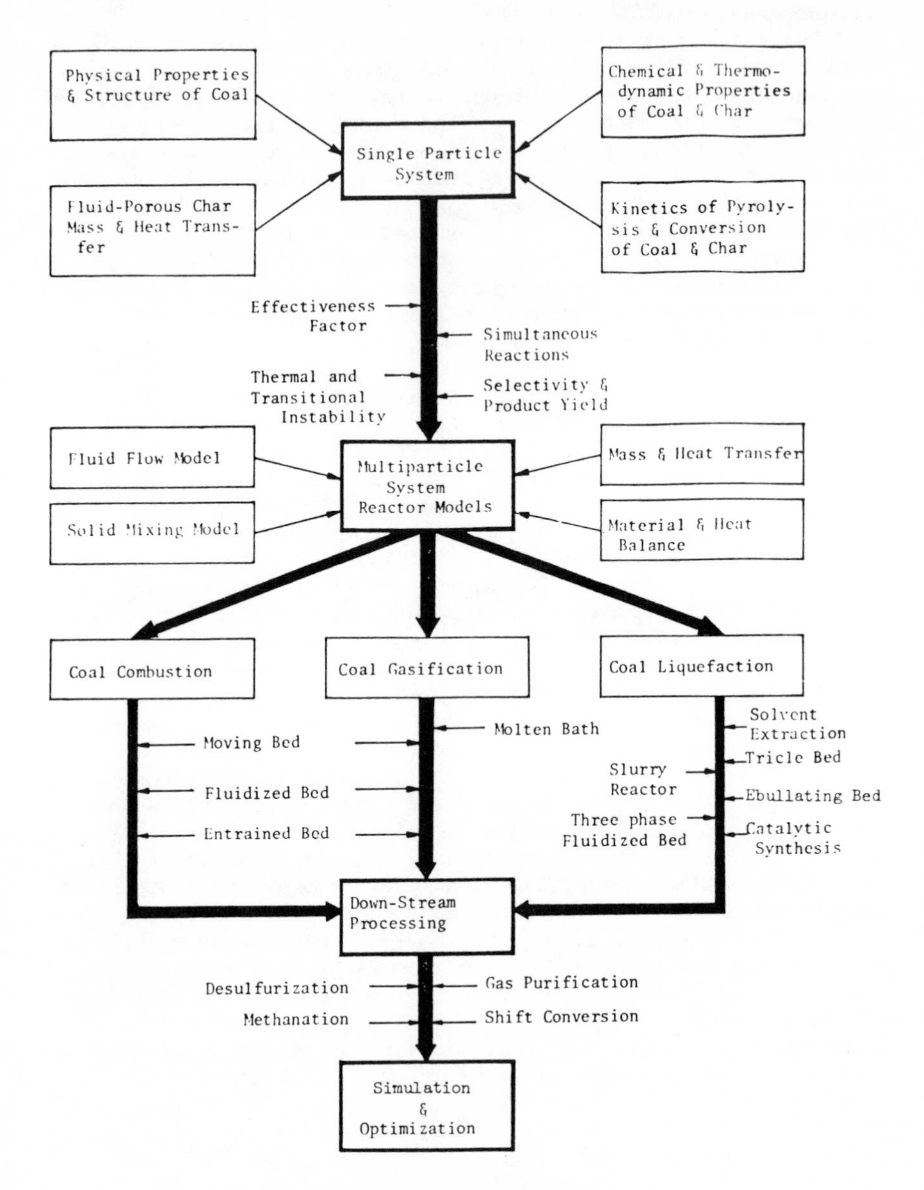

Figure 1. Coal conversion reaction engineering

nonuniform solid consisting of the metamorphosed remains of ancient vegetation. Buried and pressed by sediments with loss of water and volatile matter, the earliest stage of coal lignite was formed. As the lignite was buried deeper and compressed further, the heat associated with compression increased and accelerated devolatilization. As a result, the rank of coal became progressively higher, rising from lignite, sub-bituminous, bituminous, semi-bituminous and semi-anthracite to anthracite. The heating values of coals as a function of carbon contents are shown in Fig. 2.

Lignite and sub-bituminous coals are non-agglomerating; have higher oxygen, alkali minerals, and moisture contents; and are in general more reactive than bituminous coals, which are caking coals. Anthracite coals contain less oxygen, less moisture and less volatile matter and are much less reactive than other coals.

Mineral matter and sulfur contents in coal depend on the seam from which coal is mined. Chemically, coals contain C, H, O, N, S, and minerals in varying portions. The approximate range of the H/C atomic ratio and O/C atomic ratio for various rank coals which seems to affect the reactivity is shown in Fig. 3.

(1) PYROLYSIS AND HYDROPYROLYSIS OF COAL

The pyrolysis takes place for all coal conversion reactions when coal is heated above the "pyrolysis temperature." The behavior of coal during pyrolysis is governed by coal type and experimental conditions such as particle size, heating rate, reaction temperature and pressure, and species of gas (inert, H_2, O_2, etc.) in which it is pyrolyzed.

During pyrolysis the bituminous coal softens to form a metaplast. Before the center reaches softening temperature, the devolatilization starts and particles swell to become cenospheres and, with further thermosetting, to produce coke or char. The volatiles tend to come off in concentrated and randomly distributed jets at different points on the particle surface as shown in Fig. 4. Pyrolysis produces a range of products from hydrogen gas to heavy tar of widely varying molecular weights. The products of coal pyrolysis depend mainly on temperature, heating rate, and vapor phase residence time. High temperatures and a long vapor phase residence time tend to favor production of gases. The process of rapid pyrolysis with heating rate substantially greater than 500°C/sec has a potential of developing to one of the most effective ways of utilizing hydrocarbons contained in coal. To achieve a rapid rate of pyrolysis, pulverized coal burners, fluidized bed, free-fall, entrained bed and cyclone bed reactors are often employed. The problem of predicting product distribution from coal pyrolysis is more difficult than predicting the total yield. The products like higher hydrocarbons, liquids and tar are apparently not the result of a single decomposition reaction. Further systematic investigations with a

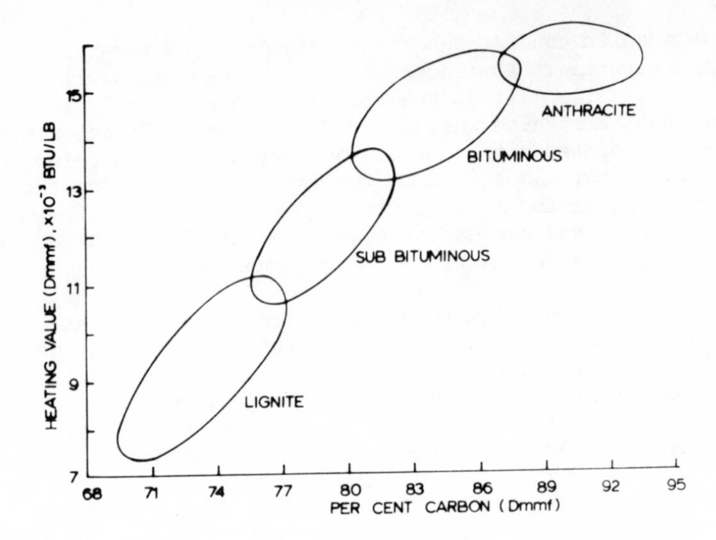

Figure 2. Heating values of various coals as a function of carbon content

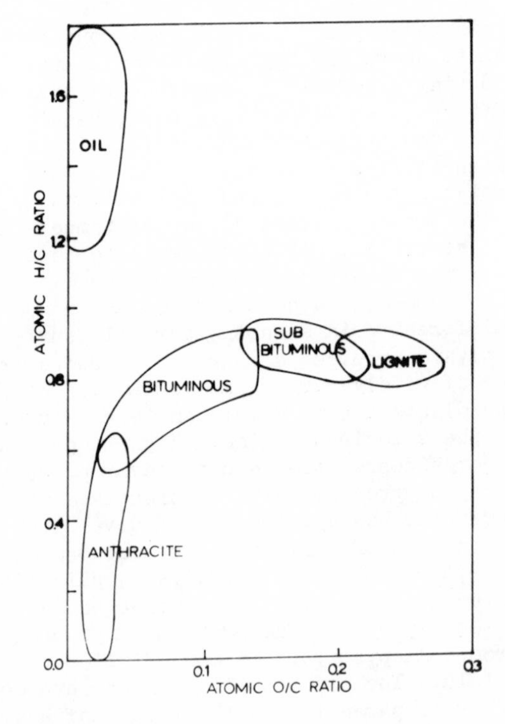

Figure 3. H/C and O/C ratios of fossil fuels

wider variety of coals, heating rates and temperature levels are needed to increase understanding of this subject.

Rates of coal pyrolysis in an inert atmosphere have been investigated by many researchers. Rate of volatile release is apparently dependent on temperature and particle size above 100 microns and probably independent on particle size below 50 microns with a transition between 50 to 100 microns. Table I lists the major rate correlations for coal pyrolysis. Equations due to Badzioch and Hawksley (1), Anthony and Howard (2,3), and Wen et al. (4) are based essentially on the concept that the rate of pyrolysis is proportional to the amount of volatile content remaining in the coal:

$$\frac{dV}{dt} = k\ (V^* - V) \qquad \text{where } k = k_o \exp\ (-E/R_g T)$$

For Badzioch and Hawksley (1) and for Wen et al. (4), the activation energy, E, is a constant. Following the idea of Pitt (8), Anthony and Howard (2) introduced Gaussian distribution of activation energy, E, with a mean value of E_o and standard deviation of σ.

The presence of a large quantity of hydrogen greatly affects the phenomena of pyrolysis, which is often referred to as hydropyrolysis. Therefore, it is logical to discuss the mechanism of pyrolysis in conjunction with the phenomenon of hydropyrolysis.

Hydropyrolysis or hydrocarbonization refers to the process in which pulverized coal particles are mixed with hydrogen at elevated temperature and pressure for a short time. The process appears attractive because it has been demonstrated that it is possible to obtain yields significantly greater than the proximate volatiles' content of coal. The reaction products include distillate oils, benzene, toluene, xylene and light gases such as methane, ethane and oxides of carbon along with a desulfurized combustible char.

Several experimental studies (3,9,10,11,12,13) have qualitatively identified the operating conditions for maximizing the hydrocarbon yields that appear to be sensitive to temperature, total pressure, hydrogen partial pressure, particle size, char and vapor residence time.

Their findings can be summarized qualitatively in the following:

Effect of Pressure

(A) Total Pressure. When coal is pyrolyzed in an inert atmosphere (under low hydrogen or low oxygen partial pressure), the total conversion (or total yield including all products) decreases as the pressure is increased. Typically, bituminous coal pyrolyzed at 1000°C yields 50-55% of the weight of coal at 10^{-4} atm but only 35-40% at 100 atm. Liquid hydrocarbons including tar also decrease from about 32% to 10%, whereas the gaseous

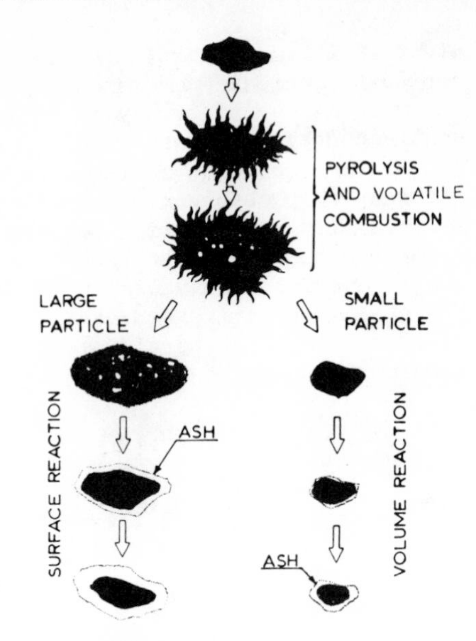

Figure 4. Pyrolysis and combustion of coal

Table I. Correlations for Pyrolysis of Coal/Char

Author and Year	Heating Rate	Temperature Range and Pressure	Total Volatiles Yield or Rate
Gregory and Littlejohn (5) (1965)	Slow, Intermediate	500 - 1100°C Atmospheric (N_2)	$V = VM - R - W$ $R = 10^A$, $A = 11.47 - 3.961 \log_{10} T + 0.05VM$ $W = 0.20\ (VM - 10.9)$
Howard and Essenhigh (6) (1967)	Rapid	200 - 1550°C Atmospheric (air)	$\frac{dV}{dt} = k_o \exp(-E/R_gT)(V_o - V)$
Jüntgen et al. (7) (1968)	Slow, Intermediate	Up to 1000°C Atmospheric (N_2)	$\frac{dv}{dT} = \frac{k_o v_o}{m} \exp\{-\frac{E}{R_gT} - \frac{k_o R_g T^2}{mE} \exp(-\frac{E}{R_gT})\}$ $m = \frac{dT}{dt}$
Badzioch and Hawksley (1) (1970)	Rapid (25000-5000°C/S)	up to 1000°C Atmospheric (N_2)	$\frac{dV}{dt} = k_o \exp(-E/R_gT)(V^* - V)$ $V^* = VM\ (1-C)Q$
Wen et al. (4) (1974)	Slow, Intermediate or Rapid	550-1500°C Atmospheric (N_2)	$\frac{dX}{dt} = k_o \exp(-E/R_gT)(f-X)$
Anthony and Howard (2,3) (1976)	Slow, Intermediate or Rapid	up to 1000°C 0.001 to 100 atm. (He and H_2)	$V = V^* \{1 - \int_0^\infty \exp(-\int_0^t k dt) f(E) dE\}$ $k = k_o \exp(-E/R_gT)$ $V^* = V^*_{nr} + V^{**}_r [1 + K_1/(K_c/P + K_2 p_{H_2}]^{-1} + K_3 p_{H_2}$ $f(E) = [\sigma(2\pi)^{1/2}]^{-1} \exp[-(E-E_o)^2/2\sigma^2]$

hydrocarbons increases from about 4% to 7% when the pressure is increased from 10^{-4} atm to 100 atm.

(B) Hydrogen Pressure. A substantially higher product yield in hydropyrolysis is clearly indicated (3,10). The presence of hydrogen significantly improves the yields of the desirable liquid and gaseous products such as methane and benzene. Increasing hydrogen pressure from 10, 100 to 150 atm typically increases the yield from 50, 60, to 70% of the weight of bituminous coal at 1000°C, respectively.

Effect of Temperature

(A) Inert Atmosphere. Fig. 5 schematically represents the effect of heating rate on relative yields and product distribution. Fig. 5 qualitatively indicates that the faster the heating rate to a given temperature, the greater the total yield up to a limit.

(B) Hydrogen Atmosphere. Hydropyrolysis in hydrogen at 100 atm shows that an increase in heating rate from 20 to 650°C/sec increases the methane yield by a factor of 1.5 and the benzene yield by a factor of more than 3 and decreases significantly (almost 1/10) in the heavy-product yield.

However, when increasing heating rate from 650°C/sec to 1400°C/sec the product yield remains essentially the same. Apparently a heating rate of 650°C/sec or greater is adequate to ensure the fragmentation of coal molecules before repolymerization takes place to form large molecules.

Effect of Solids and Gas Residence Time

(A) Hydrogen Atmosphere. Too short solid residence time does not permit the heavy species devolatilized from coal to be hydrocracked to the lighter product. Methane yield seems to increase 1.5 times when solid contact time is increased from 2 to 30 sec. At a heating rate of 650°C/sec, solid contact times of 10 sec are sufficient for particles smaller than 43 microns.

Similarly, increasing the residence time of vapor leads to increased thermal decomposition of the reaction products yielding more methane. For example, increase of vapor residence time from 0.2 to 23 sec increases the methane yield by a factor of 2.7 but decreases other hydrocarbons significantly.

Following closely the mechanism of hydropyrolysis proposed by Anthony et al. (3), Russel et al. (12) recently proposed an interesting theory describing the combined effect of chemical reactions and mass transfer occurring in a single coal particle during hydropyrolysis. Their work is briefly summarized below. The kinetic model for hydropyrolysis consists of the following steps:

(a) Devolatilization: $C \xrightarrow{k_o} (1-\nu)V + \nu V^* + C^*$

where C = reactive coal, C* = activated coal, V = unreactive volatile, V* = reactive volatile, ν = fraction of reactive volatile

rate of devolatilization, $R_o = k_o C_c$ and

$$k_o = \begin{cases} A_o \exp(-E_o/R_gT) & \text{for single reaction} \\ A_o \int_0^\infty f(E) \exp(-E/R_gT)dE & \text{for multiple reactions} \end{cases}$$

Here, f(E) is a distribution function.

(b) Deposition: $V^* \xrightarrow{k_1} S$

where S = inert char

rate of deposition, $R_1 = k_1 C_{v^*}$

(c) Stabilization: $V^* + H_2 \xrightarrow{k_2} V$

rate of stabilization, $R_2 = k_2 C_{H_2} C_{v^*}$

(d) Direct Hydrogenation: $C^* + H_2 \xrightarrow{k_3} V + C^*$

rate of direct hydrogenation, $R_3 = k_3 C_{c^*} C_{H_2}$

(e) Polymerization: $C^* \xrightarrow{k_4} S$

rate of polymerization, $R_4 = k_4 C_{c^*}$

They postulated that for slow rates of devolatilization hydrogen permeates the entire particle, immediately stabilizing all reactive volatiles and preventing deposition. Increases in the devolatilization rate reduce the hydrogen concentration within the particle with the concentration at the center eventually falling to zero. A further increase in devolatilization rate produces a core of no hydrogen and the reaction interface of hydrogen at core surface. Under this condition the product yield is reduced due to deposition of reactive volatiles. At extremely rapid rates bulk flow of volatiles effectively excludes hydrogen from the particle, and the reaction interface is at the external surface of the particle.

Assuming the coal particle to remain a porous sphere and instantaneous reaction of hydrogen with reactive volatiles at reaction interface, they formulated the conservation equation under isothermal conditions for the four gaseous species:

reactive and unreactive volatiles, hydrogen and inert gas. They numerically solved the mass balance equation similar to Eq. (9) assuming isothermal particle, no external mass transfer resistance and a pseudo steady state condition. The total yields obtained by integrating the mass balance equation over the time-temperature history were shown to agree well with the experimental data of Anthony et al. (3) for various total pressure, hydrogen partial pressure and particle size.

Volatile Combustion

Volatiles produced on pyrolysis burn with oxygen in coal combustion processes either immediately as the volatiles leave the particles or after the volatile jets break through a distance from the particle. The rate of the volatile combustion is controlled in the immediate combustion by rate of pyrolysis while in the delayed combustion by the rate of mixing with oxygen. Information on the kinetics volatile combustion is very limited.

Two hypotheses for the combustion during pyrolysis have been proposed regarding whether or not the burning occurs in the coal particles. Howard and Essenhigh (6) assumed that the burning of volatiles occurs both in the interior of the solid as well as within the laminar layer of gas surrounding the particles. Field et al. (14), on the other hand, assumed that because volatiles mix with oxygen at the particle surface and the burning rate is extremely fast, the overall rate is controlled by the boundary layer diffusion. Dobner et al. (15) argued that combustion of volatiles proceeds in the laminar layer outside of the particle and that oxygen cannot reach the interior of the particle until the combustion of volatiles outside of the particle is completed.

An alternative model is based on the common assumption that volatiles react rapidly to form CO and H_2 with the subsequent CO oxidation as the rate determining step. According to Hottel et al. (16), the following equation can be used to calculate CO oxidation rate:

$$-\frac{dC_{CO}}{dt} = A\, C_{CO} \cdot C_{O_2}^{0.3} \cdot C_{H_2O}^{0.5} \cdot \exp\,(-16{,}000/R_gT)$$

where C_i is the concentration of gaseous component i, A has values from $3\text{x}10^{10}$ to $18\text{x}10^{10}$ (units in mole, cm^3, sec). At combustion temperature of about 1500°C, CO combustion rate is about 10^5 times greater than the subsequent burning rate of char and oxygen.

(2) CHAR-GAS REACTIONS

The char that is formed as the result of the first stage reaction, namely pyrolysis and combustion of volatiles, is very different from its parent coal in size, shape and pore structure. The char-gas reactions occurring in the second stage following the pyrolysis reaction are heterogeneous reactions and take place on the surface of the solid reactants. The phenomena can be classified into two distinct modes of reactions: volumetric reaction and surface reaction. In the case of volumetric reaction, the reacting gas diffuses into the interior of the particles. As the reaction proceeds, porous char and ash layer are built up around the outer layer of the particles as the "reacting zone" continues to shrink. In the case of surface reaction, the reacting gas does not penetrate into the interior of the solid particles but is confined to react at the surface of the "shrinking core of unreacted solid" (17). Generally, surface reaction occurs when the chemical reaction is very fast, such as combustion reaction, and diffusion is the rate controlling step. Volumetric reaction, on the other hand, is the characteristic of slow reaction in a porous solid, such as gasification reaction of char by CO_2 or H_2O.

Although the rate of heterogeneous reactions is usually expressed according to the Langmuir-Hinshelwood mechanisms (Walker et al. (18)), a simpler power law expression is recommended for most of the char-gas reactions. This is to reduce the mathematical complexity in reactor modelling and the number of parameters needed to be determined by experimentation. Accordingly, the rate expression for a volumetric reaction can be described in the following forms:

$$- \frac{dC_s}{dt} = k_v \cdot \alpha_v \, (X,T) C_A{}^n \, C_s{}^m \tag{1}$$

where k_v is the volumetric reaction rate constant, and α_v (X,T) is a term representing available pore surface area of particles and is a function of carbon conversion, X, and temperature, T.

In the case of surface reaction, on the other hand, the rate is proportional to the surface area of the reaction interface and is expressed by

$$\frac{dX}{dt} = k_s \cdot S_{g_{ex}} \cdot C_A{}^n \cdot C_{so}{}^m \tag{2}$$

where Sg_{ex} (= S_{ex}/w_o) is the geometric surface area of the shrinking interface per unit original weight of a particle. k_s is the surface reaction rate constant. The solid reactant concentration is constant and is equal to the original solid reactant concentration of the char in the surface reaction expression.

Various forms of rate expressions have appeared in the literature. It is essential that a proper form is used when comparing the experimental data of different investigators. The conversion of one form of the rate expression to another is listed in Table II. A pictorial representation of pyrolysis and the subsequent char-gas reaction is shown in Fig. 4 for a large particle and a small particle, which may behave differently.

Char-Oxygen Reaction

The mechanism of char-oxygen reaction is better understood than either pyrolysis or volatile combustion. Thring and Essenhigh (19) showed that the burning rate of the char-oxygen reaction is zero order with respect to oxygen concentration below 1200°K and is first order between 1200°K and 2200°K. Glassman (20) argued that the burning rate of coal particles in either a quiescent or convective atmosphere is directly proportional to the oxygen concentration, and for coal particles surrounded by an ash layer the burning rate is proportional to the square root of oxygen concentration. The rate determining step in the combustion of char varies depending on the range of temperature, particle size and specific surface area of the char. Field (21) reported the burning rate of pulverized coal of various particle sizes and showed that for small particles (below 50 μm) the combustion is chemical reaction controlled and for large particles (above 100 μm) combustion is diffusion controlled. Mulcahy and Smith (23) reported that the burning rate at temperatures higher than 1200°K and for particles larger than 100 microns is determined by diffusion rate of oxygen to the surface.

For the temperature range above 1000°K, Field et al. (14) presented a combustion rate expression combining both the rate of chemical reaction and that of diffusion as follows:

$$-\frac{1}{S_{ex}}\frac{dn_c}{dt} = \frac{P_{O_2}}{1/k_S + 1/k_D} \qquad (3)$$

where n_c is mass of char, S_{ex} is external surface of particle, P_{O_2} is partial pressure of oxygen, k_S and k_D can be expressed as:

$$k_S = 8710 \exp(-17{,}967/T) \qquad [\text{gm/cm}^2 \cdot \text{sec} \cdot \text{atm}]$$

$$k_D = 0.292\,\psi\, D_v/T\, d_p \qquad [\text{gm/cm}^2 \cdot \text{sec} \cdot \text{atm}]$$

The mechanism factor, ψ, is a function of coal type, the ratio of CO to CO_2 formed and the particle size. ψ takes a value of 2 when CO is the direct product of char-O_2 reaction and a value of 1 when CO_2 is the direct product. The following correlations are suggested for rough estimation of ψ:

$$\psi = (2Z + 2)/(Z + 2) \text{ for } d_p \leq 50\ \mu\text{m}$$

$$\psi = [(2Z + 2) - Z(d_p - 50)/950]/(Z+2) \text{ for } 50\ \mu\text{m} < d_p \leq 1000\ \mu\text{m}$$

and $\psi = 1.0$ for $d_p > 1000\ \mu$m

where $Z = [CO]/[CO_2] = 2500 \exp(-6249/T)$

d_p in μm and $T = (T_S + T_g)/2$ in °K

The subject of mechanism of CO formation during char combustion is discussed in a later section.

Smith and co-workers (23,24) measured the burning rate of bituminous coal char for particle sizes of 18, 35, and 70 microns. They concluded that for these particles the combustion rate is slower than the rate of diffusion of oxygen to the reaction surface. The activation energy for chemical reaction controlling regime is evaluated to be 27 Kcal/mol. They reported that in the range of temperatures from 800 to 1700°K, the combustion occurred in the intermediate regime between that controlled by chemical reaction and that controlled by the pore diffusion. The Field et al. (14) rate expression of Eq. (3), however, does not agree with the data of Smith and co-workers (23,24) at the temperature range below 900°K. Their data indicate that at lower temperature combustion takes place throughout the pore surface within the char rather than at a sharp interface as implied by Eq. (3).

Hamor et al. (25) and Smith and Tyler (26) measured combustion rate of pulverized Brown coal char in an entrainment reactor and in a fixed bed reactor having a size range of 89, 49, and 22 microns. They found that below 760°K combustion of both the 89 and 49 micron particles is controlled by chemical reaction alone and shows an activation energy of 32 Kcal/mol. When temperature is raised to above 900°K, combustion of these particles is controlled by both diffusion and chemical reaction and shows an activation energy of 16.2 Kcal/mol, which is one-half of the "true" activation energy in chemical reaction controlled regime. However, for a 22 micron particle, the rate at this temperature is apparently still controlled by chemical reaction alone. Above 1550°K, combustion of an 89 micron particle is controlled by oxygen diffusion rate. For the "intrinsic" reaction rate, they proposed an empirical correlation for a temperature range from 630 to 1812°K:

$$\text{Rate} = 1.34 \exp\,[-32{,}600/R_g T_s] \qquad g/(cm^2 sec) \tag{5}$$

Dutta and Wen (27) found that the burning rate of char of a particle size from 35 to 60 mesh at low temperature is reaction controlled and obeys the volume reaction model. Their rate expression can be written as follows:

$$\frac{dX}{dt} = k_v\, \alpha_v\, C_{O_2}\, (1-X) \tag{6}$$

and obtained an activation energy of 31 Kcal/mol. According to the volume reaction model, they expressed the burning rate under the influence of intraparticle diffusion as

$$\frac{dX}{dt} = \eta\, k_v\, \alpha_v\, C_{O_2}\, (1-X) \tag{7}$$

where effectiveness factor, η, is defined as:

$$\eta = \frac{3}{\phi}\left(\frac{1}{\tanh\phi} - \frac{1}{\phi}\right)$$

Here $\phi = \phi_o [(1-X)\alpha_v']^{1/2}$, $\alpha_v' = \alpha_v/g(x)$, $g(x)$ is given by $D_e = D_{eo}g(x)$ and is a function of conversion X, $\phi_o = r_o (k_v C_{so}^{\theta}/D_{eo})^{1/2}$, and D_{eo} is the effective diffusivity of char at zero conversion.

By using Eq. (7), they showed that in the combustion of char resistance due to pore diffusion is negligible for temperatures below 576°C.

In spite of a great number of studies available on coal combustion rate, the understanding of the phenomenon is far from complete. In fact the combustion rate data available up to now are very confusing even for relatively small particles. As shown in Fig. 6 (21,22,23,24,25), the rates of combustion seem to be affected by the types of coal but the quantitative effect of the temperature and particle size as well as the rate determining factors are not yet clearly understood. This is primarily due to the difficulty in experimental evaluation of the particle temperature and the measurement of changes in physical properties of coal during the course of combustion.

Mechanism of Char Burning and CO Formation

The combustion of residual char produces various ratios of CO and CO_2 via char-O_2 reaction. Authur (28) presented an empirical correlation for CO to CO_2 ratio as indicated by Eq. (4). From Eq. (4) it is apparent that CO is the dominant product at high temperature.

The burning mechanism of char and product gas concentration distributions around the burning char are very complex, and many researchers have proposed different models. When the combustion is controlled by diffusion alone, Borghi et al. (29) maintained that for large particles it is possible for the rate of the reaction $2\ CO + O_2 \rightarrow 2\ CO_2$ to be fast enough to consume all the oxygen before it reaches the carbon surface, and the CO then is supplied by the reaction $CO_2 + C \rightarrow 2\ CO$. As the reaction becomes kinetically controlled, the atmosphere surrounding the particle will be approximately uniform, and CO_2 and O_2 will have equal opportunity to reach the surface. The $C + CO_2$ reaction then is too slow to compete with the oxidation by O_2.

Wicke and Wurzbache (30) measured the concentration profiles of CO, CO_2 and O_2 in the thin film surrounding a burning carbon rod and found evidence of the existence of a maximum in the concentration of CO_2. Degraaf (31) and Kish (32) found a temperature maxima of gas surrounding the particle which is several hundred degrees above the solid surface temperature.

On the other hand, Avedesian and Davidson (33) suggested that O_2 and CO burn rapidly in a very thin reaction zone surrounding the particle. Carbon monoxide produced at the surface diffuses out toward the reaction zone while O_2 from the main stream

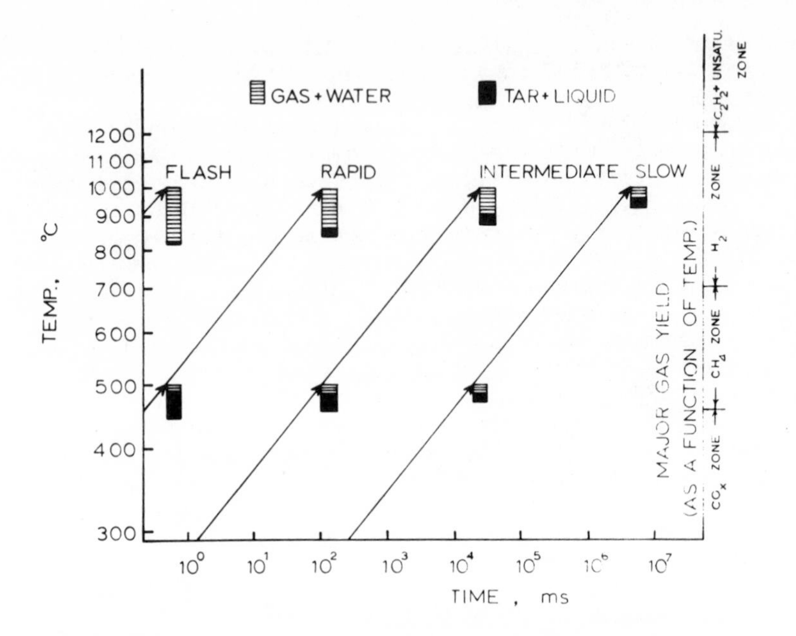

Figure 5. Relative yields and product distributions of pyrolysis in inert atmosphere as functions of temperature, time, and heating rate

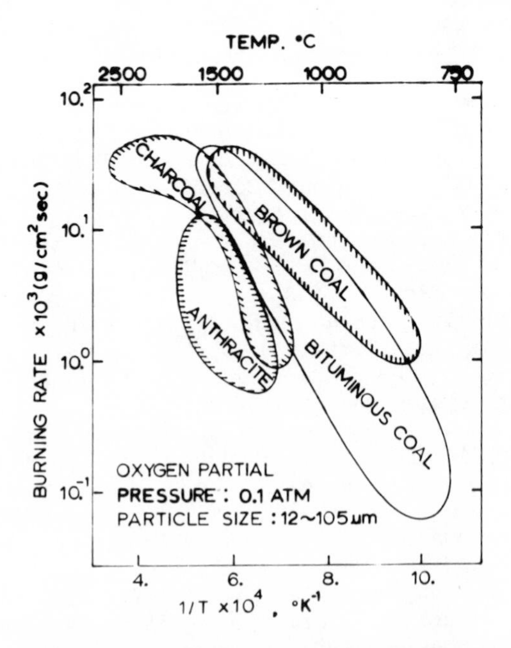

Figure 6. Combustion rates for various coals near chemical reaction controlling regime (21, 22, 23, 24, 25)

diffuses in and burns in a diffusion flame to produce CO_2 as shown in Fig. 7. According to their model, no CO appears in the main stream when there is an abundant supply of O_2.

Essenhigh (34) presented a physical model as shown in Fig. 7 that includes part of a porous solid with an adjacent diffusion boundary layer in the gas phase. Reaction between oxygen and carbon occurs heterogeneously at all available surfaces, exterior and interior. In his model the CO/CO_2 ratio rises with temperature, and CO becomes the principal product at about 1000°C and above (Authur (28)). The CO also reacts in the gas phase with oxygen to produce CO_2, partly in the particle pores and partly in the boundary layer of the char. As the oxygen concentration in the main stream is enriched, there is more CO burn-up inside the solid.

Caram and Amundson (35) suggested that large particles (> 2 mm) burn according to the double film theory (36) shown in Fig. 7, whereas small particles (< 100 μm) burn according to the single film model. In analyzing the homogeneous combustion of CO and the heterogeneous reaction of carbon with oxygen and with carbon dioxide according to double film models, they concluded that large particles (5 mm) tend to reach an upper steady state in which the particle is surrounded by a CO flame. For very small particles (50 μm) such a flame does not develop. Thus, it is evident that the char and oxygen reaction occurs in the interior surface of smaller particles at lower temperature because oxygen does not get consumed near the external surface while enough is supplied to the interior by the pore diffusion from the bulk phase.

Char-Hydrogen Reaction

The reaction of char and hydrogen is quite exothermic and produces mainly methane. This reaction is very slow when hydrogen partial pressure is low and temperature is low. But at high hydrogen partial pressure and temperature above 700°C, the rate of this reaction becomes appreciable. The mechanism of this reaction is rather complicated and has been studied by a number of investigators (37,38,39,40). The initial phase of reaction between hydrogen and coal, or hydropyrolysis, is very rapid and has been discussed in detail in the previous section. Depending on the operating condition, it is possible to convert more than 40% of coal during the first stage of hydropyrolysis. The reaction of hydrogen with the remaining char is much slower and takes place mostly on the solid surface. Wen and Huebler (41) proposed the following empirical equation for the rate of first and second stage hydrogasification.

First Stage: $$\frac{dX}{dt} = k_v\,(f-X)(C_{H_2} - C^*_{H_2})$$

where X is carbon conversion and f is the fraction of carbon that can be converted in the first stage. k_v is approximately equal to

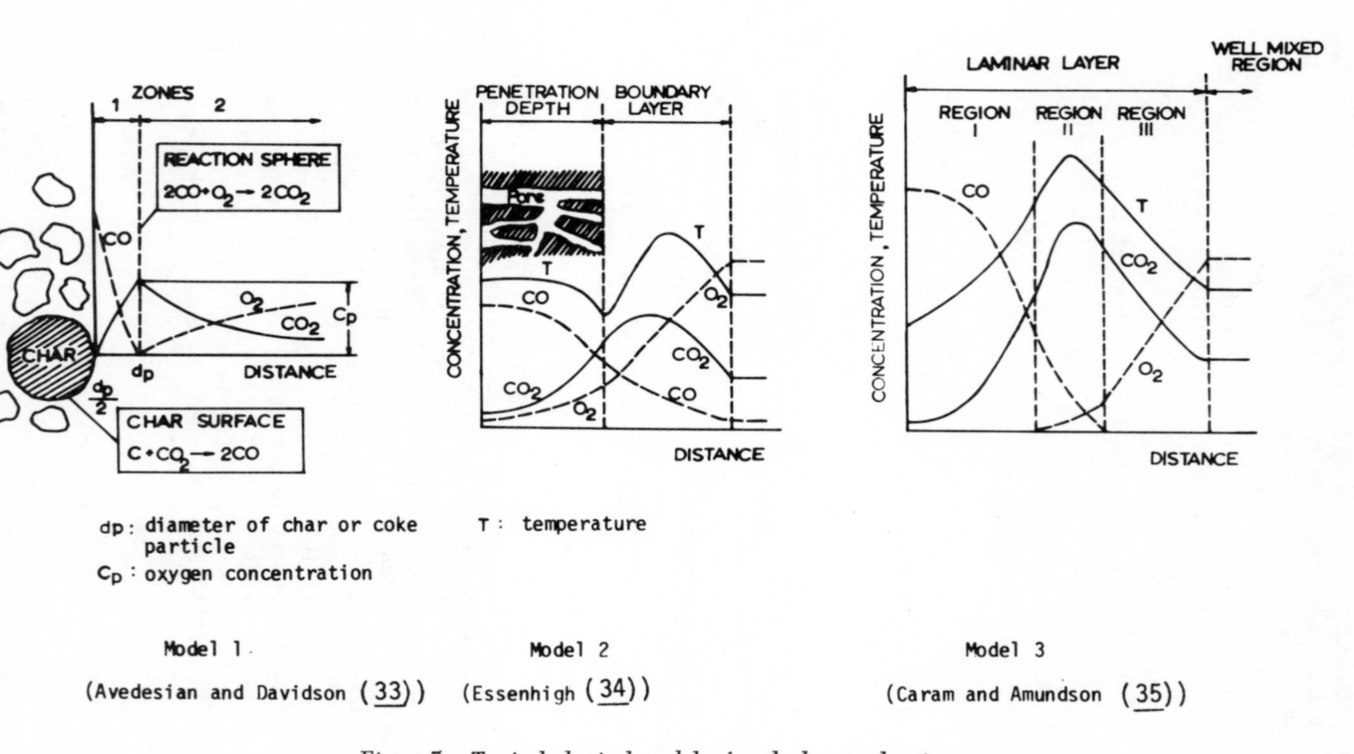

Figure 7. Typical physical models of coal–char combustion

9.5×10^{-4} and 9.0×10^{-5} (m^3/mol C-sec) for raw bituminous coal and pretreated bituminous coal, respectively. $C^*_{H_2}$, the hydrogen concentration in equilibrium with coal at various conversion, must be evaluated for different coals at different conversion and temperatures (41). The value of $C^*_{H_2}$ is much smaller compared to that for the β-graphite-H_2-CH_4 system.

$$\text{Second Stage:} \quad \frac{dx}{dt} = k'_v\,(1-x)(P_{H_2} - P^*_{H_2})$$

where $x = \dfrac{X - f}{1 - f}$ and is the carbon conversion in the second stage reaction. k'_v, which varies with the type of coal (39,44,146,147,148), has the following values at 800°C (38,40,42):

Coal Type	k'_v, (atm.sec)$^{-1}$
Lignite	$0.185 \times 10^{-5} \sim 0.42 \times 10^{-5}$
Sub-bituminous	$0.196 \times 10^{-5} \sim 0.28 \times 10^{-5}$
Bituminous	$0.097 \times 10^{-5} \sim 0.159 \times 10^{-5}$

The apparent activation energy for the second stage hydrogasification of char has been reported to vary between 30 to 41 Kcal/mol (40,43,44). Zahradnik and Glenn (45) discussed the mechanism of methane formation in hydrogasification reaction and proposed an empirical rate expression for Pittsburgh seam coal as follows:

$$MY = \frac{a + A\,e^{-E/R_gT} \cdot P_{H_2}}{1 + A\,e^{-E/R_gT} \cdot P_{H_2}} \quad \text{(for residence time 14~17 sec)}$$

where MY is the yield of methane expressed as the fraction of carbon in coal appearing as methane, a = 0.08, E = 15.42 Kcal/mol, A = 7.005 atm^{-1}, P_{H_2} in atm and T in °K.

Johnson (44) also presented an empirical correlation of hydrogasification rates for the first stage and the second stage reactions based primarily on the data obtained from a thermobalance.

Char-Carbon Dioxide Reaction

The rate of char-CO_2 reaction is relatively slow and is comparable with that of char-steam reaction. Dutta et al. (46) measured the rate of char-CO_2 reaction and concluded that for particles smaller than 300 microns and when the temperature is lower than 1000°C, the reaction is controlled by the rate of chemical reaction and takes place nearly uniformly throughout the interior of the char particle. The rate of reaction under such conditions can be expressed as (46):

$$\frac{dX}{dt} = \alpha_v\,k_v\,C^n_{CO_2}\,(1-X)$$

where α_V is the available surface area per unit weight divided by the initial available pore surface per unit weight of particle. n is the order of reaction and varies depending on the experimental conditions. Below 1300°C and with CO_2 concentration between 10^{-2} to 10 mol/m^3, n is unity. Based on a number of studies (18, 46-52), the reaction rate seems to obey the Langmuir type adsorption relation and is the first order reaction with respect to CO_2 at low CO_2 partial pressure and is a zero order reaction for high CO_2 partial pressure. The activation energy lies between 43 and 86 Kcal/mol. Dutta et al. (46) indicated that the reactivity of char increases with an increase in oxygen content of char. Recently Muralidhara (53) correlated the initial reaction rate of char and CO_2 in terms of CaO and O_2 content in the char as follows:

$$\left(\frac{dX}{dt}\right)_{X=0} = 151.4x10^6 e^{-35,000/T} + 63.1x10^6 e^{-27,000/T} C_{CaO}(1+241.4\ C_{O_2})$$

where $(dX/dt)_{X=0}$ is in $(sec)^{-1}$, C_{CaO} and C_{O_2} are weight fraction of C_{aO} and oxygen in char, respectively. This equation was formulated using data of Dutta et al. (46), Walker et al. (18) and Muralidhara (53) for lignite, bituminous coal, sub-bituminous and anthracite having particle size between 50~75 microns and CaO and oxygen contents up to 4% and 3.5%, respectively. The CO_2 partial pressure was held at 1 atm, and temperature was varied 850°C to 930°C. Apparently, the presence of alkali minerals and oxygen functional groups enhances the rate of CO_2 reaction. There is also an indication that CO may hinder the rate of CO_2 reaction for temperatures below 1100°C. The catalytic effect of alkali minerals present in the char is discussed later in the section on Catalytic Reactions.

Char-Steam Reaction

Char-steam reaction is one of the most important reactions in industrial practice for generation of CO and H_2. Most of the earlier investigators, (18,54), used Langmuir-type adsorption equations to express the rate of this reaction. This reaction is apparently controlled by chemical reaction between 1000°C and 1200°C for particles smaller than 500 microns and is affected by diffusion through the pore in the char above 1200°C (55,56,57). There is an indication that the reaction is inhibited by the presence of hydrogen. The order of char-steam reaction varies with steam concentration in much the same way as that of char-CO_2 reaction with CO_2 concentration. The order of reaction for char-steam reaction is approximately unity up to unit partial pressure of steam but tends to become zero as the steam partial pressure rises significantly (18).

An empirical equation in the chemical reaction controlling regime developed by Wen (58) based on the volume reaction model has the form:

$$\frac{dX}{dt} = k_v \left[C_{H_2O} - \frac{C_{H_2} \cdot C_{CO} R_g T}{K_{C-H_2O}}\right](1-X)$$

where $k_v = \exp(24.30 - 25120/T)$ $(cm^3/mol \cdot sec)$

$K_{C-H_2O} = \exp(17.64 - 16810/T)$ and T is in °K.

The range of the apparent activation energy has been reported to vary from 35~45 Kcal/mol (58) to 60~80 Kcal/mol (59,60,56). Johnson (44) observed that presence of steam has little effect during the rapid stage of pyrolysis. He proposed a rate expression for the second state reaction similar to that for his char-hydrogen reaction in the second stage but with a different set of rate constants.

Catalytic Reactions

It has been well-known for over half a century that char-gas reactions are catalyzed by metal salts, particularly alkali, alkali earth and transitional metals. Some of these metal salts are present in coal ash.

The catalysts found to be effective for gasification of coal are listed below in order of strength (from strong to weak).

For Production of CH_4: Li_2CO_3, Pb_3O_4, Fe_3O_4, MgO

For Production of H_2: K_2CO_3, Li_2CO_3, Pb_3O_4, CuO

For Production of CO: K_2CO_3, Li_2CO_3, Fe_3O_4, Cr_2O_3

For Gasification of Carbon: K_2CO_3, Li_2CO_3, Pb_3O_4, Cr_2O_3

Exxon Research and Engineering Co. (61) gasified Illinois coal that was treated with Na_2CO_3 and/or K_2CO_3 (up to 15% K in C) at 700°C and found that these salts catalyzed steam gasification. They also found that these salts reduced the agglomerating tendency of caking coals during gasification significantly. The rate of gasification is essentially proportional to the concentration of the catalyst. For K_2CO_3 the rate of gasification of Illinois char in fraction of carbon gasified per hour at 34 atm is roughly 20·(K/C) and 60·(K/C) at 650°C and 760°C, respectively. Here K/C is the atomic ratio of potassium and carbon in char. For Illinois seam char, K/C is approximately 0.01.

The work at Battelle's Columbus Laboratories (62) also demonstrated that impregnation of CaO into coal before gasification can prevent agglomeration of coal and greatly increase the reactivity and hydrocarbon yields in the gasifier even for large coal particles. The reaction rate for production of methane from devolatilized chars in hydrogen can be expressed as:

$$- \left[\frac{dX_c}{dt}\right]_{CH_4} = 88(1-X_c)P_{H_2} \exp(-13,800/T)$$

where t is the time in minutes, P_{H_2} is in psia, T in °K, X_c is total carbon conversion, and $[X_c]_{CH_4}$ is the fraction of carbon converted to methane. The above equation is valid for CaO treated coal in a solution of NaOH using CaO/coal ratio of 0.13 and hydrogen partial pressure of 15.3 atm.

In general, it is very difficult to evaluate the activities of catalysts present in coal whether they are added and/or impregnated prior to the gasification or are present in coal ashes. For the surface reaction and the volume reaction models, Wen and Dutta (63) modified k_v and k_s in Eqs. (1) and (2) as $k_v = Z_v k_{vt}$ and $k_s = Z_s k_{st}$ where k_{vt} and k_{st} are the rate constant without catalyst, and Z_v and Z_s represent the effect of catalyst. Z_v and Z_s depend on factors such as type and quantity of catalysts and reaction temperature.

Two reactions that seem to be catalyzed by the minerals present in ashes are (a) water-gas shift reaction and (b) methane-steam reforming reaction. The kinetics of water-gas shift reaction have been studied by various investigators (18,64,65, 66). Since the rate of reaction is very rapid, this reaction may be considered to be in equilibrium at the exit of the gasifier in most cases. However, the reaction may not have reached equilibrium within the reactor, particularly near the gas entrance. The rate expression using industrial iron oxide catalyst but correcting it by Z_v (roughly between 0.001 to 0.010) can be used to account for the catalystic effect of ashes.

The methane-steam reforming reaction, the reverse reaction of methanation reaction, is believed to be catalyzed by the minerals present in coal. Zahradnik and Grace (67) proposed the following expression for Pittsburgh seam coal:

$$- \frac{d\, C_{CH_4}}{dt} = k\, C_{CH_4}$$

where k = 312 exp $(-30,000/R_g T)$ in sec^{-1} and T in °K.

Since a number of simultaneous reactions are taking place in a coal conversion reactor, it is necessary to have a proper perspective of the relative rates of these reactions. This is essential in identifying the dominant reactions and the zones of combustion, gasification and pyrolysis reactions within the reactor. In Fig. 8, rates of pyrolysis, char-oxygen, char-hydrogen, char-carbon dioxide and char-steam reactions are plotted as a function of temperature. In this plot, the partial pressures of the reacting gases are held at one atmosphere. At such a low pressure, it is interesting to observe that the rates of char-steam and char-CO_2 reactions are moderate and roughly the same order of magnitude but are greater than char-hydrogen reactions. When the partial pressures of the reacting gases, H_2, H_2O and CO_2, are raised from one atm to the range of about 35 to 100 atm, the rates of gasification reactions increase significantly as shown by the black patch in Fig. 8. When coal contains large amounts of calcium and organic oxygen, such as in lignite, the rates of gasification reactions are also significantly greater as discussed in the section on Catalytic Reaction. If, for example, coal contains 4% CaO and 3.5% oxygen, the rate of char reacting with CO_2 at one atmosphere increases to the level represented by the black patch in Fig. 8. The rates of reactions shown in Fig. 8 are mostly in the chemical reaction controlling regime.

In solid-gas reaction, when temperature is low and the overall rate is controlled by the chemical reaction rate, the reacting gases penetrate into the interior of the particle resulting in the volume reaction. As the temperature is raised and chemical reaction rate becomes faster, the effect of diffusion becomes appreciable. When the overall rate is controlled by the diffusion rate, the reaction is confined at the surface of unreacted core and the surface reaction prevails. The criteria of reaction regime, i.e. the volume reaction prevails when the chemical reaction is rate controlling and the surface reaction prevails when the diffusion is rate controlling, have been discussed in detail by Wen and his co-workers (17,68,69).

Various reaction models for noncatalytic gas-solid reactions have been proposed and have been summarized by Szekely et al. (70). The effect of diffusion and heat transfer on the chemical reaction rate for a single particle is rather complicated, especially when multiple reactions are occurring simultaneously. This subject will be discussed in the following section.

(3) BASIC EQUATIONS FOR SINGLE PARTICLE COAL-GAS REACTIONS

We shall now attempt to present a set of governing equations for mass and heat balances around a single coal or char particle exposed to different gaseous atmospheres. The rate expressions presented in the previous section are so-called "intrinsic rates" and therefore do not include the effects of physical processes such as heat and mass transfer and bulk flow. The combined effects are formulated in this section for a single particle system.

We shall express such a system by the following stoichiometric equation:

$$\sum_i \nu_{ij} A_i + \sum_s \nu_{sj} A_s = 0 \qquad (8)$$

where i = 1, 2,⋅⋅(gaseous component), s = n + 1, n + 2,⋅⋅(solid component), j = 1, 2,⋅⋅(j-th reaction).

ν_{ij} and ν_{sj} are stoichiometric coefficients of i-th gaseous component and s-th solid component for the j-th reaction, respectively. These stoichiometric coefficients are negative if they are referred to the reactants and are positive if they are referred to the products.

We shall define the reaction rate of a single particle, R_j, such that $\nu_{ij}\ R_j$ and $\nu_{sj}\ R_j$ are moles of gaseous component i reacted per unit volume of the char particle per unit time and moles of solid component s reacted per unit volume of the char particle per unit time, respectively. The relationships between R_j and the rate expressions presented in the previous section are listed in Table II.

A general mass balance for gaseous component i can be written as:

$$\underbrace{\frac{\partial(\varepsilon C_i)}{\partial t}}_{\text{accumulation}} = \underbrace{\nabla D_{ei} \nabla C_i}_{\substack{\text{diffusion}\\ \text{through}\\ \text{porous solid}}} - \underbrace{\nabla \frac{R_g T_s}{P} C_i \sum_i N_i}_{\substack{\text{bulk flow}\\ \text{through}\\ \text{porous solid}}} + \underbrace{\sum_j \nu_{ij}\ R_j}_{\substack{\text{generation or}\\ \text{disappearance due}\\ \text{to chemical reaction}}} \tag{9}$$

and the molar flux, N_i, is defined by:

$$N_i = -\ D_{ei}\ \nabla C_i + \frac{R_g T_s}{P}\ C_i \sum_i N_i \tag{10}$$

For a system involving chemical reactions, we note that $\sum_i N_i/\nu_{ij} = 0$ and for a non-reacting system $N_I = 0$, where I is the inert component.

A mass balance for solid component s is given as:

$$\frac{\partial C_s}{\partial t} = \sum_j \nu_{sj}\ R_j \tag{11}$$

Heat balance for both solid and gas within the particle can be written as:

$$\underbrace{C_p\ \rho_b \frac{\partial T_s}{\partial t}}_{\text{accumulation}} = \underbrace{\nabla k_e \nabla T_s}_{\text{heat conduction}} + \underbrace{(\sum_i D_{ei}\ C_{pi}\ \nabla C_i)\nabla T_s}_{\substack{\text{heat transferred as the result of}\\ \text{gas diffusion in porous solid}}}$$

$$- \underbrace{\frac{R_g T_s}{P}[\sum_i C_{pi}\ C_i\ (\sum_i N_i)]\nabla T_s}_{\substack{\text{heat transferred due to bulk}\\ \text{flow through porous solid}}} + \underbrace{\sum_j(-\Delta H_j)R_j}_{\substack{\text{heat generated or absorbed due}\\ \text{to chemical reaction}}} \tag{12}$$

where $C_p\ \rho_b = \sum_i \varepsilon\ C_{pi}\ \rho_i + \sum_s C_{ps}\ \rho_s$

Here D_{ei} and ε are related through solid conversion. For convenience, an approximation may be made as (Wen, (68)):

$$D_{ei} = D_{oi}\ \varepsilon^{\beta},\ \varepsilon = \varepsilon_o + \gamma(1-X) \text{ and } X = 1 - (\sum_s C_s/\sum_s C_{so})$$

Table II. Relationship of R_j and Corresponding Rate Expression Commonly Used

Reactions	Definition of Rate of Reaction	Remarks
General reaction	$R_j = \frac{1}{\nu_{sj}} \cdot \frac{1}{V_p} \cdot \frac{dn_s}{dt} = \frac{1}{\nu_{sj}} \cdot \frac{dC_s}{dt}$	n_s = mole of solid component V_p = volume of particle [cm^3] t = time [sec]
Pyrolysis	$-\frac{dV}{dt}\ [\frac{\text{gm of VM}}{\text{gm of d.a.f. coal} \cdot \text{sec}}]$ $R_j = \frac{1}{\nu_{VMj}} \cdot \frac{\rho_{mp}}{M_{VM}} \cdot (-\frac{dV}{dt})$	V = volatile matter lost from particle to time t [gm of VM/gm of d.a.f. coal] ρ_{mp} = d.a.f. coal density [gm/cm^3] M_{VM} = molecular weight of volatile matter
Char combustion	$-\frac{1}{S_{ex}} \cdot \frac{dn_c}{dt}\ [\frac{\text{gm}}{cm^2 \cdot \text{sec}}]$ $R_j = \frac{1}{\nu_{cj}} \cdot \frac{S_{ex}}{V_p M_c} \cdot (\frac{dn_c}{S_{ex} dt})$	surface reaction n_c = gm of carbon/particle S_{ex} = external surface of particle [cm^2] M_c = molecular weight of carbon
Char gasification Char combustion	$\frac{dX}{dt}\ [\frac{\text{wt. fraction}}{\text{sec}}]$ $R_j = \frac{C_{so}}{\nu_{sj}} \cdot (-\frac{dX}{dt})$	volume reaction X = solid conversion [wt. fr.] C_{so} = initial solid concentration [wt./cm^3 of particle]
Catalytic reaction	$-\frac{1}{w_{ash}} \cdot \frac{dn_{co}}{dt}\ [\frac{\text{mole of CO}}{\text{gm of ash} \cdot \text{sec}}]$ $R_j = \frac{1}{\nu_{coj}} \cdot \varepsilon_{ash} \cdot \rho_{ash} (\frac{1}{w_{ash}} \cdot \frac{dn_{co}}{dt})$	water-gas shift reaction etc. w_{ash} = weight of ash [gm] n_{co} = mole of carbon monoxide ε_{ash} = fraction of ash in particle [-] ρ_{ash} = density of ash [gm/cm^3]

The boundary and initial conditions are:

B.C.

at $r = 0$, $D_{ei}\ \nabla C_i = 0$, $k_e\ \nabla T_s = 0$, and $\nabla N_i = 0$

at $r = r_o$, $- D_{ei}\ \nabla C_i = k_{gi}\ (C_i - C_{io})$, and

$$- k_e\ \nabla T_s = h_c\ (T_s - T_g) + h_r\ (T_s^4 - T_g^4)$$

at $t = 0$, $C_i = 0$, $C_s = C_{so}$, and $T_s = T_{so}$

It is obvious that these equations cannot be solved easily, particularly when a number of simultaneous reactions are involved. Therefore, many simplifying assumptions must be made to reduce the complexity of the equations. This becomes more essential when applied to the modelling of a coal conversion reactor because of additional mathematical complexity resulting from intraparticle phenomena and hydrodynamics of solids and gas in the reactor.

Some of the terms, for example, in the mass and heat balance equations can be neglected without serious errors, depending on the condition. The terms relating to the bulk flow are not important except during pyrolysis or hydropyrolysis. The accumulation term for gaseous species in the mass balance equation can usually be ignored, and a pseudo steady state assumption can be applied without serious errors.

To what extent these basic equations can be simplified depends on the accuracy of experimental data used in generating reaction rate, R_j, and the accuracy required in the simulation or design of an integral reactor for coal conversion. This topic is the subject of discussion in the next section.

DESIGN AND MODELLING OF COAL CONVERSION REACTORS

(1) CHARACTERISTICS OF VARIOUS COAL CONVERSION REACTORS

Depending on the process and the final product desired, coal conversions are carried out in various types of reactors. For gasification and combustion of coal, moving bed (or fixed bed) reactor, fluidized bed reactor, and entrained bed (or transport, or suspension) reactor are employed. For coal liquefaction, three phase reactors such as slurry reactor, fixed bed reactor and ebullating bed reactor are used. The operating conditions, temperature, pressure, flow rates of gas and solids, residence times and mixing of solids and gases, and direction of flows are different in these reactors.

What sets one type of coal conversion reactor from another is the reliability of performance, which depends above all on the simplicity of design. Simplicity in design would mean ease of maintenance and high availability. Other desirable characteristics in coal conversion reactors, for example in gasifier, include a capability for processing a wide variety of coals, production of gas free from tars and of low dust loadings, and ability for quick shutdowns and restarts.

The advantages and disadvantages of fixed bed, fluidized bed and entrained bed gasifiers are presented in Table III. An excellent report concerning the status of coal gasifier technology based on EPRI's workshop by Yerushalmi is available (71).

In the design of a coal conversion reactor, it is necessary to consider not only the reaction kinetics of a single particle but also the hydrodynamics of gas and mixing of solids and the accompanying heat and mass transfer occurring in the reactor.

The mathematical models for coal conversion reactors, whether they are combustor, gasifier, or liquefier, are invariably complicated, containing a set of basic equations describing the system and a number of model parameters. The model must represent the actual reactor closely enough to yield useful information for design and analysis. However, such a model can never represent a complete picture of reality. Depending on the purpose, a simple model may be quite adequate in some instances. A much more refined and elaborate model, however, may be necessary in other circumstances. Obviously, a more complicated and rigorous model is more costly to develop. A good model, therefore, must recognize its own inadequacies so that it can serve as a means to develop a more complete picture of reality. Hence, in developing a coal reactor model it is imperative that we differentiate the major factors that are significantly important from the minor factors that may be safely neglected.

By analyzing the behavior of the coal conversion reactor model and comparing it with the actual reactor performance, one can learn how and in which direction the improvement of the model should be attempted.

(2) FIXED BED (MOVING BED) GASIFIER MODELS

In a fixed bed gasifier, coals move downward while coming into contact with gases flowing upward countercurrently. The bed consists of a preheating zone at the top followed by a pyrolysis zone, a gasification zone, a combustion zone and an ash zone at the bottom. A schematic diagram of temperature and concentration profiles is presented in Fig. 9. The maximum temperature (about 1300°C) is usually located at the lower part of the bed and depends on oxygen to steam ratio of the feeding gas. The advantages and disadvantages of fixed bed gasifiers are summarized Table III.

There have been several fixed bed gasifier models developed based on some simplified assumptions (72,73,74,75). Yoon et al. (76) proposed a model assuming gas and solids to be at the same temperature and no heat loss from the wall of the gasifier.

The simplifying assumptions made in most of these studies were to reduce the mathematical complexity, but in many instances they may have resulted in a misleading temperature and

Table III. Characteristics of Various Types of Gasifiers

(I) Fixed Bed Gasifier (Dry Ash)
(Lurgi, Woodall-Duckham, Wellman-Galusha, etc.)

Advantages

- High thermal efficiency and carbon conversion.
- Large residence time of solids (1 to 3 hours).
- Low contamination of gas with solids when compared with fluidized bed and entrained bed.
- Capable of operating at elevated pressures.

Disadvantages

- Caking coals cannot be used without pretreatment to render them nonagglomerating or without modifying the mechanical design.
- Uniformly sized coal containing a minimum of fines (1~5 cm) having reasonable mechanical strength is needed.
- Ash-fusion temperature imposes an upper temperature limit, and a large amount of steam is needed to control the temperature at the bottom of the bed. Much of the steam passes without reacting, contributing to heat losses and large volume of a dilute liquor.
- Gas leaving contains a large amount of tars necessitating expensive treatment.
- In spite of pressure, capacity is small requiring a large number of gasifiers.
- Poor adaptability to changing fuel. Minimum temperature operable, (lignite 690°C, Sub-bituminous coal 750°C, Semi-anthracite, 800°C) depends on coal reactivity.

(II) Slagging Fixed Bed Gasifier (Slagger)
(Lurgi Slagger, Secord-Grate, etc.)

Advantages

- Steam requirement is about a fifth of that needed for dry ash fixed bed gasifier, and nearly all the steam is reacted.
- Lower production of liquor and higher thermal efficiency.
- Slagger is capable of processing 3 to 4 times more coal/unit area than a dry ash gasifier.
- Fines and tars may be disposed by injecting into the slagging zone.

Table III (Continued). Characteristics of Various Types of Gasifiers

Disadvantages

- Tars do escape and caking coals cannot be processed.
- Although a wider range of coals can be processed, coals that are mechanically weak can cause fines to be blown by a high velocity blast of steam and oxygen.
- Materials of construction, containment and withdrawal of slag and formation of molten iron in reducing conditions are major problems.

(III) Fluidized Bed Gasifier
(Winkler, Hygas, Cogas, CO_2 accepter, Synthane, Battelle/Union Carbide, Westinghouse, U-gas, EXXON Catalytic, etc.)

Advantages

- Good temperature control, easy solids handling, capability for bringing cold solids or gas feed instantaneously to bed temperature.
- Ability to tolerate variation in quality of fuel during operation.
- Capable of operation at part load, and can be stopped and restarted rather easily.

Disadvantages

- Operation temperature is limited. The upper temperature is the clinkering temperature (around 1040°C) and the lower temperature is indicated by coal reactivity and the escape of tars.
- Build-up of micron-size carbon fines in the bed and loss of this carbon and ash entrainment can be a serious problem. Recycle of fines does not improve carbon utilization very much because of low reactivity of fines.
- Appreciable amount of carbon is contained in the ash withdrawn due to need for maintaining sufficient carbon-inventory in the bed.
- Unless a burn-up cell or second stage fluidized bed is provided, complete conversion can not be achieved in one stage fluid bed.
- Feeding of caking coal without pretreatment or of wet non-caking coal is still a problem.
- Formation of clinkers near the oxygen inlet point may disrupt operation.
- Mixing of solids and gas and the number of feeding points required are still not well understood for scale-up of the reactor.

Table III (Continued). Characteristics of Various Types of Gasifiers

(IV) Entrained Bed (Suspension) Gasifier (Koppers-Totzek, Texaco, Brand W, Foster Wheeler, Combustion Engineering, etc.)

Advantages

- Ability to utilize any type of coal irrespective of swelling and caking including fines, and with slight modification coal-oil mixture can be processed.
- High coal throughput capacity particularly at high pressure.
- Produces gas free of tars, phenols and very little methane.
- High carbon utilization due to high reaction rates.
- Simple, flexible and easy to scale-up.

Disadvantages

- Refractories and materials of construction are problems in slagging zone.
- Low heat-recovery efficiency resulting from co-current operation. Outlet gas temperature is high and needs sensible heat recovery.
- Continuous feeding of coal into pressurized gasifier and slag withdrawal at high pressure are some of the problems. Changing coal feed rate to follow load change may be difficult.
- Dust loading in product gas could be high requiring expensive collection equipment. Char recycle is needed and would be difficult at high pressure.
- Low fuel inventory and oxygen is required.

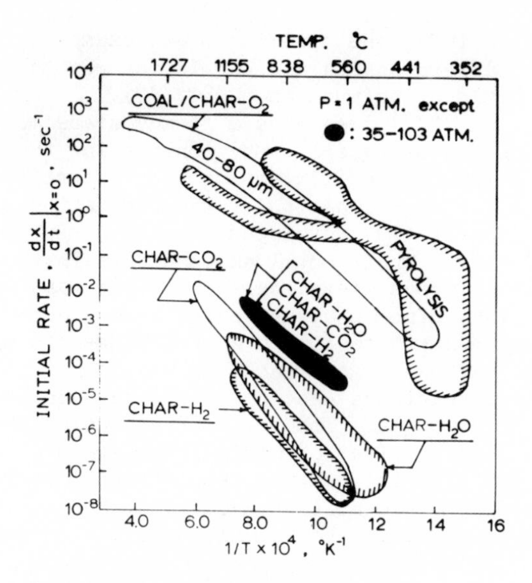

Figure 8. Comparison of initial rates of pyrolysis, combustion, and gasification of coal–char

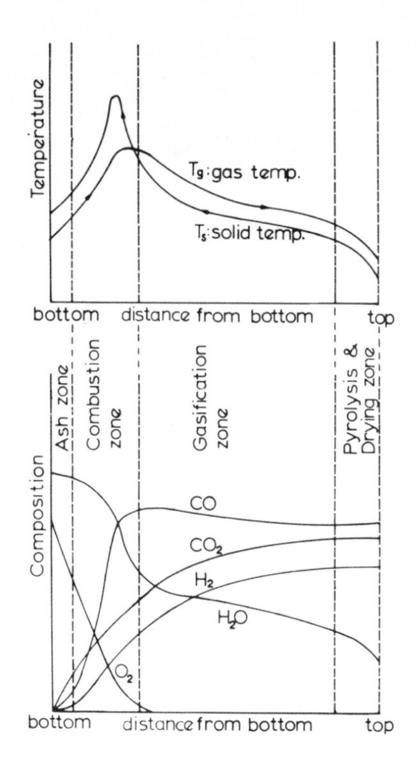

Figure 9. Representation of temperature and concentration profiles in a moving bed gasifier

concentration profile of the gasifier. For example, the temperature difference between solids and gas in the combustion zone of Lurgi gasifier could be as much as 600°C (72).

Borowiec et al. (77) presented a steady state model for a countercurrent moving-bed gasifier by considering the effect of interphase heat transfer coefficient on temperature and composition profiles, and on the location of the combustion zone.

Amundson and Arri (78) developed a model of Lurgi type moving bed gasifier of char. Their model assumed that in the upper gasification zone, carbon-steam, carbon-hydrogen and water gas shift reactions take place whereas in the lower combustion zone, the particles are assumed to follow a shrinking-core model dominated by the carbon-oxygen reaction. Although within the core, gasification reactions also occur but only carbon dioxide emerges as the product gas. The parametric studies showed that radiation had a marked effect on maximum temperature in the bed. Their study showed that the maximum temperature occurred at the bottom of the bed if residual carbon emerged, but the maximum temperature could wander in the bed if there was an ash layer in the bed.

Axial and radial dispersions of mass and heat for both gas and solids may not be negligible, particularly for large diameter gasifiers.

A general material balance based on a unit volume of a fixed bed gasifier can be written as:

For gas phase:

$$E_{zg} \frac{\partial^2 \hat{C}_i}{\partial z^2} + E_{rg} \frac{1}{r} \frac{\partial}{\partial r} \left(r \frac{\partial \hat{C}_i}{\partial r}\right) - \frac{\partial u_g \hat{C}_i}{\partial z} + (1-\varepsilon_B) \sum_j \nu_{ij} R_j = 0 \quad (13)$$

For solid phase:

$$E_{zs} \frac{\partial^2 \hat{C}_s}{\partial z^2} + E_{rs} \frac{1}{r} \frac{\partial}{\partial r} \left(r \frac{\partial \hat{C}_s}{\partial r}\right) \mp \frac{\partial u_s \hat{C}_s}{\partial z} + (1-\varepsilon_B) \sum_j \nu_{sj} R_j = 0 \quad (14)$$

where for countercurrent flow the sign of the third term is positive, and for co-current flow it is negative. Here $\hat{C}_i = F_i/A_T u_g$, $\hat{C}_s = F_s/A_T u_s$, $u_g = u_{go} \sum_i F_i / \sum_i F_{io}$ and $u_s = u_{so} \sum_s F_s / \sum_s F_{so}$.

A heat balance for a moving bed gasifier can be similarly written as:

For gas phase:

$$k^g_{ez} \frac{\partial^2 T_g}{\partial z^2} + k^g_{er} \frac{1}{r} \frac{\partial}{\partial r} \left(r \frac{\partial T_g}{\partial r}\right) - \frac{[(\sum_i u_g \hat{C}_i C_{pi}) T_g]}{\partial z}$$

$$= h_p a(T_g - T_s) - (1-\varepsilon_B) \sum_j (-\Delta H_j R_j)(1-\delta_j) \quad (15)$$

For solid phase:

$$k_{ez}^{s} \frac{\partial^2 T_s}{\partial z^2} + k_{er}^{s} \frac{1}{r} \frac{\partial}{\partial r} \left(r \frac{\partial T_s}{\partial r}\right) \mp \frac{\partial[(\sum_s u_s \hat{C}_s C_{ps}) T_s]}{\partial z}$$

$$= h_p a(T_s - T_g) - (1-\varepsilon_B) \sum_j (-\Delta H_j R_j) \delta_j \qquad (16)$$

where δ_j is unity when reactions take place on the surface of the solid and is zero when they take place in the gas phase. For example, for $O_2 + H_2$, $CO + O_2$, etc., $\delta_j = 0$.

A set of boundary conditions normally used at the entrance and at the exit of the reactor for the mass balance equations and the heat balance equations are applicable. The conditions of the symmetry about the reactor axis and the imperviousness at the reactor wall also apply for the mass balance equation. For the heat balance equation, at the wall ($r = r_o$) the following conditions are imposed:

$$- k_{er}^{g} \frac{\partial T_g}{\partial r} = h_w^g (T_g - T_w) \qquad (17)$$

$$- k_{er}^{s} \frac{\partial T_s}{\partial r} = h_w^s (T_s - T_w) \qquad (18)$$

The concentration profiles of gaseous species and temperature profiles of solids and gases can be obtained in theory by integration of the above equations.

Although correlations of axial and radial dispersion coefficients for gases through a fixed bed are available (79), the corresponding dispersion coefficients for solids are difficult to estimate. The temperature distributions of solids are indeed affected by these dispersion coefficients as well as other factors such as heat transfer coefficients.

A number of numerical methods to solve the above sets of equations are available. Most of them, however, suffer from problems and slow convergence. The subject matter deserves a separate discussion but is beyond the scope of this paper.

Thoma and Vortmeyer (80) analyzed a moving bed catalytic reactor and showed that the ratio of "flow capacities", $\Sigma F_i C_{pi}/\Sigma F_s C_{ps}$, is the most important parameter besides inlet conditions. The same conclusion was drawn by Luss and Amundson (81) when they analyzed a countercurrent liquid-liquid spray column. Thoma and Vortmeyer then performed an experiment and confirmed the range of multiplicity of the steady state solutions, which they obtained from their moving bed model.

The typical temperature and concentration profiles in a moving bed gasifier are shown in Fig. 9. The gas temperature usually intersects the solid temperature at the combustion zone near the bottom, and both solid and gas temperatures reach maximum values at some distances from the bottom of the gasifier.

Schaefer et al. (82) theoretically examined the stability of a countercurrent shaft furnace by treating a simple step change in heat generation rate and reported the multiplicity of solution depends on model parameters and boundary conditions. Mori and Muchi (83) treated the case of a first order reaction occurring in a catalytic moving bed and examined the reactor stability.

For noncatalytic gas-solid moving bed reactors, Ishida and Wen (84) examined the stability of operations by a numerical method and a graphical method for co-current and countercurrent operations and pointed out that a transitional instability of the rate controlling regime could exist. In such a case, a sudden shift from one controlling regime to another can occur in an exothermic reaction system depending on the system parameters and the initial temperatures of the solids and gas feeds. They showed that the pseudo steady-state analysis may become misleading when a sudden shift in controlling regime occurs.

This type of phenomena implies that the moving bed gasifier can exhibit ignition, extinction and hysteresis. For this purpose we shall use simple schematic diagrams shown in Fig. 10 to illustrate the phenomena under discussion.

In a noncatalytic system, the heat generation curves gradually vary as the solid reactant is consumed. Heat exchange curves also move depending on the temperature as indicated in the figure. Case A represents the situation in which solid and gas are both at the same temperature. Such situations could be encountered either when the heat transfer between the solids and the gas in an actual operation is extremely rapid or due to an assumption made in the model to simplify the mathematics. When h is infinity, there is no sudden shifting of the rate controlling regime, and only one solution can be obtained. However, as shown in Case B, when the heat transfer coefficient is small and temperatures of the gases and the solids are different, it is possible for the reaction to follow the path of A B C D E F G H in which a sudden shift occurs in the rate controlling regime from B to C. On the other hand, the reaction can also follow the path of A' B' C D D F G H displaying multiple steady state. Which of the steady states realized in a gasifier depends on the initial temperature of the system. Needless to say, the thermal instability and multiple steady states can occur not only in moving bed gasifiers but also in entrained bed gasifiers and fluidized bed gasifiers.

(3) ENTRAINED BED GASIFIER MODEL

The entrained flow coal gasifier is normally operated cocurrently, either downflow or upflow, and at temperatures significantly higher than either fixed or fluidized bed gasifiers. The characteristics of entrained bed gasifiers are listed in Table III. Pulverized coals and gasifying medium (oxygen, steam, etc.) are injected through nozzles into the gasifier where considerable

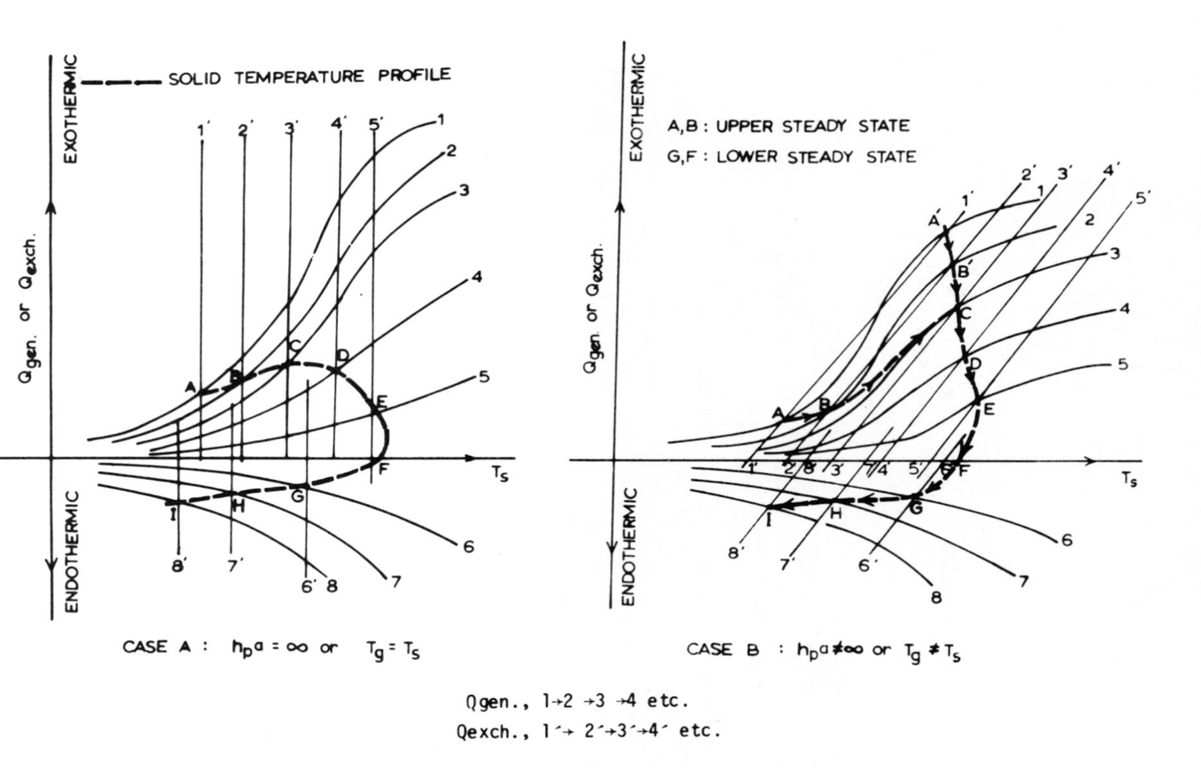

Figure 10. Temperature excursions in a moving bed gasifier

mixing takes place due to turbulence and swirling of both gases and solids. The schematic temperature and concentration profiles of an entrained gasifier are shown in Fig. 11. In some of the gasifiers, the mixing depends on axial jets from injection nozzles whereas others develop a vortex field induced by tangential firing. It is therefore very difficult to model such complex hydrodynamics in an entrained flow system. Kane and McCallister (85) recently analyzed the flow field of an entrained flow gasifier and determined the dimensionless groups that govern the scaling laws of the gasifier. Among the important dimensionless groups they identified are the swirl number, geometric scale ratio, Froude number and particle loading ratio. Existing models are not adequate to predict solid concentrations and gas velocity for such a complex flow system. Most of the models of pulverized fuel combustion systems and entrained bed gasifiers have been formulated based on simple flow patterns such as complete-mixing (86), isothermal plug flow (87,88,89) and a combination of complete-mixing and plug flow (86,90). The combination of complete-mixing and plug flow seems to be the most commonly used flow pattern because the empirically measured residence time distributions can be fitted to formulate an approximate size of the mixing zone in modelling an entrained gasifier. Ubhayakar et al. (91) also considered a combination model in which coal pyrolysis takes place in the mixed zone near the nozzle, which is followed by a plug flow zone in an entrained bed. The mixing time was used as an adjustable parameter in their model.

In a number of entrained bed models a homogeneous flow is assumed when solid particles are very small. However, in entrained coal gasifiers, the solids are subjected to a high temperature environment and are under high gas velocity. Therefore, the residence time of solids is extremely short, requiring careful analysis of both solids and gas flow patterns. Field et al. (14) reviewed the flow patterns and mixing of fuel and air in turbulent jet flow within a combustion chamber. Thring (92) proposed an empirical equation to estimate the mass rate of material recirculation. The recirculation current is set up as the coal-feed jet decreases from its nozzle velocity and entrains surrounding fluid. Zahradnik and Grace (67) examined the methane yield and operating conditions of the equipment development unit of the Bi-Gas process by evaluating the recirculation ratio and the entrainment and disentrainment distance using the correlation developed by Thring (92). Tester et al. (93) also modelled the two stage Bi-Gas process and obtained the temperature and concentration profiles assuming the temperature of solids and gas as equal. Since the entrained bed gasifier operates at high temperatures (980° to 1930°C), radiative heat transfer plays an important role. Consequently the evaluation of radiative properties of materials such as refractory, gas and cloud of particles including coal, char, soot and ash at flame temperature becomes important. Dobner (15) presented a review of entrained bed

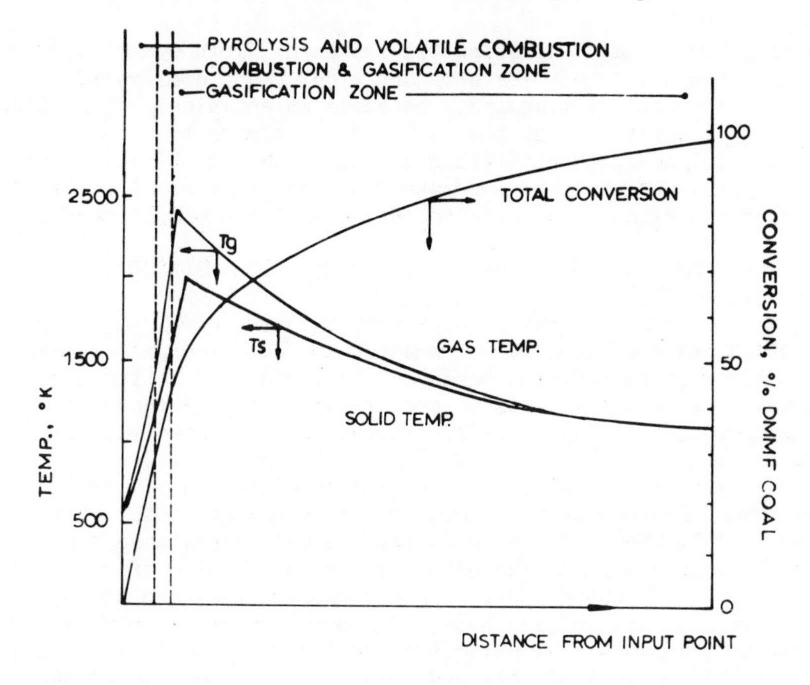

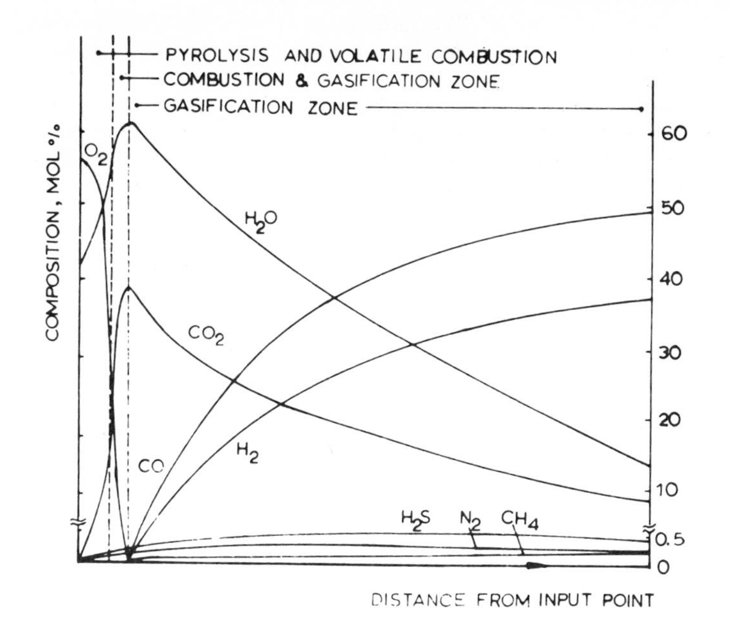

Figure 11. Temperature and concentration profiles in a typical entrained bed gasifier

gasifiers' modelling discussing the issues of fluid dynamics, heat and mass transfer, and chemistry and kinetics of coal gasification. A sensitivity analysis based on an entrained bed gasifier model (94) indicates that the effluent gas compositions are comparatively easy to simulate since they are very close to the equilibrium, whereas the temperature and velocity profiles of the gasifier are rather difficult to estimate from a simple model.

(4) FLUIDIZED BED COMBUSTOR (FBC) AND GASIFIER MODELS

There has been a far greater effort in the development of fluidized bed processes for combustion and gasification of coal than corresponding efforts based on other modes of solids-gas contacting processes. This is due primarily to the inherent advantages that fluidized beds offer (which are listed in Table III).

The direct combustion of coal in fluidized beds containing limestone/dolomite additives appears to be the most attractive scheme for burning coal in an environmentally acceptable manner. Also, from an economic standpoint, the high heat transfer coefficient and heat generation rates in fluidized bed combustors (FBC) results in a smaller boiler volume for a given duty as compared to conventional pulverized coal burning boilers. The potential advantages of this new combustion scheme have prompted considerable research in fluidized bed combustion in recent years, notably in the U.S. and U.K., and much useful information is becoming available from several bench-scale and pilot plant experiments. The fundamental and engineering aspects of fluidized bed coal combustion are discussed by Beer (95). However, only recently, attempts are being made to develop theoretical models for predicting the performance of FBC. A review of the modelling efforts in fluidized bed combustion has been recently presented by Caretto (96).

Almost all of the models proposed to date are based on the two phase theory of fluidization originally proposed by Toomey and Johnstone (97) and later modified by Davidson and Harrison (98). According to the theory, the fluidized bed is assumed to consist of two phases, viz., 1) a continuous, dense particulate phase (emulsion phase) and 2) a discontinuous, lean gas phase (bubble phase) with exchange of gas between the bubble phase and emulsion phase. The gas flow rate through the emulsion phase is assumed to be at minimum fluidization and that in excess of the minimum fluidization velocity passes through the bubble phase. This formulation of the two phase theory is based on the assumption that the voidage of the emulsion phase remains constant. However, as pointed out by Rowe (99) and Horio and Wen (100) this assumption may be an over-simplification. In particular, experiments with fine powders ($d_p < 60\ \mu m$) conducted by Rowe show that the dense phase voidage changes with gas velocity, and as much as 30 percent of the gas flow occurs interstitially. This effect can be taken into account by introducing a multiplier of minimum fluidization velocity, a coefficient greater than unity that is a function of gas velocity and bed height.

The models that have appeared in the literature deal with different aspects of fluidized bed combustion. However, the important variables that relate to the performance of FBC, viz., combustion efficiency, carbon concentration, bed temperature profile, etc.,have been the primary emphasis in all these models noted in the following section.

Avedesian and Davidson (33) developed a combustion model based on the two phase theory. The combustion was assumed to be controlled by (1) interphase transfer of oxygen from bubbles of air to surrounding ash particles and (2) diffusion of oxygen through the ash phase. (See Fig. 7). The theoretical prediction was shown to be in good agreement with experimental data. Campbell and Davidson (101) later modified this model to include the presence of CO_2 in the particulate phase of the combustion and applied the model to predict the carbon particle size distribution in a continuously operated fluidized bed combustion.

Horio et al. (102) developed a general mathematical model for the FBC, employing the modified version of the bubble assemblage model (103,104). Predictions of combustion efficiency, axial temperature profile and sulfur retention efficiency in the bed were compared with experimental data obtained from the National Coal Board and Exxon Miniplant. Fig. 12 presents typical profiles of carbon concentration and temperature in the bed. Fig. 13 indicates the SO_2 retention by limestone additives as a function of bed temperature.

Baron et al. (105) formulated a model for the FBC based on the two phase theory for predicting the combustion efficiency and carbon concentration in the bed. They accounted for the elutriation loss and attrition of particles in the bed using the Merrick and Highley correlation (106). Within the range of particle size for which the correlations are valid they found that the model predictions were reasonable. However, for larger particles, the correlation underestimated the elutriation loss. Gibbs (107) derived a mechanistic model for the combustion of coal in a fluidized bed that enables the combustion efficiency, carbon hold-up, and spatial distribution of oxygen in the bed to be calculated. The burning rate of coal was assumed to be diffusion controlled. The carbon loss due to elutriation, attrition and splashing of coal from bursting of bubbles on the bed surface was taken into account in the model formulation. The carbon loss, predicted by the model,was strongly dependent on the mean bubble diameter and excess air.

Gordon and Amundson (108) examined the influence of several operating variables on the steady state performance of a fluidized bed combustion via a mathematical model. Multiple steady state solutions were found to exist for the typical range of operation variables. In particular, it was noted that one of the key

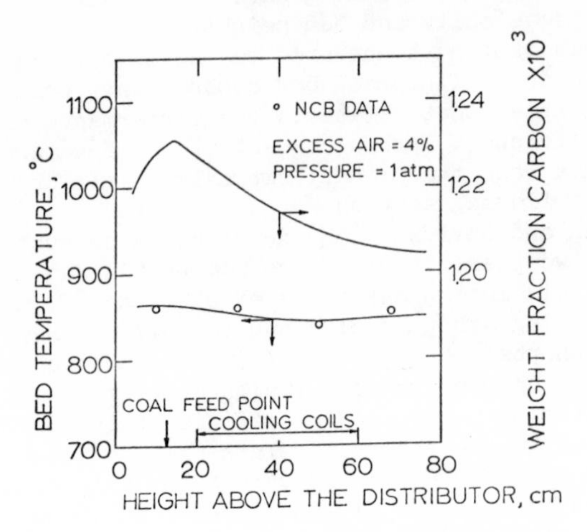

American Institute of Chemical Engineers

Figure 12. Temperature and carbon concentration profiles in FBC (113)

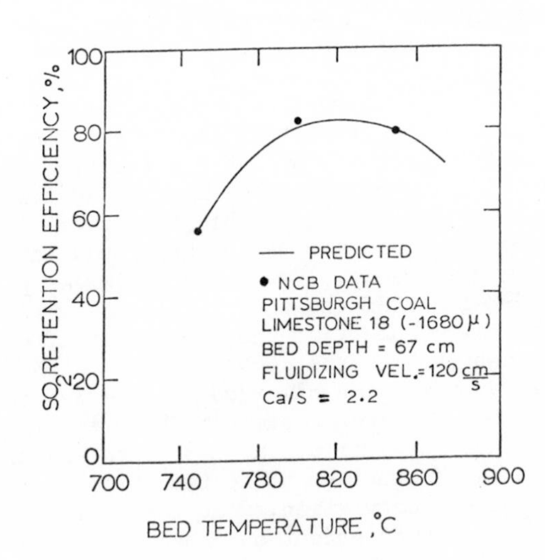

American Institute of Chemical Engineers

Figure 13. Effect of bed temperature on SO_2 retention (113)

factors in determining the state of the bed, as well as the multiplicity of the system, was the gas interchange coefficient between the bubble phase and emulsion phase.

Chen and Saxena (109) used a three phase bubbling bed model (bubble phase, cloud-wake phase and emulsion phase) for predicting the sulfur retention efficiency in a fluidized bed combustor. The size distribution of coal and limestone particles in the feed, overflow and elutriated fractions were treated by population balance. Varying bubble size along the bed height, and plug flow of gas in all the three phases were assumed in the model. However, the bed was assumed isothermal.

Wen et al. (110) developed a fluid-bed reactor model for the hydrogasification of char using the bubble assemblage concept. Solids were assumed to be completely mixed in each compartment with exchange of gas between the bubble phase and emulsion phase. Particle size distribution was not considered in the model and this may affect the predicted conversions of char.

None of the models discussed above are versatile enough for scale-up purposes because of their incomplete treatment of the coupled complex phenomena occurring in the fluidized bed. The main difference in the various models is their treatment of the gas flow in the two phases and in the solids mixing mode in the dense phase.

A classification of the fluidized bed models is presented in Table IV.

Important aspects of all the models include limestone-SO_2 kinetics, combustion kinetics and other gas-solids reaction kinetics, gas phase material balances, solid phase material balances, gas exchange between the bubble and emulsion phases, heat transfer, and bubble hydrodynamics.

The governing equations describing the physico-chemical processes occurring in the bed can be found in the individual papers cited in Table IV and are not presented here.

Future modelling efforts should be directed toward improving the existing models. The deficiencies and ways of improving the present models are discussed below.

Bubble Hydrodynamics

Bubbles coalesce and grow in size as they ascend through the bed. Models (105,107) based on constant bubble size show that the bubble size does indeed affect the combustion efficiency and carbon concentration in the bed. Furthermore, the phenomenon of jetting at the distributor surface should be taken into account because bubbles are not formed yet in this region. It has been reported (114,115) that the gas-solid mixing adjacent to the distributor is markedly different from the freely bubbling zone. This point has to be taken into consideration in modelling.

For scale-up purposes, good correlations to account for the changing bubble size in the presence of internals (cooling coils)

Table IV. Classification of Models Proposed for Fluidized Bed Combustion and Gasifiers

Model Description	Investigators	Gas Flow Pattern		Solids Mixing in the Bed	Remarks
		Bubble Phase	Emulsion Phase		
Fluidized bed combustors					
Two phase bubbling bed model	Avedesian and Davidson (33) Gibbs (107) Campbell and Davidson (101) Gordon and Amundson (108) Becker et al. (111) Baron et al. (29)	a) Plug flow b) Plug flow c) Complete mixing	Plug flow Complete mixing Complete mixing	Complete mixing	1) Bubble diameter is assumed to be uniform throughout the bed in most cases and is an adjustable parameter 2) Cloud and wake are combined in the emulsion phase 3) No combustion of carbon in the bubble phase
Two phase compartment in series models	Horio and Wen (112) Horio et al. (102) Rengarajan et al. (113)	Complete mixing in each compartment	Complete mixing in each compartment	Complete mixing in each compartment with backflow of solids from one compartment to another	1) Bubbles grow along the bed height 2) The backflow solids mixing parameter is adjustable 3) Cloud and wake are combined in the bubble phase 4) Combustion of carbon occurs in both bubble and emulsion phase
Three phase bubbling bed model	Chen and Saxena (109)	Plug flow	Plug flow	Complete mixing	1) Cloud-wake is treated as a separate phase and is in plug flow 2) Solids in the cloud-wake is at minimum fluidization 3) Bubbles grow along the bed height 4) Combustion occurs in the cloud-wake and emulsion phases 5) Isothermal condition throughout the bed
Fluidized bed gasifiers					
Two phase compartment in series model	Wen et al. (110)	Complete mixing in each compartment	Complete mixing in each compartment	Complete mixing in each compartment with backflow of solids from one compartment to another	1) Bubbles grow along the bed height 2) No solids in the bubble phase 3) Solids backflow mixing parameter is adjustable

in the combustor should be developed (113).

Chemical Reactions in Fluidized Beds

The mechanism of carbon combustion in FBC is assumed to be diffusion controlled in most of the modelling efforts as discussed in the section on Char-Oxygen Reaction. This is true only for large particles (> 300 microns) at high temperatures (> 1200°K). Feed coal contains a wide range of sizes, and assuming a diffusion controlled kinetics for all particle sizes would lead to over-estimation of the combustion rate.

As discussed earlier, pyrolysis of coal, composition and yield of volatiles are strongly dependent on coal particle size, temperature and heating rate. Instantaneous coal devolatilization at the feed point can be expected only when the particle size is small. The time needed for the devolatilization of a 1000 micron coal particle is 0.5 to 1.0 second, which is of the same magnitude as solids mixing time in the bed (95,29). This necessitates the consideration for the rate of devolatilization of coal.

A realistic model for the FBC should also include SO_2 absorption by limestone additives and NO production from fuel nitrogen and its subsequent reduction by char (95,116,117).

Attrition and Elutriation

Size distributions of solids (coal and limestone) in the feed and in the bed should be considered in the evaluation of attrition and elutriation loss. Standard correlations for elutriation rate constants have been found to be inadequate for the calculation of solids elutriation. Data obtained from large pilot-scale FBC show large disagreement from those calculated based on existing elutriation rate correlations. Recently, correlations for attrition and elutriation of bed particles have been proposed by Merrick and Highley (106). For larger particles, this correlation underestimates the carbon loss.

Solids Mixing

In most of the modelling studies, solids in the emulsion phase are assumed to be completely mixed. Though this is a reasonable assumption in many cases, it has been observed that in the presence of closely packed horizontal coils (Exxon miniplant data) solids mixing is severely hindered resulting in steeper temperature profile. In such cases the complete mixing assump-tion would be erroneous. In some models (102,110,113), the solids mixing due to bubble motion is accounted for by an adjustable backmix parameter. However, this is still not an adequate treat-ment of the mixing process.

Reactivities of coal and other solids (limestone, dolomite, etc.) in fluidized bed combustors or gasifiers decrease as they are converted. The degree of mixing of solids and gas, therefore, affects the overall conversion of solids and product gas composition. Hence, in modelling such behavior, it is necessary to apply population balance of the varying size particles having varying reactivities in the mixing process prevailing in a fluidized bed reactor.

Another area of modelling for fluidized bed coal combustors and gasifiers that needs to be developed is the construction of dynamic models that can be used to simulate and analyze the behavior during the start-up, shutdown, turn-up and slumping of the fluidized bed. This is important from the point of view of developing control and following the load demand of the bed.

Freeboard Reactions

Most of the fluidized bed coal combustion and gasification models ignore freeboard reactions of volatiles and char. For shallow beds, volatiles burn predominantly in the freeboard. In addition the char particles splashing from the bed surface can also react with the oxygen in the freeboard. Yates and Rowe (118) have proposed a model for the reactions occurring in the freeboard. Such an approach can be adopted for modelling the freeboard reactions in the FBC.

COAL LIQUEFACTION REACTIONS

There are four major types of coal liquefaction processes being developed today. They can be classified as (1) Pyrolysis; (2) Solvent Extraction; (3) Catalytic Liquefaction, and (4) Indirect Liquefaction. A comprehensive report on assessment of technology for the liquefaction of coal has been issued by the National Research Council (119).

(1) CHARACTERISTICS OF COAL LIQUEFACTION PROCESSES

The pyrolysis and hydropyrolysis process produces liquid product and char. Either fluidized beds or entrained beds are used for this process. The reaction kinetics and reactor modelling of solid-gas systems have already been discussed earlier.

The solvent extraction process involves the contacting of coal and a hydrogen donor solvent at a temperature up to 500°C to produce solid or liquid product. The extraction is carried out either directly under hydrogen pressure or without hydrogen in the dissolver but with the solvent being hydrogenated in a separate step before it is returned to the extraction step.

Catalytic liquefaction process allows a slurry of coal and oil to be hydrogenated over active catalysts in a fixed bed reactor, in an ebullating bed reactor or in a trickle bed reactor to produce a liquid hydrocarbon product.

Indirect liquefaction can be carried out in a fixed bed or

fluidized bed catalytic reactor in which synthesis gas produced from coal gasification is converted to hydrocarbons and methanol. Methanol may be further converted over zeolite catalysts to gasoline. The advantages and disadvantages of the coal liquefaction processes are listed in Table V. Since pyrolysis and hydropyrolysis processes have been discussed in the previous sections and the indirect liquefaction processes are mostly based on the conventional catalytic reaction engineering, their discussion will be excluded in this section.

(2) MECHANISM OF COAL DISSOLUTION

For successful operation, the solvent must be thermally stable at reaction conditions, and it must act either as a hydrogen donor or hydrogen transfer agent or both. Van Krevelen (120) suggested that Lewis' basicity of the solvent is an additional important parameter in successful coal extraction. Oele et al. (121) classified the solvents into five groups with respect to their effect on coal. The three groups that are of interest in liquefaction practice are the specific solvents (e.g. pyridine), degrading solvents (e.g. anthracene), and reactive solvents (e.g. tetralin). Dryden (122) suggested to use the square of the solubility parameter in correlating solvent effectiveness. The solubility parameter is a measure of the cohesive forces in a solution that has no excess entropy of mixing (123). Silver and coworker, based on Kiebler's data (124), found that solvents with a nonpolar solubility parameter of 9.5 (cal/c.c.) appeared to be most effective for coal dissolution (125).

The degree of dissolution of coal or hydrogenation is an indication of the effectiveness of the process concerned. Unfortunately, no uniform definition exists in this matter. It is a common practice to subject reactor effluent, after venting gaseous products, to extraction by an organic solvent. A variety of solvents have been used, e.g. benzene, pyridine, cresol, xylenol, etc. Benzene was commonly used in the earlier days by most investigators. Any material insoluble in benzene is assumed to be unreacted coal and does not contribute to the viscosity of the product oil. In the last few years, it has been found that this assumption needs re-examination. Part of the benzene insolubles are pyridine soluble and therefore can be considered as reacted coal. This fraction, named pre-asphaltene by Sternberg (126) and asphaltols by Farcasiu et al. (127), can be clearly distinguished from asphaltene and product oil by its high viscosity. Therefore, it is the pyridine insolubles that may be regarded as unreacted coal, after making any adjustment for mineral matter and catalyst. It is very important to distinguish the data of coal dissolution rate based on benzene wash from those obtained by pyridine wash since we are dealing with different process steps as will be discussed later.

Table V. Characteristics of Coal Liquefaction Processes (*119*)

(I) Pyrolysis and Hydropyrolysis Processes
(Lurgi-Ruhrgas, COED, Occidental, etc.)

Advantages

- Operating pressures may be low.
- Addition of hydrogen or other reactant to coal is not necessary.
- Equipment is relatively simple and low in cost due to very short residence time.

Disadvantages

- Approximately one-third of the coal can be converted to liquid.
- Separation of the heavy oil product from char and ash is difficult.
- The liquid product requires further treatment to make it acceptable as fuel.
- The char produced has limited market value.

(II) Solvent Extraction Processes
(CSF, SRC, SRL, Costeam, EDS, etc.)

Advantages

- Operating temperatures are lower than pyrolysis.
- Varying degree of extraction and hydrogenation can be applied to produce quality of product desired.

Disadvantages

- Separation of unreacted coal and ash is difficult.
- The product is a friable solid at room temperature and is difficult to transport, store and handle in conventional equipment.
- The handling and recycling of coal-oil slurry presents problems.

(III) Catalytic Liquefaction
(Bergius, H-Coal, Synthoil, CCL, etc.)

Advantages

- Recovery of catalyst from solid residues is not needed.
- Operating pressure is lower than 270 atm.
- Residence time is short, and product quality can be regulated.

Disadvantages

- Separation of unreacted coal and ash is difficult.
- Hydrogen and recycling oil are required.
- Catalysts deactivate rapidly.

(IV) Indirect Liquefaction
(Fisher-Tropsch, Methanol Synthesis, Mobile Zeolite, etc.)

Advantages

- Almost any coal can be used.
- The product quality can be controlled and made free from nitrogen and sulfur.

Disadvantages

- Coal must be gasified and product gas must be purified before being converted into liquid products.
- Thermal efficiency of the process is much lower than coal hydrogenation processes.
- The plant is complex, and capital cost is high.

Weller et al. (128) provide a classic working model for coal liquefaction. The mechanism proposed is based on reaction of coal under hydrogen pressure and stannous sulfide catalyst. The reaction path proposed is coal → asphaltene → oil; both reactions are first order with water and gas as a by-product. However, no oil vehicle is involved in this model.

Curran et al. (129) studied the mechanism of hydrogen transfer from tetralin to bituminous coal without the presence of molecular hydrogen. Coal is thermally cleavaged into free radicals, which are then stabilized by capturing hydrogen atoms from a donor solvent. The extent of extraction is independent of the solvent composition, being a primary function of the quantity of hydrogen transferred. The rate controlling step in the hydrogen transfer reaction is the rupture of covalent bonds that cannot be promoted by hydrogenating or cracking-type catalyst. It should be noted that for bituminous coal, even at 400°C, the temperature is considered to be too low for any extensive disintegration or pyrolysis. Thus, the role of a donor solvent, besides transferring hydrogen to coal, tends to promote the thermal cleavage. Curran (129) also concluded that only 0.2 weight % (based on maf coal) or less hydrogen is needed to obtain the first 50% coal conversion; but to reach 92% conversion, 1.4% hydrogen consumption is required. This is essentially the same conclusion that Neavel (131) and researchers in Mobil (130) have found.

Heredy and Fugassi (132) studied dissolution of coal in phenanthrene via thermal cracking and simultaneous hydrogen disproportionation reactions. Phenanthrene, which does not possess hydrogen donor property, probably plays the role of a free radical carrier or hydrogen transfer agent. Recently, investigators from Mobil (133) also reported a hydrogen shuttling mechanism, whereby coal fragments into smaller soluble forms with the aid of solvents that are good "shuttlers" of hydrogen:

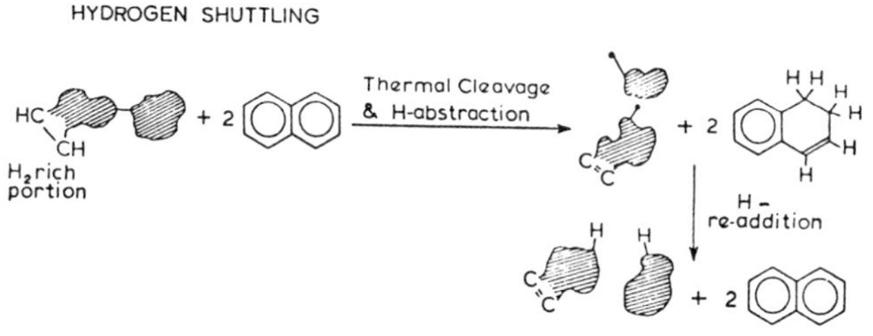

HYDROGEN SHUTTLING

HC CH + 2 → Thermal Cleavage & H-abstraction → + 2

H_2 rich portion

H-re-addition

+ 2

Neavel (131,134) measured the rate of conversion of an Illinois high volatile bituminous coal in tetralin at 400°C. Approximately 90% of the coal becomes soluble in pyridine in less than five minutes. About 40% becomes soluble in benzene. Researchers in Mobil conducted experiments using several coals and a variety of solvents at both short and long contact times (130). Coal dissolution is very fast initially, converting 70-80% of West Kentucky Coal at 427°C in about three minutes. Very little hydrogen is consumed at this stage. Forty percent of the oxygen and sulfur is removed quickly with near stoichiometric loss of hydrogen from the coaly matter. At longer time, hydrogen consumption is significantly higher than the stoichiometry required for H_2O or H_2S production.

Table VI gives an example of the reaction pathways proposed by different investigators for the production of liquid fuel from coal. It is apparent that as the mechanism becomes more and more complicated, so do the mathematical models of the reaction.

It is clear now that the role of solvent and heating is to facilitate the thermal degradation of coal resulting in the formation of free radicals of relatively low molecular weight. These free radicals are stabilized by hydrogen transfer from hydroaromatic solvent molecules. As long as a hydrogen donor or a hydrogen transfer solvent is present, dissolution occurs even when molecular hydrogen is not present. However, as the hydrogen inventory in the donor solvent or in coal becomes depleted, condensation or repolymerization reactions prevail, which tend to yield substances of high molecular weight. Kang et al. (136) speculated that coke formation results when thermal cracking gets ahead of hydrogenation reactions.

The rate of coal dissolution appears to be insensitive to the coal particle size (129,137), but it is dependent on temperature, pressure, reactor hydrodynamics and types of coal (137). The function of hydrogen and the catalyst is to subsequently rehydrogenate the vehicle solvent, although the catalyst is also responsible for desulfurization, denitrogenation and the production of lighter liquid products. Paradoxically, the ratio of hydrogen to carbon in Solvent Refined Coal is lower than that in the original feed coal. This is because hydrogen is consumed in the production of hydrocarbon gases, removal of heteroatoms, production of recycle solvents, etc., thus raising the aromatic content of the Solvent Refined Coal. A proposed working model for coal dissolution is

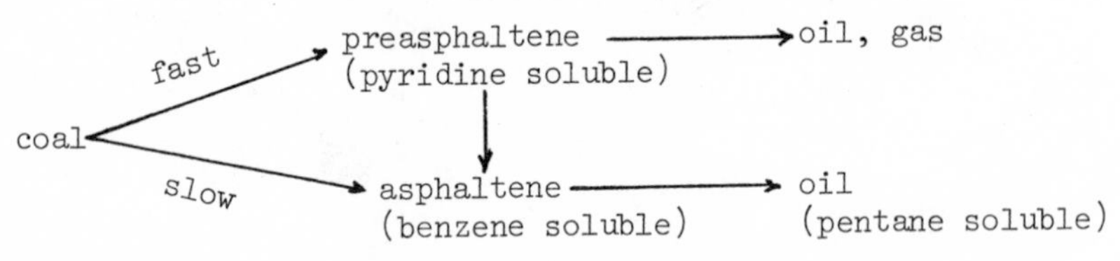

In this scheme, a fraction of the coal would undergo fast thermal reaction whereby the free radicals would be stabilized by the hydrogen inventory within the coal or solvent itself. This can take place by hydrogen shuttling or hydrogen transfer. Very little molecular hydrogen would be consumed at this stage. The rate of reaction is a strong function of temperature. The remaining portion of the coal would take the second and much slower path to form asphaltene. Since the fast reaction would deplete the available hydrogen inventory, the hydrogen for the second reaction would have to come from molecular hydrogen diffusing into the solvent and "donating" the hydrogen to the coal particles. The rate for the second reaction would depend on, besides temperature, hydrogen pressure and reactor hydrodynamics for further dissolution and hydrogenation to yield product oil.

(3) COAL LIQUEFACTION REACTOR ANALYSIS

When coal is heated in the presence of hydrogen and a catalyst, dissociative chemisorption of the hydrogen on the catalyst surface can yield active hydrogen that can stabilize the thermally-produced reactive fragments (141). In addition, the catalyst also promotes hydrogenation of the aromatic structures with subsequent ring opening and cracking reactions, thereby reducing the size of large clusters. Since most bituminous coal softens upon heating, the resulting caking and sticking problems can plug reactors, not to mention causing massive coke deposition. In order to avoid these problems, most liquefaction processes brought the coal-solvent slurry into contact with a catalyst bed (packed or ebullating) under hydrogen pressure.

Feldman et al. (142) concluded that in the hydrogenation of coal tar and a coal-coal tar slurry using a Co-Mo catalyst, the rate of hydrogenation is limited by the diffusion of hydrogen from the gas bubbles to the liquid phase rather than by inter or intraphase diffusion involving the catalyst. Thus the general assumption in a catalyzed system that the reaction rate was proportional to the mass of catalyst is not justified. In fact it is suggested (142) that catalyst activity should be reduced, to a level where there is no hydrogen starvation at the catalyst surface, in order to minimize carbon deposition.

It should be pointed out that in most liquefaction processes, the flow of the reactants is cocurrent upward. In cocurrent operation, there is no flooding limit, and greater throughput is attained compared with countercurrent column of similar size. Moreover, for the same values of gas and liquid flow rates, interfacial area, liquid mass transfer coefficient and pressure drop values in a cocurrent upward packed column are always higher than those obtained in downflow towers (143). Upflow operations also give a better performance due to larger liquid holdup and better liquid distribution throughout the catalyst bed (144).

Shah et al. (145) studied the catalytic liquefaction of a sub-bituminous coal using the axial dispersion model. They concluded that a minimum gas flow rate could be found that would eliminate possible hydrogen mass transfer resistance from the gas phase to the catalyst surface.

The effect of temperature on the coal dissolution rate is

Table VI. Summary of Liquefaction Reaction Mechanisms

1. Weller et al. (128)

   Coal ⟶ Asphaltene ⟶ Oil

2. Falkum and Glenn (138)

   Coal I ↘ / Coal II ⇄ Asphaltene ⟶ Oil

3. Ishii et al. (139)

   Coal ⟶ Oil
   Coal ↘ Asphaltene ⟶ Oil

4. Liekenberg and Potgieter (140)

   Coal ⟶ Asphalt ⟶ Heavy oil

   Coal ⟶ Asphalt

   Coal ⟶ Heavy oil

5. Squires (135)

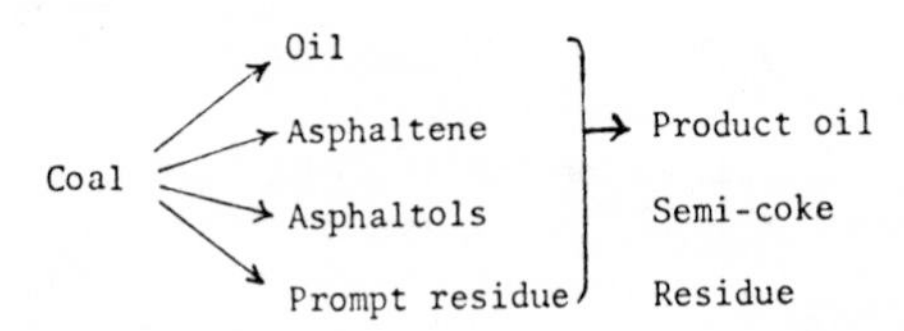

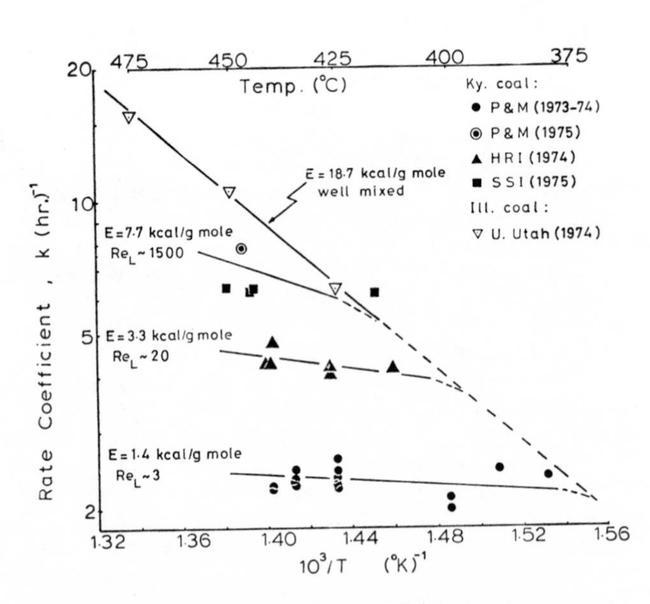

Figure 14. Effect of temperature on coal dissolution rate coefficient, indicating hydrodynamic influence

shown in Fig. 14. The rate of dissolution is based on the benzene soluble fraction. The data shown represent a variety of reaction diameter and length (137). The benzene soluble fraction gives a significantly smaller coal dissolution compared to the pyridine soluble fraction at short time (about 5 min or less) but approaches closely to that based on the pyridine soluble at long time (about 30 min. or more). As can be seen in Fig. 14, the activation energy seems to increase as the slurry flow rate is increased, indicating that the mass transfer effect becomes less as the degree of turbulence is increased at high liquid flow rates. Since bituminous coal is soften and becomes "plastic-like" at temperatures around 325-350°C, vigorous mixing is needed to disperse the viscous material to enhance heat and mass transfer in the coal slurry. It becomes apparent that a considerable hydrodynamic effect on coal dissolution exists in both the preheater and dissolver. A similar hydrodynamic effect can also be seen on rate of hydrogen absorption which appears to be controlled by the liquid side phenomena. Therefore, in design and scale-up of liquefaction reactors, we need to examine the extent of the fluid mixing for both slurry and gas and to formulate realistic flow models to account for the mass and heat transfer of the two phase flow prevailing in the reactor.

Acknowledgment

The authors wish to express their appreciation to K. W. Han, R. Krishnan, P. Rengarajan and Christy Boyle for their assistance in preparation of this paper. This work is partly supported by Department of Energy, Washington, DC.

Nomenclature

a = specific surface area of particles, cm^2/cm^3
A_T = cross-sectional area of bed, cm^2
C = constant in the Badzioch and Hawksley Equation and is related to the residual volatile matter, [-]
C_A = gas concentration of component A, mol/cm^3
C_i = concentration of gaseous component i in single particle, mol/cm^3
$\hat{C}_i$ = concentration of gaseous component i in the moving bed, mol/cm^3
C_{pi} = heat capacity of gaseous component i, cal/mol·°C
C_{ps} = heat capacity of solid component, cal/mol·°C
C_s = concentration of solid component in single particle; C_{so}, same at time t = 0, mol/cm^3
$\hat{C}_s$ = concentration of solid component in the moving bed, $mol/cm^3 \cdot bed$
d_p = particle diameter, cm
D_e = effective diffusivity of gaseous component, cm^2/sec
D_{ei} = effective diffusivity of gaseous component i, cm^2/sec
D_{oi} = molecular diffusivity of gaseous component i, cm^2/sec
E = activation energy, cal/mol
E_r = radial dispersion coefficient; E_{rg}, same of gaseous component; E_{rs}, same of solid component, cm^2/sec
E_z = axial dispersion coefficient; E_{zg}, same in the gas phase; E_{zs}, same in the solid phase, cm^2/sec
f = final fraction conversion of coal due to pyrolysis (= V*/100), [-]
F_i = mass flow rate of gaseous component i, mol/sec
F_s = mass flow rate of solid component, mol/sec
h_c = convective heat transfer coefficient, $cal/cm^2 \cdot sec \cdot °C$

h_p = heat transfer coefficient between gas and solid particle, cal/cm^2.sec·°C
h_r = radiative heat transfer coefficient, cal/cm^2·sec·°C
h_w = wall heat transfer coefficient; h_w^g, same between wall and the gas phase; h_w^s, same between wall the the solid phase, cal/cm^2·sec·°C
ΔH_j = heat of the j-th reaction, cal/mol
k, k_o = pyrolysis rate constant, sec^{-1}
k_D = diffusive mass transfer coefficient, gm/cm^2·sec·atm
k_e = effective thermal conductivity of coal, cal/cm^2·sec·°C
k_{er} = effective thermal conductivity for radial direction; k_{er}^g, same of the gas phase; k_{er}^s, same of the solid phase, cal/cm.sec·°C
k_{ez} = effective thermal conductivity for axial direction; k_{ez}^g, same of the gas phase; k_{ez}^s, same of the solid phase, cal/cm·sec·°C
k_s = surface reaction rate constant, cm$^{3(m+n-2)}$/(mol)$^{m+n-1}$·sec
k_v = volumetric rate constant, cm$^{3(m+n-1)}$/(mol)$^{m+n-1}$·sec
k_{gi} = mass transfer coefficient of gaseous component i between the gas phase and the solid phase, cm/sec
K_1, K_2 = constant in the Badzioch and Hawksley Equation in pyrolysis of coal, °K^{-1} and °K, respectively
K_c = overall mass transfer coefficient of the primary volatiles formed within the pores of coal particles, due to pyrolysis, sec^{-1}
N_i = molar flux of gaseous component i, mol/cm^2·sec
P_{H_2} = partial pressure of hydrogen; $P_{H_2}^*$, same in equilibrium with coal at various conversions
Q = constant in the Badzioch and Hawksley Equation and is a function of coal type, [-]
Q_{exch} = heat exchange rate between the gas phase and the solid phase, cal/sec
Q_{gen} = heat generation rate due to chemical reaction, cal/sec
R_g = gas constant, cal/mol·°K or atm·cm^3/mol·°K
R_j = rate of the j-th reaction based on the volume of solid particle, mol/cm^3·particle·sec
r = radial distance in the moving bed, cm
r_o = radius of the moving bed, cm
S_{ex} = geometric surface area of the shrinking interface of reacting solid particles, cm^2
t = time, sec
T = temperature; T_g, same of the gas phase; T_s, same of the solid phase, °K
T_w = wall temperature of the reactor, °K
u_g = superficial velocity of the gas phase; u_{go}, same at inlet of the moving bed, cm/sec
v = volume of any particular component of gaseous volatiles evolved due to pyrolysis, cm^3
v_o = volume of any particular component of gaseous volatiles released due to pyrolysis at time t = ∞, cm^3
V = yield of total volatiles, percent of moisture and ash-free coal, [-]
V* = ultimate yield of total volatiles (percent) from coal due to pyrolysis, in hydrogen or inert atmosphere
VM = proximate volatile matter content of coal, (percent)(moisture and ash-free coal basis)
V_{nr}^* = ultimate yield of nonreactive volatiles (percent) from coal due to pyrolysis in inert atmosphere
V_p = volume of solid particle, cm^3
V_r^{**} = reactive volatiles (percent) formed up to t = ∞ (potential ultimate yield of reactive volatiles)
W_o = original weight of solid particle, gm
X = fraction conversion due to pyrolysis, or due to the first stages of hydrogasification and char-steam reactions (= V/100), [-]
Z_s, Z_v = catalytic activity factor, [-]

Greek Letters

α_v = relative pore surface area function, [-]
β = positive constant, a value that depends on structure of porous particle and physical properties of diffusing gas, [-]
γ = positive or negative constant, and depends on density of the solid product, [-]
δ_j = distributed fraction of heat of the j-th reaction to the solid phase, [-]
ε = porosity of solid particle; ε_o, same at time t = 0, [-]
ε_B = voidage of the moving bed, [-]
η = effectiveness factor based on volume, [-]
ν_{ij} = stoichiometric coefficient of gaseous component i in the j-th reaction, [-]
ν_{sj} = stoichiometric coefficient of solid component in the j-th reaction, [-]
ϕ = Thiele modulus, [-]

Literature Cited

1. Badzioch, S. and Hawksley, P. B. W., Ind. Eng. Chem., Process Des. Dev. (1970) 9, 521.
2. Anthony, D. B. and Howard, J. B., AIChE J. (1976) 22, 625.
3. Anthony, D. B., Howard, J. B., Hottel, H. C. and Meissner, H. P., Fuel (1976) 55, 121.
4. Wen, C. Y., Bailie, R. C., Lin, C. Y. and O'Brien, W. S., "Coal Gasification", Adv. Chem. Ser. (1974) 131, 9.
5. Gregory, D. R. and Littlejohn, R. F., The BCURA Monthly Bulletin (1965) 29 (6), 173.
6. Howard, J. B. and Essenhigh, R. H., Ind. Eng. Chem., Process Des. Dev. (1967) 6, 74.
7. Juntgen, H. and van Heek, K. H., Fuel (1968) 47, 103.
8. Pitt, G. J., Fuel (1962) 41, 267.
9. Suuberg, E. M., Peters, W. A. and Howard, J. B., Ind. Eng. Chem., Process Des. Dev. (1978) 17, 37.
10. Graff, R. A., Dobner, S. and Squires, A. M., Fuel (1976) 55, 109-112.
11. Steinberg, M., and Fallon, P., "Coal Liquefaction by Rapid Gas Phase Hydrogenation", (Nov., 1974), Report BNL-19507.
12. Russel, W. B., Saville, D. A. and Greene, M. I., Ann. Meetg., AIChE., 70th, New York, No. 10f, Nov. 14-17, 1977.
13. Bush, T. W., Howard, J. B., Kenda, S., Mead, D., Peters, W. A., and Suuberg, F. M., "Basic Studies of Coal Pyrolysis and Hydrogasification", (Jan-June, 1977), Report FE-MIT-2295T26-2.
14. Field, M. A., Gill, D. W., Morgan, B. B. and Hawksley, P. G. W., "Combustion of Pulverized Coal", 179, BCURA, Leatherhead, 1967.
15. Dobner, S., "Modelling of Entrained Bed Gasification", The Issues The Clean Fuels Institute, The City College of the City University of New York, EPRI, Palo Alto, Cal. Jan. 15, 1976.
16. Hottel, H. C., Williams, G. C., Nerheim, N. M.. and Schneider, G. R., 10th Intern. Symp. on Combustion, 111-121, (1965).
17. Ishida, M., and Wen, C. Y., AIChE J. (1968), 14, 311.
18. Walker, P. L., Jr., Rusinko, F., Jr. and Austin, L. G., Advan. Catalysis, (1959), 11, 133.
19. Thring, M. W.. Essenhigh, R. H., "Chemistry of Coal Utilization-Supplementary Volume", Lowry, H. H., Ed., Ch. 17, John Wiley and Sons, New York, 1963.
20. Glassman, I., Ann. Meetg., AIChE., 70th, New York, Nov. 13-14, 1977.
21. Field, M. A., Combust. Flame (1969), 13, 237.
22. Field, M. A., Combust. Flame (1970), 14, 237.
23. Mulcahy, M. F. R. and Smith, I. W., Rev. Pure and Appl. Chem. (1969) 19, 81.
24. Sergent, G. D. and Smith, I. W., Fuel (1973) 52, 52.
25. Hamor, R. J., Smith, I. W. and Tyler, R. J., Combust. Flame (1973) 21, 153.
26. Smith, I W. and Tyler, R. J., Combust. Sci. Techn. (1974) 9, 87.
27. Dutta, S. and Wen, C. Y., Ind. Eng. Chem., Process Des. Dev. (1977), 16, 31.
28. Authur, J. R., Trans. Faraday Soc. (1951), 47, 164.
29. Borghi, G., Sarofim, A. F. and Beer, J. M., Ann. Meetg., AIChE, 70th, New York, No. 34C, Nov. 14-17, (1977).
30. Wicke, E. and Wurzbacher, G., Int. J. Heat Mass Transfer (1962), 5, 277.
31. DeGraaf, J. G. A., Brennst-Warme-Kraft (1965) 17, 227.
32. Kish, D., Ber. Bunsengas Phys. Chem. (1967), 71, 60.
33. Avedesian, M. M. and Davidson, J. F., Trans. Instn. Chem. Engrs. (1973), 51, 121.
34. Essenhigh, R. H., "Combustion of Coal" in Coal Conversion Technology ed. by E. S. Lee and C. Y. Wen, Addison-Wesley Publishing Co. in print, 1977.
35. Caram, H. S. and Amundson, N. R., Ind. Eng. Chem. (1977), 16, 171.
36. Burke, S. P. and Schuman, T. E. W., Proc. 3rd Int. Conf. Bituminous Coal (1931), 2, 485.
37. Zielke, C. W., and Gorin, E., Ind. Eng. Chem. (1955) 47, 820.
38. Hiteshue, R. W., Anderson, R. B., and Schlesinger, M. D., Ind. Eng. Chem. (1957) 49, 2008.
39. Hiteshue, R. W., Anderson, R. B. and Schlesinger, M. D., Ind. Eng. Chem. (1960) 52, 577.
40. Pyrcioch, E. J., and Linden, H. R., Ind. Eng. Chem. (1960) 52, 590.
41. Wen, C. Y., and Huebler, J., Ind. Eng. Chem., Process Des. Dev. (1965) 4, 142.
42. Feldkirchner, H. L., and Linden, H. R., Ind. Eng. Chem., Process Des. Dev. (1963) 2, 153.
43. Birch, T. J., Hall, K. R., and Urie, R. W., J. Inst. Fuel (1960), 33, 422.
44. Johnson, J. L., "Coal Gasification", Adv. Chem. Ser. (1974), 131, 145.
45. Zahradnik, R. L., and Glenn, R. A., Am. Chem. Soc., Div. Fuel Chem. Preprints, New York, Sept. 1969.
46. Dutta, S.. Wen, C. Y., and Belt, R. J., Ind. Eng. Chem., Process Des. Dev. (1977), 16, 20.
47. Turkdogan, E. T., Koump, V., Vinters, J. V., and Perzak, T. F., Carbon (1968) 6, 467.
48. Turkdogan, E. T., and Vinters, J. V., Carbon (1969) 7, 101.
49. Wen, C. Y., and Wu, N. T., AIChE J. (1976) 22, 1012.
50. Fucks, W. E., and Yavorsky, P. M., Am. Chem. Soc., Div. Fuel Chem. Preprint (1975), 20, (3), 115.
51. Yoshida, K., and Kunii, D., J. Chem. Eng. Japan, (1969) 2, 170.
52. Austin, L. G., and Walker, P. L. Jr., AIChE J. (1963) 9, 303.
53. Muralidhara, H. S., "The Effect of Solid Chemical Composition on Coal Char Reactivity in a Carbon Dioxide Atmosphere", Ph. D. dissertation, West Virginia University, (1978).
54. Gadsby, J. Hinshelwood, C. N., and Sykes, K. W., Proc. Royal Soc. (London) (1946), A187, 129.
55. Jensen, G. A., Ind. Eng. Chem., Process Des. Dev. (1975) 14, 314.
56. Klei, H. E., Sahagian, J., and Sundstrom, D. W., Ind. Eng. Chem., Process Des. Dev. (1975) 14, 470.

57. Kayembe, N. and Pulsifer, A. H.. Fuel (1976), 55, 211.
58. Wen, C. Y., (Project Director), "Optimization of Coal Gasification Processes", R&D Research Report No. 66. (1972), 1, Ch. 4, 74.
59. Rossberg, M., Z. Elektrochem (1956) 60, 952.
60. Montet, G. L. and Myers, G. E., Carbon (1968), 6, 627.
61. Kalina, T., Exxon Catalytic Coal Gasification Process-Predevelopment Program, Tech. Prog. Rep. (1976), FE-2369-4-5-6-7.
62. Chauhan, S. P., Feldmann, H. F., Stambaugh, E. P., and Oxley, J. H., Am. Chem. Soc., Div. Fuel Preprint, (1975) 20 (4) 207-18.
63. Wen, C. Y. and Dutta, S., "Coal Conversion Technology (ed. by Wen & Lee) - Rates of Coal Pyrolysis and Gasification Reactions" Addison Wesley Publishing Co. (1977), 71.
64. Lowry, H. H., "Chemistry of Coal Utilization", Supplementary Volume 913, John Wiley, New York, 1963.
65. Bohlbro, H., Acta Chem. Scandinivica (1961) 15, 502.
66. Moe, J. M., Chem. Eng. Progr. (1962) 58, 33.
67. Zahradnik, R. L., and Grace, R. J., Adv. Chem. Ser. (1974), 131, 126.
68. Wen, C. Y., Ind. Eng. Chem. (1968), 60 (9), 33.
69. Wen, C. Y., and Wang, S. C., Ind. Eng. Chem. (1970), 62, 30.
70. Szekely, J.. Evans. J. W. and Sohn, H. Y., "Gas-Solid Reactions", Academic Press, New York, 1976.
71. Yerushalmi, J., "Report on EPRI's Workshop on Clean Gaseous Fuels from Coal", prepared for EPRI, Jan. 15, 1977.
72. Rudolph, P. F. H., presented at the 4th Synthetic Gas Symp., Chicago, October 30-31, 1972.
73. Woodmansee, D. E., Energy Communications (1976), 2 (1), 13.
74. Lewis, P. S., Liberafore, A. J., Rahfuse, R. V., MERC/RI-75/1 (Morgantown Energy Research Center).
75. Desai, P. and Wen. C. Y., "Computer Modelling of MERC's Fixed Bed Gasifier", Report prepared for Morgantown, Energy Research Center, Morgantown, West Virginia, February 1978.
76. Yoon, H. Wei, J. and Denn, M. M., Ann. Meetg., AIChE, 69th, Chicago, Nov. 28, 1976.
77. Borowiec, S. P., Lanza, C. A., Schulz, H. W. and Spencer, J. L., Ann. Meetg. AIChE, 70th, New York, No. 10d, Nov. 14-17, 1977.
78. Amundson, N. R. and Arri, L. E., AIChE J. (1978), 24, 87.
79. Wen, C. Y. and Fan, L. T., "Models for Flow Systems and Chemical Reactors", p. 171, p. 188, Marcel Dekker Inc., New York, 1975.
80. Thoma, K. and Vortmeyer, D., "Chemical Reaction Engineering-Houston", Weekman, V. W., Jr. and Luss, D., Ed., Am. Chem. Soc. Sym. Ser. (1978), 65, 539-549.
81. Luss, D. L. and Amundson, N. R., Ind. Eng. Chem., Fundam. (1967), 6, 437.
82. Schaefer, R. J., Vortmeyer, D. and Watson, C. C., Chem. Eng. Sci. (1974), 29, 119.
83. Mori, S. and Muchi, I., "Heat Process Engineering" (Adv. Chem. Eng., 9, in Japanese) p. 120, Maki-Shotten, Tokyo (1975).
84. Ishida, M. and Wen, C. Y., Ind. Eng. Chem., Process Des. Dev. (1971), 10 (2), 164.
85. Kane, R. S. and McCallister, R. A., AIChE J., (1978) 24, 55.
86. Beer, J. M. and Lee, K. B., 10th Intern. Sym. Combust. (1965) 1187-1202.
87. Nusselt, W., Z. Ver. Dt. Ing., (1924), 68, 124-128.
88. Hottel, H. C. and Stewart, I. M., Ind. Eng. Chem. (1940), 32, 719-730.
89. Essenhigh, R. H., J. Inst. Fuel (1961), 34, 239-244.
90. Mehta, A. K., Report presented to ERDA by Combustion Engineering, FE-1545-26, August (1976).
91. Ubhayaker, S. K., Stickler, D. B. and Gannon, R. E., Fuel, (1977) 56, 281.
92. Thring, M. W., "The Science of Flames and Furnaces", 2nd Ed., John Wiley, New York, 1962, p. 625.
93. Tester, R. J., Wei, J. and Denn, M. M., Ann. Meetg., AIChE, 70th, New York, No. 10e, Nov. 14-17, 1977.
94. Wen, C. Y. and Chaung, T. Z., "Entrained Bed Gasifier Model for Texaco Partial Oxidation Unit", paper in preparation.
95. Beer, J. M., Paper presented at 16th International Symposium on Combustion, The Combustion Institute, 439-460, 1977.
96. Caretto, L. S., Paper presented at the 1977 Fall Meeting, Western States Section - The Combustion Institute, Stanford University, October 1977.
97. Toomey, R. D. and Johnstone, H. F., Chem. Eng. Prog. (1952) 48, 220.
98. Davidson, J. F., and Harrison, D., "Fluidized Particles", Cambridge University Press, Cambridge 1963.
99. Rowe, P. N., "Chemical Reaction Engineering-Houston" Am. Chem. Soc. Symp. Ser. (1978) 65, 436.
100. Horio, M., and Wen, C. Y., AIChE Symp. Ser. (1977), 73, (161), 9.
101. Campbell, E. K., and Davidson J. F., Paper. A-2 in Institute of Fuel Symposium Series, No.1, 1975.
102. Horio, M., Rengarajan, P., Krishnan, R., and Wen, C. Y., "Fluidized Bed Combustor Modeling", Report to NASA Lewis Research Center, Cleveland. Jan. 1977.
103. Kato, K., and Wen, C. Y., Chem. Eng. Sci. (1969), 24, 135.
104. Mori, S. and Wen, C. Y., AIChE J. (1975) 21, 109.
105. Baron, R. E., Hodges, J. L., and Sarofim, A. F., Ann. Meetg., 70th, AIChE, New York, 1977.
106. Merrick, D., and Highley, J., AIChE Symp. Ser. (1974) 70, (137), 366.
107. Gibbs, B. M., Paper A-5 Institute of Fuel Symposium Series, 1975.
108. Gordon, A. L., and Amundson, N. R., Chem. Eng. Sci. (1976) 31, 1163.
109. Chen, T. P., and Saxena, S., Fuel (1977) 56, 401.

110. Wen, C. Y., Mori, S., Gray, J.A., and Yavorsky, P. M., AIChE Symp. Ser. (1977) 73 (161), 86.
111. Becker, H. A., Beer, J. M., and Gibbs, B. M., Paper A-1, Institute of Fuel Symposium Series 1975.
112. Horio, M., Mori, S. and Wen, C. Y., Preprints of Int. Fluidization Conf., Pacific Grove, California, Engg Foundation Conference, Page vii-5 1975.
113. Rengarajan, P., Krishnan, R., and Wen, C. Y., Paper Ann. Meetg., 70th, AIChE, New York, 1977.
114. Wen, C. Y., and Dutta, S., AIChE. Symp. Ser. (1977) 73 (161), 1.
115. Behie, L. A., and Kehoe, P., AIChE J. (1973) 19, 1070.
116. Honda, T., Furusawa, T., and Kunii, D., Proc. Second PACHEC Meeting, Denver, Colorado, 2, 1236, 1977.
117. Horio, M., Mori, S. and Muchi, I., Proc. 5th FBC Conference, Washington, D.C., Dec. 1977.
118. Yates, J. G. and Rowe, P. N., Trans. Instn. Chem. Engrs. (1977) 55 (2), 137.
119. Assessment of Technology for the Liquefaction of Coal prepared by Ad Hoc Panel on Liquefaction of Coal of Fuels, National Research Council, National Academy of Science, Washington, D.C., 1977.
120. Van Krevelen, D. W., "Coal: Topology, Chemistry, Physico and Constitution", Elsevier, New York, 1961.
121. Oele, A. P., Waterman, H. L., Goedkoop, M. L. and Van Krevelen, D. W., Fuel (1951) 30, 169.
122. Dryden, I. G. C., "Chemistry of Coal Utilization-Supplementary Volume", Lowry, H. H., Ed., Ch. 6, John Wiley and Sons, New York, 1963.
123. Hildebrand, J. H. and Scott, R. L., "Regular Solutions", 166, Prentice-Hall, Englewood Cliffs, N.J., 1962.
124. Kiebler, M. W., Ind. Eng. Chem. (1940), 32, 1389.
125. Angelovich, J. M., Pastor, G. R. and Silver, H. F., Ind. Eng. Chem., Process Des. Dev. (1970) 9, 106.
126. Steinberg, H. W., "A Second Look at the Reductive Alkylation of Coal and at the Nature of Asphaltenes", Am. Chem. Soc., Div. Fuel Chem. Preprints, (1976), 21, (7), 1.
127. Farcasiu, M., Mitchell, T. O., and Whitehurst, D. D., "On the Chemical Nature of the Benzene Components of Solvent Refined Coals", Am. Chem. Soc., Div. Fuel Chemistry, preprints (1976) 21, (7), 11.
128. Weller, S., Pelipetz, M. G. and Friedman, S., Ind. Eng. Chem., Process Des. Dev. (1951), 43, 1572, 1575.
129. Curran, G. P., Struck, R. T. and Gorin, E., Ind. Eng. Chem., Process Des. Dev. (1967), 6, 166.
130. Mobil, R. and Corp, D., "The Nature and Origin of Asphaltenes in Processed Coals", EPRI-AF-252, Annual Report, Feb. 1976.
131. Neavel, R. C., Fuel (1976), 55, 237.
132. Heredy, L. A. and Fugassi, P., "Coal Science", Adv. Chem. Ser. (1966) 55, 448.
133. Mobil, R. and Corp, D., "The Nature and Origin of Asphaltenes in Processed Coals", EPRI AF-480, Annual Report, July 1977.
134. Neavel, R. C., "Coal Plasticity Mechanism: Inferences from Liquefaction Studies", Proc. Symp. Agglomeration and Conversion of Coal, West Virginia University, Morgantown, WV, 1975.
135. Squires, A. M., "Reaction Paths in Donor Solvent Coal Liquefaction", Conf. Coal Gasification/Liquefaction, Moscow, USSR., Oct. 1976.
136. Kang, C. C., Nongbri, G. and Stewart, N., "The Role of Solvent in the Solvent Refined Coal Process", Am. Chem. Soc., Div. Fuel Chemistry, preprints, (1976) 21, (65), 19.
137. Han, K. W., Dixit, V. B. and Wen, C. Y., Ind. Eng. Chem., Process Des. Dev. (1978) 17, 16.
138. Falkum, E. and Glenn, R. A., Fuel, 31, 133 (1952).
139. Ishii, T., Maekawa, Y., and Takeya, G., Kagaku Kogaku (Chem. Eng. Japan) (1965) 29, (12), 988.
140. Liebenberg, B. J. and Potgieter, H. G. J., Fuel (1973) 52, 130.
141. Wiser, W. H., Anderson, L. L., Oader, S. A. and Hill, G. R., J. Appl. Chem. Biotechnol. (1971) 21, 82.
142. Feldman, H. F., Williams, D. A. and Simons, W. H., "Role of Hydrogen Diffusion in the hydrogenation of coal tar over a Co-Mo Catalyst", Nat. Meetg., 68th, AIChE, Houston, Texas,1971.
143. Specchia, V., Sicardi, S. and Giametto, A., AIChE J. (1974) 20, 646.
144. Montagna, A. and Shah, Y. T., Chem. Eng. J. (1975) 10, 99.
145. Shah, Y. T., Cronauer, D. C., McIlvries, H. G. and Paraskoo, J. A., "Chemical Reaction Engineering-Houston", Am. Chem. Soc. Symp. Ser. (1978), 65, 303.
146. Goring, G. E., Curran, G. P., Tarbox, R. P., and Gorin, E., Ind. Eng. Chem. (1952), 44, 1057.
147. Zielke, C. W. and Gorin, E., Ind. Eng. Chem., (1955), 47, 820.
148. Blackwood, J. D., Austral. J. Chem. (1959), 12, 14.

RECEIVED March 30, 1978

4

Transport Phenomena in Packed Bed Reactors

E. U. SCHLÜNDER

Institut für Thermische Verfahrenstechnik, Universität Karlsruhe, F.R.G., West Germany

I. Preface

The performance of packed bed reactors may be influenced and sometimes even controlled by transport phenomena in the voids between the particles as well as inside the particles. This paper is restricted to transport phenomena mainly between the particles. There occurs heat, mass and momentum transfer between the fluid and the particles. Moreover, transport of heat and mass also occurs between the voids themselves. While mass transport through the particles is negligible, heat conduction through the particles is always involved with transport of heat between the voids and therefore must be considered simultaneously.

In this paper emphasis is put on recent developments rather than on a more or less complete literature survey. However, experimental data will be thoroughly reported, because they are the basis of any theoretical analysis.

II. Outline

Since transport of heat and mass in the voids of a packed bed cannot be understood without knowing the movement of the fluid, hydraulic phenomena will be treated first. Second the heat and mass transfer between the fluid and the particles will be discussed. Third the radial transport of heat and mass through the voids (and the particles) will be treated.

In the past these transport phenomena usually have been treated assuming that the fluid passes through the reactor in plug flow. This assumption then has been revised after careful studies have revealed, that there is not only a microscopic but also a macroscopic void fraction distribution. Consequently,there is a certain

0-8412-0432-2/78/47-072-110$12.80/0

macroscopic flow rate distribution in a tubular reactor. But only recently the consequences of this macroscopic non-uniform distribution of flow rate have been evaluated with regard to heat and mass transfer between the fluid and the particles. This effect will be treated in this paper to some extent.

Another problem, which still has not been solved is the heat transfer between particles and rigid walls. However, some achievements can be reported already now.

III. Flow rate distribution and pressure drop in tubular reactors

It is well known that the void fraction is larger near the walls of a tubular reactor than in the central parts. Fig. 1 shows experimental data of Benenate and Brosilow (1). The void fraction ψ decreases from 1 at the wall to a minimum of 0,23 and follows a damped oscillation function towards the center of the tube. The abscissa y/d is the distance from the wall divided by the particle diameter. A semitheoretical analysis of this distribution has been given by Ridgeway and Tarbuck (2). For practical purposes Martin (3) gave an empirical expression as shown in Fig. 1. The average void fraction remains practically constant for $y/d > 0,5$ and is about 0,40 for a random packed bed of equal sized spheres. However, near the retaining wall the average void fraction is about 0,50. This means that models of packed bed reactors introducing an average hydraulic diameter must be based at least on two average diameters. In this case one obtains a higher flow rate $\dot{V}_2$ near the wall and a lower one $\dot{V}_1$ in the central parts as shown in Fig. 2.

Applying Ergun's law (4) for calculating the pressure drop in each of these two parallel average flow channels

$$\frac{\Delta P}{L} = C \frac{(1-\psi_1)^2}{\psi_1^3} \frac{\eta u_1}{d^2} + B \frac{1-\psi_1}{\psi_1^3} \frac{\rho u_1^2}{d} \tag{1}$$

$$\frac{\Delta P}{L} = C \frac{(1-\psi_2)^2}{\psi_2^3} \frac{\eta u_2}{d^2} + B \frac{1-\psi_2}{\psi_2^3} \frac{\rho u_2^2}{d} \tag{2}$$

yields the velocity ratio $\omega = u_2/u_1$

$$\omega = \frac{\varphi(1+K)-1+\sqrt{[\varphi(1+K)-1]^2+4(\varphi+MZ)(1-\varphi+Z)K}}{2(\varphi+MZ)} \tag{3}$$

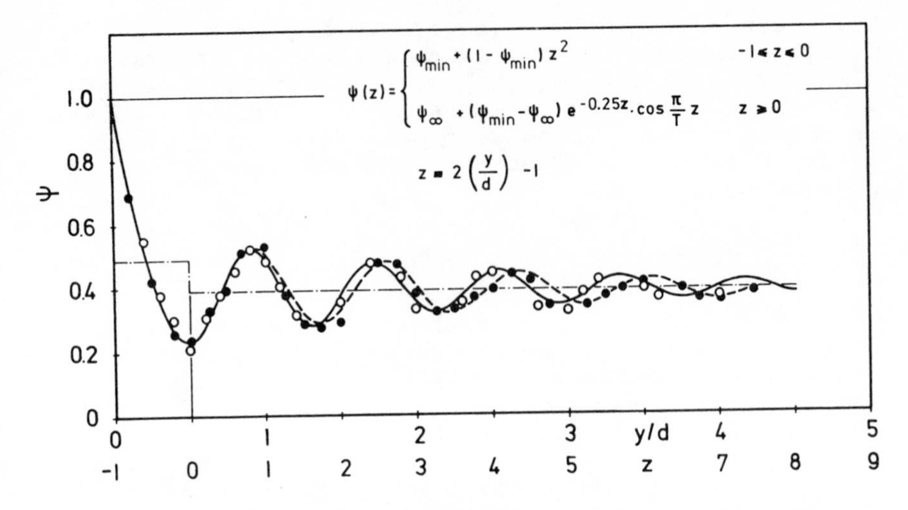

Figure 1. Void fraction distribution near the wall of a packed bed of spheres. Data points from Ref. 1.

D/d	T	*Symbol*	*Empirical Correlation*
∞	$\sqrt{2/3}$	○	————
20,3	*0,876*	●	------------

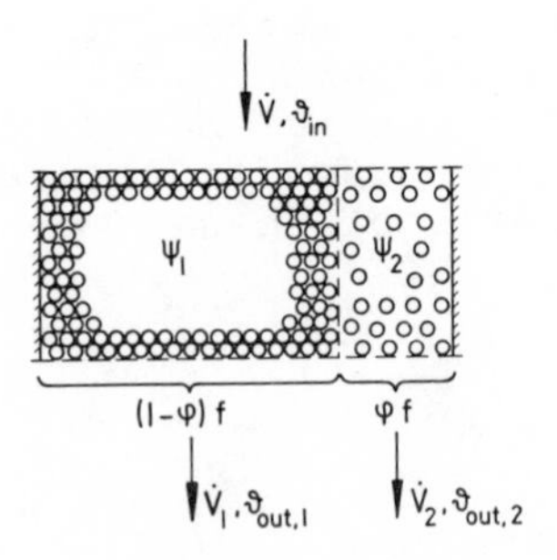

Figure 2. Model for non-uniformly packed beds

where

$$Z = \frac{B}{C} \frac{Re}{1-\psi_1}$$

$$K = \left(\frac{1-\psi_1}{1-\psi_2}\right)^2 \left(\frac{\psi_2}{\psi_1}\right)^3$$

$$M = \frac{1-\psi_1}{1-\psi_2}$$

$$Re = \frac{\rho\ u\ d}{\eta}$$

$$u = (1-\varphi)\ u_1 + \varphi u_2 \qquad \text{and}$$

$$\psi = (1-\varphi)\ \psi_1 + \varphi \psi_2 \quad .$$

In case of low Reynoldsnumbers Eq. 3 reduces to

$$\omega = \left(\frac{1-\psi_1}{1-\psi_2}\right)^2 \cdot \left(\frac{\psi_2}{\psi_1}\right)^3 \tag{3 a}$$

and incase of high Reynoldsnumbers to

$$\omega = \sqrt{\frac{1-\psi_1}{1-\psi_2} \cdot \left(\frac{\psi_2}{\psi_1}\right)^2} \tag{3 b}$$

In a packed bed reactor with $\psi_1 = 0,40$ and $\psi_2 = 0,50$ Eq. 3 a gives $\omega = 2,81$ which means that near the retaining walls the fluid velocity is considerably higher.
The flow rate ratio $\dot{\nu} = \dot{V}_2/(\dot{V}_1+\dot{V}_2)$ is

$$\dot{\nu} = \frac{1}{1 + \frac{1-\varphi}{\varphi\omega}} \tag{4}$$

The fraction φ of the cross sectional area with the larger voids ψ_2 near the wall can be estimated by

$$\varphi \cong 2\ \frac{d}{D} \tag{5}$$

which means that usually φ is between 0,01 and 0,20. Take e.g. $\varphi = 0,10$ and $\omega = 2,81$ then $\dot{\nu}$ is 0,238, which means that the flow rate near the wall is more than twice as much as it would be in case of plug flow.
Fig. 3 shows experimental data of Schwartz and Smith

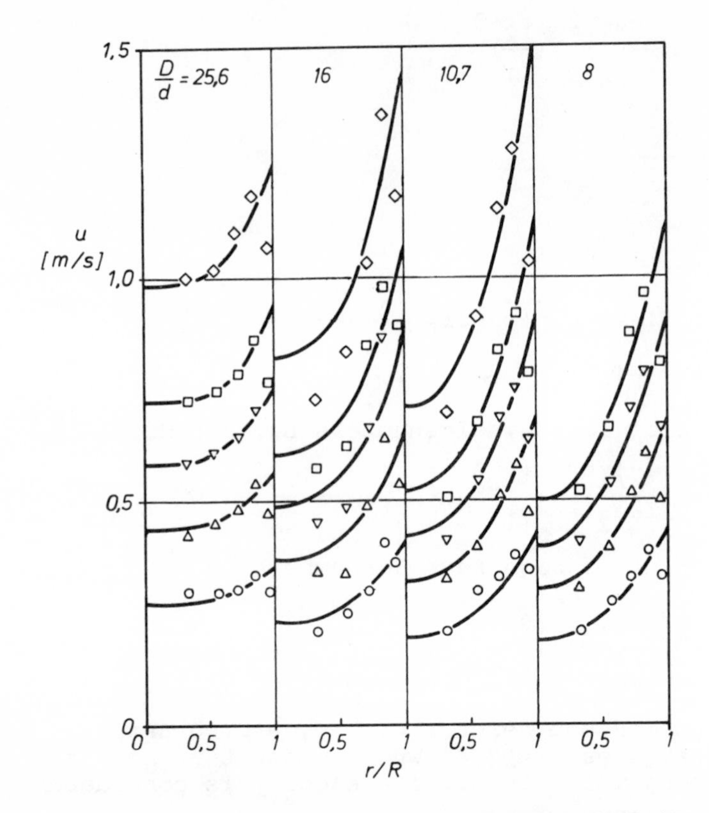

Industrial Engineering Chemistry

Figure 3. Velocity distribution in a tubular packed bed reactor obtained by Schwartz and Smith (5)

(5) which are in good agreement with this estimation. One might object that the constants C and B in Ergun's law have been derived from experiments assuming plug flow. However, it can be shown that re-evaluation of these data in terms of Eq. 1 and 2 gives only slightly lower constants (10 % to 15 %).

IV. Particle-to-fluid heat transfer in tubular reactors

The heat transfer between a single particle surrounded by an extended fluid flow is well known and can be predicted by a dimensionless correlation

$$Nu_{\text{Single Particle}} = 2 + F \sqrt[3]{Pr} \sqrt{Re} \tag{6}$$

$$\text{with} \quad F = 0{,}664 \sqrt{1 + \left[\frac{0{,}0557\ Re^{0,3}\ Pr^{0,67}}{1+2{,}44(Pr^{2/3}-1)Re^{-0,1}}\right]^2}$$

The dimensionless groups are

$Nu = \alpha d/\lambda$ the Nusselt number

$Re = u_\infty d/\nu$ the Reynolds number

$Pr = \nu/\kappa$ the Prandtl number

α is the heat transfer coefficient,

λ the heat conductivity,

κ the thermal diffusivity,

ν the kinematic viscosity,

u_∞ the free stream fluid velocity and

d the particle diameter.

Eq. 6 has been presented by Martin (3) based on experimental data correlated by Gnielinski (6). The same equation also holds for packed beds if it is multiplied with an enhancement factor depending on the void fraction. According to Martin (3) this factor is $f_\psi = 1+1{,}5(1-\psi)$ and therefore

$$Nu_{\text{Packed Bed}} = (1+1{,}5(1-\psi))\ Nu_{\text{Single Particle}} \tag{7}$$

where the Nusselt number for the single particle follows from Eq. 6 by calculating the Reynolds number with an average fluid velocity u_∞/ψ, which means that Re has to be replaced by Re/ψ. Applied to a randomly packed bed of spheres with $\psi = 0{,}4$ Eq. 7 predicts the packed bed heat transfer coefficient for intermediate Re numbers to be about three times ($[1+1{,}5(1-0{,}4)]/\sqrt{0{,}4} = 3{,}0$) those of the single sphere at the same superficial velocity u_∞. This is in good agreement with Ranz's suggestion (7) to use the single sphere equation with a modified velocity in Re, which should be 9,1 times the superficial velocity and therefore gives $\sqrt{9,1} \cong 3$ times greater Nu numbers. Recently Gnielinski (8) has collected quite a number of experimental data and compared them with the predictions of Eq. 6 and 7. Fig. 4 a, b and c show the Nusselt number Nu versus the Reynolds number Re_ψ with the Prandtl number Pr as Parameter. The void fraction has been varied between 0,26 and 0,935. The correlation contains data from heat and mass transfer experiments to cover a wide range of Prandtl (or Schmidt) numbers from 0,6 to 10 000. The agreement seems to be rather good. However, there is a remarkable restriction as to the application of Eq. 6 and 7, since the Eq. 7 is based on the assumption that the flow rate of the fluid throughout the packed bed is uniform, which means that the fluid velocity profile equals the plug flow profile. One recognises that in Fig. 4a, b and c all the experimental data are in a range beyond Re Pr = Pe $\sim$ 100. For low Peclet numbers Pe Soerensen and Stewart (31) solved the equations of motion and energy for creeping flow through a duct formed by a cubic array of spheres. Under the assumption of constant average fluid flow rate throughout the whole crossectional area they found - as to be expected - that the Nu number becomes constant for vanishing Pe number. The limiting value for a large number of particle layers is Nu = 3,89. For a randomly packed bed with ψ = 0,4 Eq. 6 and 7 yield Nu = 3,8 which is very close to the theoretical value. However, experimental data in the low Peclet number regime (Pe < 100), collected by Kunii and Suzuki (32) fall far below this limiting value 3,8 by some orders of magnitude, see Fig. 5. Moreover, the slope of Nu versus Pe partly is greater than one in the log-log plot and eventually, the Nu number is lower for smaller particle diameters. Based on a model recently presented by Schlünder (33), Martin (3) has shown that these discrepencies could be explained by the non uniform flow rate distribution in packed beds due to the larger void fraction near the retaining walls. Since the analysis given by Schlünder and Martin has fare reaching consequences, it shall be treated in some detail.

Assume the particles of the packed bed to have a constant temperature T_S. Then the outlet temperature of fluid flow rate $\dot{V}_2$ near the wall is given by

$$\frac{T_{out,2}-T_S}{T_{in}-T_S} = \exp(-NTU_2) \tag{8}$$

and for the flow rate $\dot{V}_1$ one obtains

$$\frac{T_{out,1}-T_S}{T_{in}-T_S} = \exp(-NTU_1) \tag{9}$$

where $NTU_1 = \frac{\alpha_1 A_1}{\rho c_p \dot{V}_1}$ and $NTU_2 = \frac{\alpha_2 A_2}{\rho c_p \dot{V}_2}$

are the number of transfer units of the respective flow rates. A_1 and A_2 resp. are the particle surface areas, ρ is the density and c_p the specific enthalpy of the fluid.

The average number of transfer units of the system is defined by

$$NTU = \ln \frac{T_{out}-T_S}{T_{in}-T_S} \tag{10}$$

where T_{out} is the mixing temperature of the two streams $\dot{V}_1$ and $\dot{V}_2$, respectively

$$T_{out} = \frac{\dot{V}_2}{\dot{V}_1+\dot{V}_2} T_{out,2} + \frac{\dot{V}_1}{\dot{V}_1+\dot{V}_2} T_{out,1} \tag{11}$$

Combining the Eq.'s 8 to 11 yields

$$NTU = -\ln \left[(1-\dot{\nu})\exp(-NTU_1)+\dot{\nu}\exp(-NTU_2)\right] \tag{12}$$

where $\dot{\nu} = \dot{V}_2/(\dot{V}_1+\dot{V}_2)$.

By definition the number of transfer units is connected with the Nusselt number and the Peclet number

$$NTU \equiv \frac{Nu}{Pe/a_v L} \tag{13}$$

where a_v is the particle surface area per unit volume and L is the height of the packed bed. Thus one obtains for the average Nusselt number of the system

$$Nu = -\frac{Pe}{a_v L} \ln \left[(1-\dot{\nu})\exp(-NTU_1)+\dot{\nu}\exp(-NTU_2)\right] \tag{14}$$

The individual number of transfer units NTU_1 and NTU_2 can be calculated with the help of Eq. 6, 7 and 13, which yield

$$NTU_1 = \frac{Nu_1}{Pe_1 a_v/L} \text{ and } NTU_2 = \frac{Nu_2}{Pe_2 a_v/L} \tag{15}$$

The average Peclet number Pe is correlated with the individual ones:

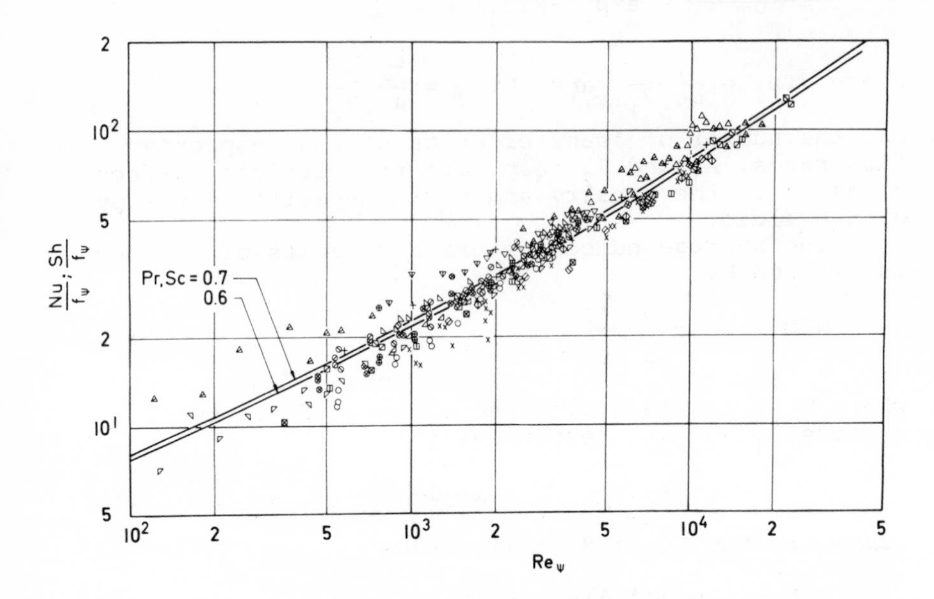

Figure 4a. Heat transfer in packed beds of spheres with different void fraction ψ. Nu as a function of Re with Pr parameter according Gnielinski (8). The authors of the experimental data are given in the legend.

Legend to Figure 4 a

Symbol	Author	Diameter d/mm/	Ψ	Pr; Sc
◁	Hobson, M. a. G. Thodos /10/	9,4	0,475	0,61
▽ ▼	Gupta, A.S.a. G. Thodos /11/	15,9 15,9	0,444 0,576	0,61 0,61
⊕ ⊘ ⦸ ⊗ ○	Rowe,P.N. a. K.T.Claxton/12/	38,1 38,1 38,1 38,1 38,1	0,26 0,365 0,488 0,632 0,476	0,73 0,73 0,73 0,73 0,73
◸	Glaser, M.B. a.G.Thodos /13/	7,9	0,453	0,7
⧄ ⊞ ⊠ ⧅	Malling,G.F. a.G.Thodos /14/	15,9 15,9 15,9 15,9	0,366 0,545 0,788 0,386	0,6 0,6 0,6 0,6
▲ △ ◬	Gupta, A.S. /15/	15,9 16,1 15,9	0,778 0,444 0,576	0,6 0,6 0,6
+	Thoenes, D.a.H.Kramers /16/	15,0	0,260	0,84
◺ ◿	Glaser, H. /17/	12,1 10,3	0,42 0,392	0,72 0,72
◇ ◇ ◈ ◇	Jaeschke, L. /18/	22,5 22,5 22,5 22,5	0,668 0,779 0,935 0,476	0,56 0,56 0,56 0,56
×	Bradshaw,R.D.a.J.E.Myers/19/	8,7	0,400	0,6
○ ⦶ ⊖	Gaffney,B.J.a.T.B. Drew /20/	12,2 9,4 6,3	0,519 0,504 0,513	160 170 160-180

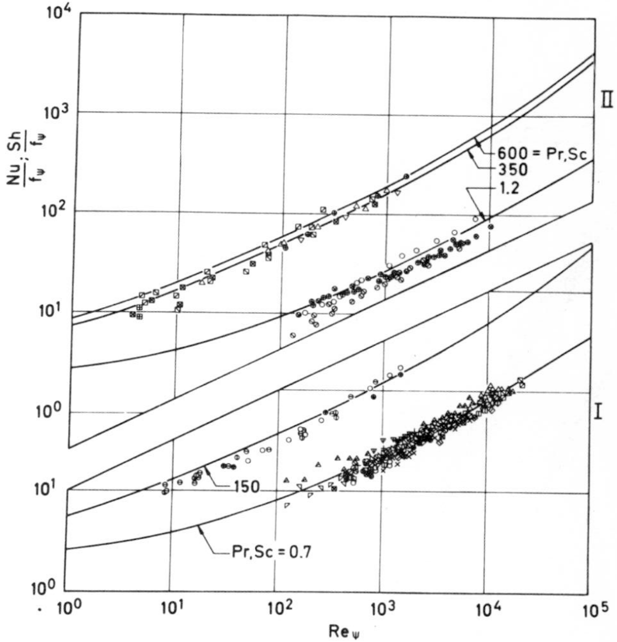

Figure 4b. *Heat transfer in packed beds of spheres with different void fraction ψ. Nu as a function of Re with Pr as parameter according Gnielinski (8). The authors of the experimental data are given in the legend.*

Legend to Figure 4 b (Part I)

Symbol	Author	Diameter d/mm/	Ψ	Pr; Sc
◁	Hobson, M. a. G. Thodos /10/	9,4	0,475	0,61
▽ ▽	Gupta, A.S. a. G. Thodos /11/	15,9 15,9	0,444 0,576	0,61 0,61
⊕ ⊘ ⊘ ⊗ ○	Rowe, P.N. a. K.T. Claxton/12/	38,1 38,1 38,1 38,1 38,1	0,26 0,365 0,488 0,632 0,476	0,73 0,73 0,73 0,73 0,73
◸	Glaser, M.B. a.G.Thodos /13/	7,9	0,453	0,7
⊠ ⊞ ⊠ ⊠	Malling, G.F. a.G.Thodos /14/	15,9 15,9 15,9 15,9	0,366 0,545 0,788 0,386	0,6 0,6 0,6 0,6
△ △ △	Gupta, A.S. /15/	15,9 16,1 15,9	0,778 0,444 0,576	0,6 0,6 0,6
+	Thoenes, D.a.H.Kramers /16/	15,0	0,260	0,84
◺ ◿	Glaser, H. /17/	12,1 10,3	0,42 0,392	0,72 0,72
◇ ◇ ◇ ◇	Jaeschke, L. /18/	22,5 22,5 22,5 22,5	0,668 0,779 0,935 0,476	0,56 0,56 0,56 0,56
×	Bradshaw, R.D.a.J.E.Myers/19/	8,7	0,400	0,6
○ ⊖ ⊖	Gaffney, B.J.a.T.B. Drew /20/	12,2 9,4 6,3	0,519 0,504 0,513	160 170 160-180

Legend to Figure 4 b (Part II)

Symbol	Author	Diameter d/mm/	Ψ	Pr; Sc
⊕ ○ ⊗ ⊘ ⊘	Thoenes, D.a.H.Kramers /16/	15,0 15,0 15,0 15,0 15,0	0,260 0,476 0,320 0,480 0,480	1,24 1,24 1,24 1,24 1,14
⊠ ⊞ ⊠ ⊠	Gaffney, B.J. a. T.B.Drew/20/	12,9 9,5 9,6 6,3	0,504 0,497 0,5 0,524	340-430 420 380 370-420
△ ▽	Venkateswaran, S.D. a. G.S. Laddha /26/	10,0 10,0	0,47 0,555	640-680 670
⊕	Thoenes, D. a. H.Kramers /16/	15,0	0,26	450-605

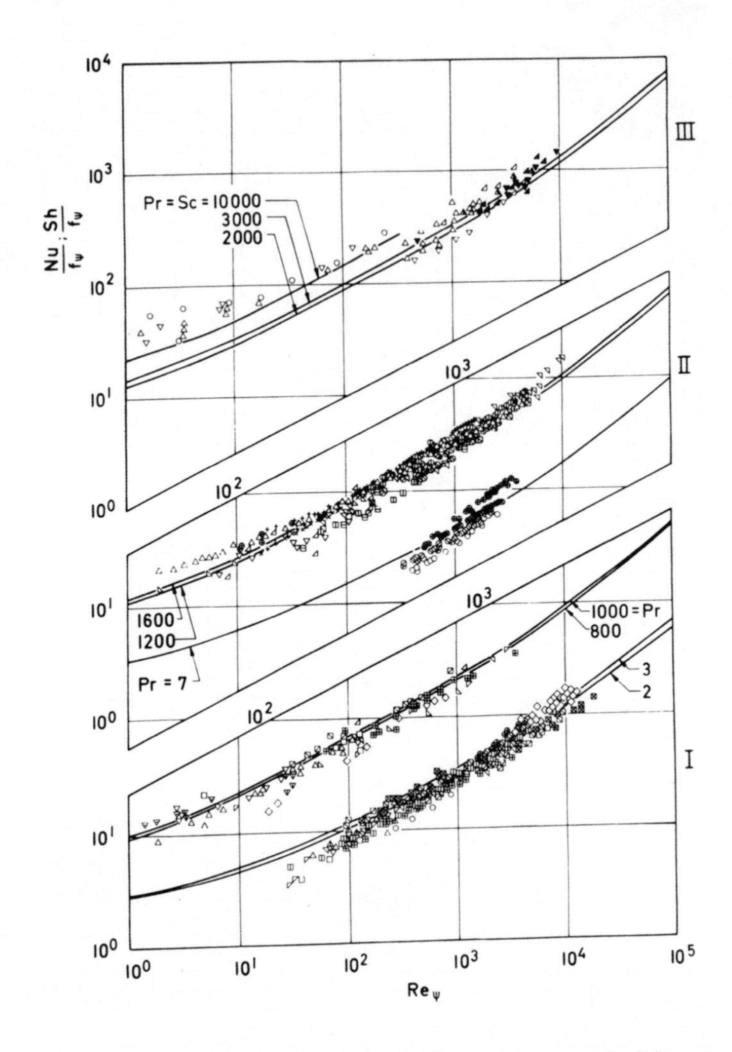

Figure 4c. Heat transfer in packed beds of spheres with different void fraction ψ. Nu as a function of Re with Pr as parameter according Gnielinski (8). The authors of the experimental data are given in the legend.

Legend to Figure 4 c (Part I)

Symbol	Author	Diameter d/mm/	ψ	Pr;Sc
⊠	Thoenes,D. *a.* H.Kramers /16/	15,0	0,26	2,06
◫		15,0	0,476	2,06
⧅		13,8	0,32	2,06
⧄		15,0	0,48	2,06
⊞		15,0	0,26	2,8
□		15,0	0,476	2,8
◹		13,8	0,32	2,8
◸		15,0	0,48	2,8
◿	Wilkins,G.S.*a.*G.Thodos /21/	2,6	0,423	3,72
◺		3,1	0,421	3,72
◇	v.d.Decken *et al.* /22/	30,0	0,380	2,57
△	Hobson,M.*a.*G.Thodos /10/	9,4	0,475	1,87
▲		9,4	0,475	1,7
▽		9,4	0,475	1,76
⊕	Rowe,P.N.*a.*K.T.Claxton /12/	38,1	0,26	2,54
○		38,1	0,476	2,54
⊖		38,1	0,365	2,54
⊘		38,1	0,488	2,54
⊗		38,1	0,632	2,54
⊞	Thoenes,D. *a.* H.Kramers /16/	15,0	0,26	780-1070
⧅		15,0	0,395	910-1060
□		15,0	0,476	800-1090
⧄		13,8	0,320	880-1040
▽	Wilson, E.J. *a.* C.J.Geanko-polis /23/	6,2	0,436	876-1107
▽		6,3	0,401	760-1000
▽		6,3	0,441	950-1070
◸	Bhattacharya, S.N. *a.* M. Raja Rao /24/	23,4	0,53	794
◹		23,4	0,49	794
◺		7,9	0,39	794
◿		17,6	0,53	930
△	Evans,G.C. *a.* C.F.Gerald/25/	2,0	0,51	1051-1104
◇	Venkateswaran, S.D. *a.* G.S. Laddha /26/	19,0	0,483	867-892

Legend to Figure 4 c (Part II)

Symbol	Author	Diameter d/mm/	Ψ	Pr; Sc
⊕	Rowe,P.N.*a.*K.T.Claxton /12/	38,1	0,260	6,0-7,2
⊗		38,1	0,365	6,7-7,0
⊖		38,1	0,488	6,7-7,0
⊘		38,1	0,632	6,6-7,0
○		38,1	0,476	6,4-7,0
◹	Thoenes,D. *a.* H.Kramers/16/	15,0	0,26	1175-1840
◸		15,0	0,395	1100-1320
◿		15,0	0,476	1100-1430
◺	Wilson,E.J. *a.* C.J. Geanko-polis /23/	6,3	0,403	1171-1231
▽	Williamson, J.E. *et al.* /27/	6,1	0,431	1103-1140
⧄	McCune,L.K.*a.*R.H.Wilhelm /28/	6,4	0,375	1200-1450
◫		4,8	0,369	1240-1405
⊟		3,2	0,355	1330-1350
⧅		6,4	0,375	1340
□		1,3	0,445	1340
⊠		2,1	0,433	1300
△	Karabelas, A.J. *et al.* /29/	12,7;25,4;76,2	0,26	1600
⊕	Rowe,P.N.*a.*K.T.Claxton /12/	15,9	0,26	1360-1670
⦶		6,4; 12,7	0,37	1380-1440
⊗		38,1	0,26	1360-1400
○		38,1	0,365	1310-1410
⊖		38,1	0,488	1280-1410
⊘		38,1	0,632	1320-1370
+	Jolls,K.R.*a.*T.J.Hanratty /30/	25,4	0,410	1695
+		25,4	0,410	1740
x		25,4	0,410	1780

Legend to Figure 4 c (Part III)

Symbol	Author	Diameter d/mm/	Ψ	Pr; Sc
▼	Thoenes,D. *a.* H.Kramers /16/	15,0	0,26	1960-2300
▲		15,0	0,26	3300-3650
◿		15,0	0,395	2400-2800
▽		15,0	0,476	1960-2800
△		15,0	0,476	2870-3540
▲		13,8	0,32	3380-3530
○	Gaffney,B.J. *a.* T.B.Drew /20/	12,2	0,52	10100-11000
△		9,2	0,497	10400-11000
▽		6,0	0,507	10700-11500

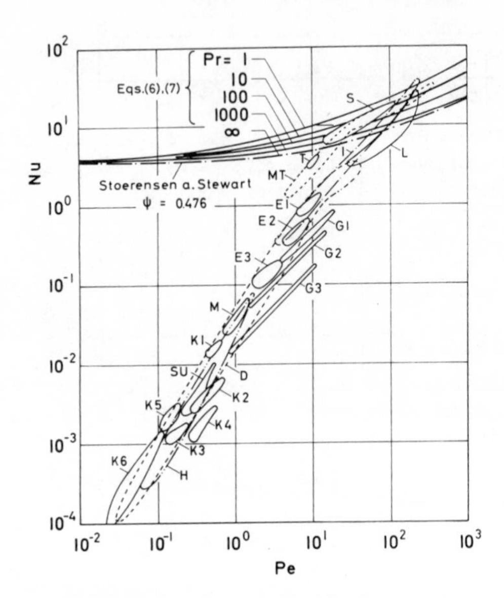

Figure 5. Heat transfer in packed beds of spherical particles. Full lines calculated from the empirical Equations 6 and 7. Dotted line from theoretical calculations by Soerensen and Stewart (31). Data of various authors as collected by Kunii and Suzuki (32).

KEY	Authors	Particle diameter (mm)	Fluid
L:	Löf and Hawley [37]	32·8	air
G1: G2: G3:	Grootenhuis [34]	0·39 0·21 0·064	air
E1: E2: E3:	Eichhorn and White [35]	0·66 0·52 0·28	air
S:	Satterfield and Resnick [39]	5·1	
K1: K2: K3: K4:	Kunii and Smith [36]	1·02 0·57 0·40 0·11	air
D:	Dannadieu [38]	0·13 ~ 1·1	air
I:	Kunii and Ito [40]	0·45~5·0	air
M:	Mimura [41]	3·7	air
SU:	Suzuki [42]	1·10	air
T:	Tokutomi [43]	1·0	air
K5: K6:	Kunii and Smith [36]	1·02 0·57	water
H:	Harada [44]	0·55~3·1	water
MT:	Mitsumori [45]	1·18 ~11·0	water

$$Pe_1 = \frac{1-\dot{\nu}}{1-\varphi} Pe \text{ and } Pe_2 = \frac{\dot{\nu}}{\varphi} Pe \tag{16}$$

It is of interest to study the asymptotic behavior of Eq. 14. For large Pe numbers the individual NTU's go to zero. Then Eq. 14 may be expanded to give

$$\lim_{NTU \to O} Nu = (1-\varphi) \frac{a_{v1}}{a_v} Nu_1 + \varphi \frac{a_{v2}}{a_v} Nu_2 \tag{17}$$

Eq. 17 says that in the limiting case NTU→O the average Nusselt number is equal to the artihmetic mean value of the individual Nusselt numbers with respect to the individual crossectional and surface areas.

For low Pe numbers or large NTU, respectively the asymptotic solution of Eq. 14 becomes

$$\lim_{NTU \to \infty} Nu = \frac{a_{v2}}{a_v} \frac{\varphi}{\dot{\nu}} Nu_{single\ sphere} \cdot f_\psi \tag{18}$$

This value may be far below $Nu_{single\ sphere} \cdot f_\psi$. In case of $NTU_{1,2} \to O$ (large Pe) the outlet temperatures of both streams $T_{out,1}$ and $T_{out,2}$ are close to the inlet temperature T_{in}. If, however, $NTU_{1,2} \to \infty$ (low Pe) the outlet temperatures T_{out1} and T_{out2} are close to the particle temperature T_S.

For intermediate Pe numbers it may happen that NTU_2 is still very low while NTU_1 is rather large, provided that there is a sufficiently large variation in flow. In this case the outlet temperature $T_{out,1}$ is already close to the wall temperature T_S, while the outlet temperature $T_{out,2}$ is still close to the inlet temperature T_{in}. In this case Eq. 14 reduces to

$$Nu = \frac{-Pe}{a_v L} \ln \dot{\nu} = \frac{-\ln \dot{\nu}}{6(1-\psi)} \frac{d}{L} Pe \tag{18 a}$$

or $$NTU = -\ln \dot{\nu} \tag{18 b}$$

Eq. 18 b shows clearly that the number of transfer units only depends on the flow rate ratio $\dot{\nu}$ but not on the flow rate itself nor on the surface area and the thermal properties of the fluid. This is because the average outlet temperature is nothing but the result of a mixing process between some fluid portions having the inlet temperature and others having the particle temperature or in other words between some fluid unchanged in temperature and the rest having reached thermal equilibrium. Since the average NTU is a purely hydraulic property of the non uniform system, this regime of intermediate Peclet numbers shall be called the hydraulically controlled regime. Within this regime the basic laws of heat conduction and heat convection only do apply to the local transfer processes but not at all to the average transfer rate. Therefore the average heat transfer coefficient in the hydraulically controlled regime has no longer a true physical meaning. In particular, one cannot compare the average heat or mass transfer coefficient with the rate of a chemical reaction in order to decide, whether the whole process is either transfer or reaction controlled. One has to compare the local transfer rates with the local reaction rates. Otherwise the result might be wrong by orders of magnitude.

Eq. 18 a, b also hold for low Peclet numbers in case that the surface area a_{v2} does not contribute to the transfer process. All experimental data realizing the boundary condition of constant particle temperature have been mass transfer experiments (evaporation, sublimation). Since the retaining wall itself does not contribute to the mass transfer, see Fig. 6, one might expect Eq. 18 a, b to be valid not only for intermediate but also for low Peclet numbers (Pe→O).

All heat transfer experiments have been carried out with the boundary condition of constant heat flux $\dot{q}$ at the interface of the particles. In this case the average Nusselt number of the non uniform system is given by

$$\frac{1}{Nu} = \frac{1}{Nu_1} + \varphi^*\left(\frac{1}{Nu_2} - \frac{1}{Nu_1}\right) + \frac{a_v L}{Pe}\,\frac{\dot{\nu}}{1-\dot{\nu}}\left(1 - \frac{\varphi^*}{\dot{\nu}}\right)^2 \qquad (19)$$

with

$$\varphi^* = \varphi\,\frac{1-\psi_2}{1-\psi} = \varphi\,\frac{a_{v2}}{a_v}$$

as Martin (3) has shown. For large Pe numbers this equation reduces to

$$\lim_{Pe\to\infty} \frac{1}{Nu} = (1-\varphi)\,\frac{a_{v1}}{Nu_1} + \varphi\,\frac{a_{v2}}{Nu_2} \qquad (19\ a)$$

while for low Pe numbers the asymptote becomes

$$\lim_{Pe\to 0} Nu = \frac{1}{6(1-\psi)\left(1-\frac{\varphi^*}{\dot{\nu}}\right)\frac{2\,\dot{\nu}}{1-\dot{\nu}}} \cdot \frac{d}{L}\,Pe \qquad (19\ b)$$

According Eq. 19 b Nu goes to zero when Pe goes to zero. Furthermore Nu decreases with decreasing particle diameter d and increasing bed height. This is basicly the same result as for the boundary condition of constant particle temperature in the range of intermediate Pe numbers and also for low Pe numbers if a_{v2} is assumed to be zero, see Eq. 18 a. In Fig. 7 the Nusselt number Nu is plotted against the Peclet number Pe according to equation 19 with the ratio of bed

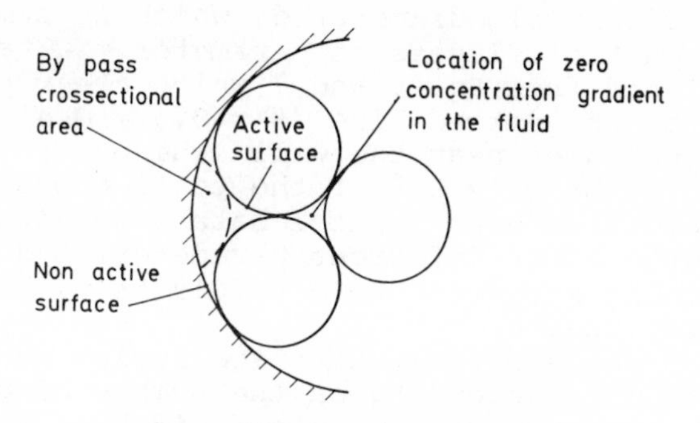

Figure 6. Illustration of the bypass effect

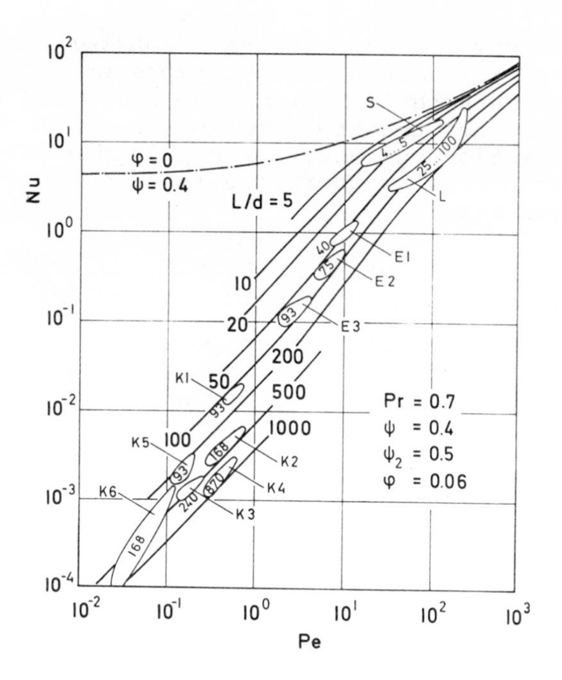

Figure 7. Heat transfer in packed beds. Curves calculated from Equation 19 with $Nu_{1,2} = Nu\ (Pe_{1,2}, \psi_{1,2}, Pr)$ from Equations 6 and 7 and $\mathring{v}$ from Equation 4, ω from Equation 3. Data with actual L/d from Figure 5. Dotted line ($\varphi = 0, \psi = 0{,}4$) uniform packed bed based on experimental data at high Pe numbers and on calculations by Soerensen and Stewart (31) at low Pe numbers.

height L to particle diameter d, which is simply the number of particle layers as parameter. Nu_1 and Nu_2 are calculated from Eq. 6 and 7 which are represented in Fig. 7 by the dotted line ($\varphi = 0, \psi = 0{,}4$). The crossectional area near the wall (where $\psi = \psi_2 = 0{,}5$) is assumed to be only 6 % of the total crossectional area. In addition experimental data according to various authors for different parameters L/d are shown in Fig. 7. The agreement with Eq. 19 is within the experimental scatter.

Fig. 7 shows clearly that for Peclet numbers less than 10^3 to 10^2, depending on the number of particle layers, the average particle-to-fluid heat transfer is entirely hydraulically controlled, which means that the number of transfer units is independent of the suface area per unit volume, independent of the fluid flow rate as well as of the thermal properties of the fluid.

Therefore, the average heat transfer coefficient (Nusselt number) does no longer represent a heat transfer process but rather a pure hydraulic phenomenon. If a heat or mass transfer process is combined with a chemical reaction one always has to apply the local heat or mass transfer coefficients, e.g. to form a Hatta number etc.

V. Heat Transfer, Energy Transport and Temperature Profiles in Tubular Packed Bed Reactors

Fig. 8 shows a typical temperatur profile in a tubular packed bed reactor. Profiles like this have been measured for example by Seidel (46). Near the wall there is a very steep temperature gradient. This observation suggests the introduction of a so called wall heat transfer coefficient

$$\alpha_W = \frac{\dot{q}}{T_o - T_{Wall}} \tag{20}$$

where $\dot{q}$ is the heat flux.

The temperature profiles for the fluid phase $T_G(r)$ and for the solid phase $T_S(r)$ in the central parts of the reactor can be predicted if an adequate model for the energy transport both in the fluid as well as in the solid phase can be developed and if the parameters in this model are available. In the following section V.1 such a model and its parameters will be presented. In the section V.2 the wall heat transfer coefficient will be discussed.

V.1 Energy Transport in Packed Beds

The most widely used model, which will be described below, is based on the following hypothesis:
1. The heat flux both in the solid as well as in the gas phase follows Fourier's law

$$\dot{q}_S = - \lambda_{SO} \frac{\partial T_S}{\partial r} \tag{20}$$

$$\dot{q}_G = - \Lambda \frac{\partial T_G}{\partial r} \tag{21}$$

where λ_{SO} and Λ are the apparent heat conductivities in the solid and in the gas phase, respectively.
2. Both heat fluxes $\dot{q}_S$ and $\dot{q}_G$ penetrate through the packed bed along the space variable r in parallel and independently. Consequently, the total heat flux is

$$\dot{q} = - \lambda_{SO} \frac{\partial T_S}{\partial r} - \Lambda \frac{\partial T_G}{\partial r} \tag{22}$$

In the abscence of a chemical reaction T_S is equal to T_G and one obtains

$$\dot{q} = - (\lambda_{SO} + \Lambda) \frac{\partial T}{\partial r} \tag{22 a}$$

The first hypothesis is arbitrary and remains subject to experimental verification. The second hypothesis is to be justified by physical reasons. Schlünder (47) has calculated the local heat flux penetrating a gap formed by a plane and an adjacent sphere filled with a stagnant gas, see Fig. 9. taking into account that the heat conductivity of the gas λ is affected by the gap width, which near the contacting point is always smaller than the mean free path of the gas molecules. Therefore, the heat conductivity λ goes to zero, when the gap width goes to zero. However, the amount of heat transferred locally $d\dot{Q}/d\rho$ shows a maximum value very close to the contacting point, see Fig. 9. Consequently, most of the heat is transferred near the contacting point, say between $0 < \rho < 0{,}5$ r. The same result would be obtained for two adjacent spheres.

If in addition a gas is blown through the gap between the plane and the sphere, the local flow rate will vary with the gap width. In laminar flow the local flow rate is proportional to the third power of the gap width. At a position $\rho = 0{,}5$ r the local flow rate ratio $\dot{V}_G(\rho)/\dot{V}_G(r)$ is equal to

$$\left(\frac{S}{d/2}\right)^3 = \left(1 - \sqrt{1 - \left(\frac{\rho}{d/2}\right)^2}\right)^3 = 0{,}008$$

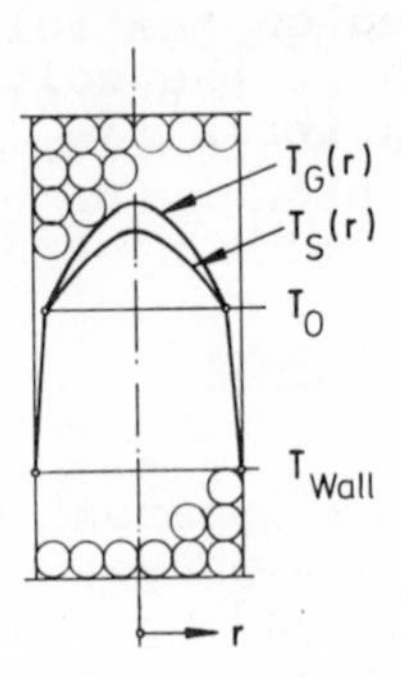

Figure 8. Typical temperature profile in a tubular packed bed reactor.

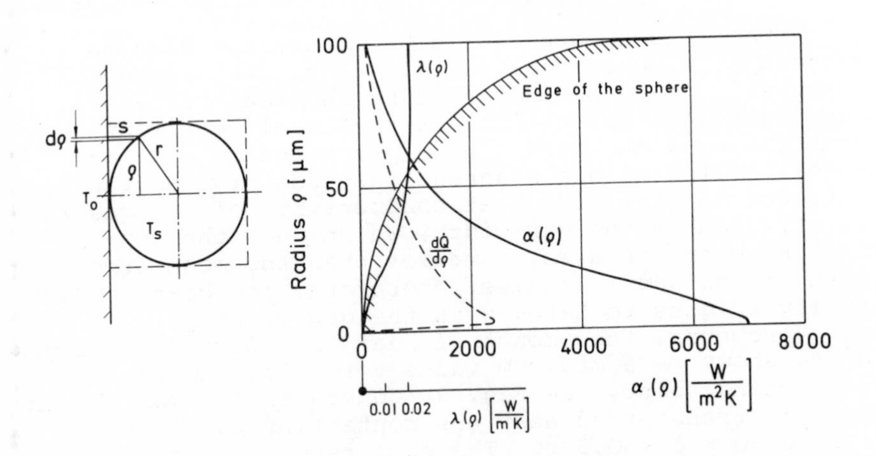

Figure 9 Heat transfer between a plane surface and a sphere through a stagnant gas. λ = local heat conductivity of the gas in the gap between plane and sphere; α (ρ) = local heat transfer coefficient defined by α(ρ) = $d\dot{Q}/2\pi\rho\, d\rho(T_o - T_s)$. Particle diameter 200 μm. Air at 300°K and 1 bar. $d\dot{Q}/d\rho$ without numbers.

and is less than 1 %. Consequently, the gas flow does not disturb the stagnant gas film near the contacting point. This is the justification for the second hypothesis that the energy transport through the solid particles separated by small amounts of stagnant gas by conduction and through the gas phase in the interstices by turbulent motion occur independently and parallel in the packed bed.

A. Energy transport by conduction in the solid phase

Since the last 50 years quite a number models have been developed by various authors to predict the heat conductivity λ_{SO} of packed beds of spheres filled with a stagnant gas. Recently Zehner (48) has compared the models presented by Krischer (49), Yagi and Kunii (50), Schumann and Voss (51), Kunii and Smith (52), Schlünder (53), Woodside (54), Deissler and Boegli (55), Wakao and Kato (56) and Krupitzka (57). Based on a detailed examination of the various models Zehner developed a new one, which seems to agree with a large number of experimental data in a wide range of temperature and pressure better than the previous ones. In 1977 Bauer (58) has extended the Zehner-model to describe packed beds of non spherical particles like cylinders, raschig rings etc. as well as to packed beds with a particle size distribution. The Zehner-Bauer-model yields to following equations for the prediction of the apparent heat conductivity of a packed bed filled with a stagnant gas λ_{SO}:

$$\frac{\lambda_{SO}}{\lambda} = (1-\sqrt{1-\psi})\left[\frac{\psi}{\psi-1+\lambda/\lambda_D}+\psi\frac{\lambda_R}{\lambda}\right]+\sqrt{1-\psi}\left[\varphi\frac{\lambda_S^*}{\lambda}+(1-\varphi)\frac{\lambda_{SO}^*}{\lambda}\right] \tag{23}$$

where

$$\frac{\lambda_{SO}^*}{\lambda} = \frac{2}{P}\left[\frac{B\left(\frac{\lambda_S^*}{\lambda}+\frac{\lambda_R}{\lambda}-1\right)\frac{\lambda}{\lambda_D}\frac{\lambda}{\lambda_S^*}}{P^2}\ln\frac{\left(\frac{\lambda_S^*}{\lambda}+\frac{\lambda_R}{\lambda}\right)\frac{\lambda}{\lambda_D}}{B\left(1+\left(\frac{\lambda}{\lambda_D}-1\right)\left(\frac{\lambda_S^*}{\lambda}+\frac{\lambda_R}{\lambda}\right)\right)} - \frac{B-1}{P}\frac{\lambda}{\lambda_D}+\frac{B+1}{2B}\left[\frac{\lambda_R}{\lambda}\frac{\lambda}{\lambda_D}-B\left(1+\left(\frac{\lambda}{\lambda_D}-1\right)\frac{\lambda_R}{\lambda}\right)\right]\right] \tag{23a}$$

with

$$P = \left(1+\left(\frac{\lambda_R}{\lambda}-B\frac{\lambda_D}{\lambda}\right)\frac{\lambda}{\lambda_S}\right)\frac{\lambda}{\lambda_D}-B\left(\frac{\lambda}{\lambda_D}-1\right)\left(1+\frac{\lambda_R}{\lambda}\frac{\lambda}{\lambda_S^*}\right) \tag{23b}$$

The individual heat conductivities λ, λ_S^*, λ_R and λ_D are:

a) λ is the heat conductivity of the fluid (gas).

b) $$\lambda_S^* = \lambda_S\left(1+\frac{1}{Bi_{ox}}\right) \tag{23c}$$

with

$$Bi_{ox} = \frac{S_{ox}}{\lambda_{ox}} \cdot \frac{\lambda_S}{\bar{d}}$$

For non metallic particles λ_S^* is equal to the heat conductivity of the solid particle λ_S. Only in case of metallic particles may an external oxide layer form an additional heat transfer resistance S_{ox}/λ_{ox}. $\bar{d}$ is the average diameter of the particles the reciprocal of which

$$\frac{1}{\bar{d}} = \sum_{i=1}^{i=n} \frac{\Delta f_i}{d_{V,i}}$$

is to be formed by summing up all particle size fractions Δf_i devided by the volume equivalent sphere diameter $d_{V,i}$. In the case of a monodisperse packed bed $\bar{d}$ is of course, equal to the sphere diameter d.

c) $$\lambda_R = \frac{0,04\ C_S}{2/\varepsilon - 1} \cdot \left(\frac{T}{100}\right)^3 \cdot x_R \qquad (23\ d)$$

λ_R is the equivalent heat conductivity for the energy transfer by radiation as already suggested by Damköhler (59).

x_R is an average distance between radiating surfaces within the packed bed and is to be calculated by

$$\frac{1}{x_R} = \sum_{i=1}^{i=n} \frac{\Delta f_i}{R_{Form} \cdot d_{V,i}}$$

R_{Form} is a shape factor taking into account that the apparent shape of the particle with respect to radiative energy exchange may be different from its geometrical one and must be determined from experiments.

d) $$\lambda_D = \frac{\lambda}{1 + \frac{2\sigma_0}{x_D} \frac{2-\gamma}{\gamma}} \qquad (23\ e)$$

λ_D takes into account the Smoluchowski effect. λ is the heat conductivity of the continuum gas, σ the mean free path of the molecules, γ the accomodation coefficient and x_D is the average gap width between the particles with respect to heat conduction in a rarified gas and is to be calculated from

$$\frac{1}{x_D} = \sum_{i=1}^{i=n} \frac{f_i}{D_{Form} \cdot d_{V,i}}$$

D_{Form} is a shape factor taking into account that the apparent shape of the particle with respect to Kundsen diffusion may be different from the geometrical one. D_{Form} is to be determined from experiments.

Eventually, the quantity B in Eq. 23 is a parameter by which the real shape of the particle is to be replaced by an artificial one so that the streamlines of the heat flux remain parallel throughout the packed bed. (This is one of the essentials of the Zehner-model). B follows from

$$B = C_{Form} \left(\frac{1-\psi}{\psi}\right)^{\frac{10}{9}} f(\xi_r) \tag{23 f}$$

C_{Form} is a shape factor with respect to particle-to-particle heat conduction. $f(\xi_R)$ is a particle size distribution function. In the case of monodisperse packed beds f equals unity. In general both C_{Form} and $f(\xi_r)$ must be determined from experiments.

The quantity φ in Eq. 23 is the relative contact area of adjacent particles depending on the relative radius of this area

$$\varphi = \frac{23\, \rho_K{}^2}{1+22\, \rho_K{}^{4/3}} \tag{23 g}$$

according to Wakao and Kato (56). ρ_K must be determined by experiments.

In the following figures Eq. 23 is compared with experimental results. The data of Imura and Tagegoshi (60) show the influence of the gas pressure on the heat conductivity of packed beds of equal sized spheres. In this case the various shape factors are $C_{Form}=1,25$, $x_R = d$, $x_D = d$.

Fig. 10 a, b, c and d demonstrate that all the data for poorly conducting spheres can be fitted with $\rho_K{}^2=3,5\cdot10^{-4}$. Fig. 11 a, b und c show that the data for metallic spheres can only be fitted if an oxid layer is assumed.

For packed beds of cylinders the shape factors are

$$C_{Form} = 2,5;\; x_R = d\, \sqrt[3]{31/2d},$$

which represent data at normal pressure very well, see (58).

The heat conductivity λ_{SO} of packed beds of hollow cylinders at various temperatures and normal pressure according (58) is shown in Fig. 12 a und b. The shape factors are $C_{Form} = 2,5\ (1+[d_i/d]^2)$ and $x_R = d\, \sqrt[3]{31/2d}$.

The heat conductivity of binary mixtures of ceramic spheres are shown in Fig. 13 a and b as a function of the mass fraction Δf_1 with the diameter ratio d_1/d_2 and the temperature T as parameters. The shape factors are $C_{Form} = 1,25$; $1/x_R=\Delta f_1/d_1+(1-\Delta f_1)/d_2$. The distribution function $f(\xi_R) \equiv 1+3\ \xi_1^1$ with

$$\xi_1 = \left[\frac{\Delta f_1 + (1-\Delta f_1)(d_1/d_2)^2}{\Delta f_1 + ([1-\Delta f_1] d_1/d_2)^2} - 1 \right]^{1/2}$$

according to Rumpf and Gupte (61). The influence of the pressure on the heat conductivity of binary mixtures of spheres has been measured by Imura and Takegoshi (60). Fig. 14 a and b show good agreement with Eq. 23 with the shape factor $1/x_D = \Delta f_1/d_1 + (1-\Delta f_1)/d_2$. So far Eq. 23 seems to predict experimental data for various conditions with sufficient accuracy for practical purposes.

B. Energy transport by turbulent motion of the fluid in the voids.

Fig. 15 shows schematically the flow pattern in a packed bed, where the flow separates from the particles and complete mixing occurs in the voids. From this model Schlünder (62) derived the apparent radial heat conductivity in the fluid phase due to the turbulent motion in the interstices for a two dimensional packed bed

$$\frac{\Lambda}{\lambda} = \frac{Pe}{K} \tag{24}$$

where $Pe = (\dot{m}c_p/\lambda) \cdot x_F$, with the mass flow rate $\dot{m}$, the heat conductivity of the fluid λ, the specific enthalpy of the fluid c_p and the mixing length x_F, which is nearly equal to the particle diameter. Bauer (58) has extended this model to polydispered packed beds. He derived the following formula for the computation of the average mixing length

$$x_F = \sum_{j=1}^{j=m} \Delta\psi_j \frac{1}{\sum_{i=1}^{i=n} \Delta f_i / x_{F,i}} \tag{25}$$

where $x_{F,i} = F_{F,i}\ d_{F,i}$.

$d_{F,i}$ ist the diameter of a sphere equivalent in volume, $F_{F,i}$ is a shape factor, which must be determined from experiments.

The constant K in Eq. 24 has the value 8 according (62) for an infinite packed bed. For a finite packed bed Schlünder (62) recommended

$$K = 8\ (2 - (1 - 2\frac{d}{d_{Tube}})^2\ ) \tag{26}$$

For packed beds of equal sized spheres Bauer (58) found the shape factor $F_F = 1{,}15$, while $d_F = d$. Fig. 16 shows experimental data for Λ/λ versus Pe number for

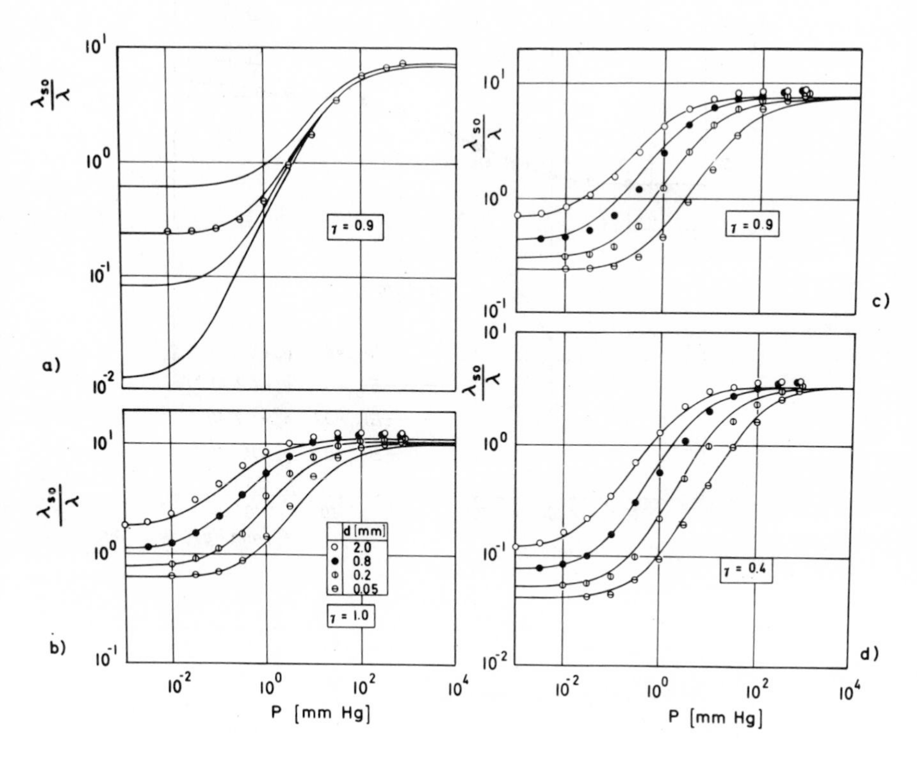

Figure 10. Heat conductivity λ_{so} of packed beds of equal-sized ceramic spheres at various pressures and room temperature according to Ref. 60 and Equation 23, $Bi_{ox} = 0$, $\psi = 0,39$, $\rho_K^2 = 3{,}5 \cdot 10^{-4}$. (a) Nitrogen $\gamma = 0.9$; (b) R 12 $\gamma = 1.0$; (c) nitrogen $\gamma = 0.9$; (d) helium $\gamma = 0{,}4$.

(a)

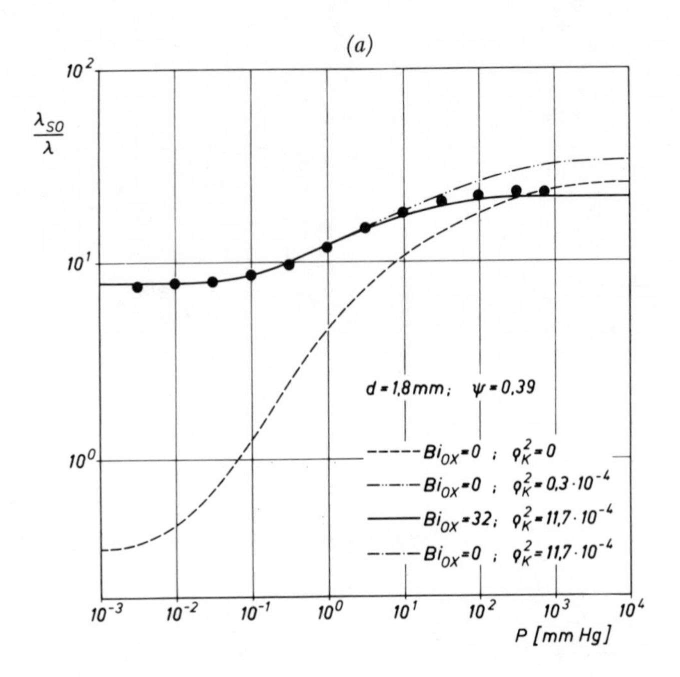

λSO/λ
d = 1,8 mm ; ψ = 0,39
Bi_OX = 0 ; φ_K^2 = 0
Bi_OX = 0 ; φ_K^2 = 0,3 · 10^-4
Bi_OX = 32 ; φ_K^2 = 11,7 · 10^-4
Bi_OX = 0 ; φ_K^2 = 11,7 · 10^-4
P [mm Hg]

(b)

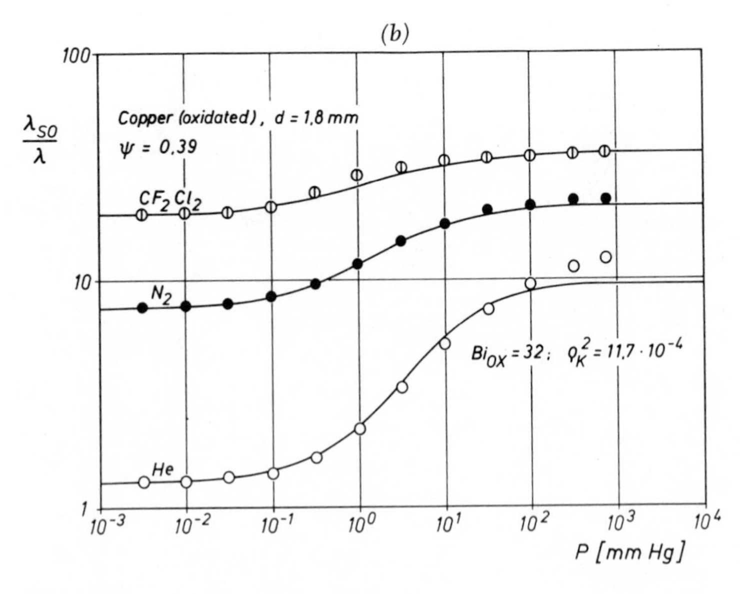

λSO/λ
Copper (oxidated), d = 1,8 mm
ψ = 0,39
CF2 Cl2
N2
He
Bi_OX = 32 ; φ_K^2 = 11,7 · 10^-4
P [mm Hg]

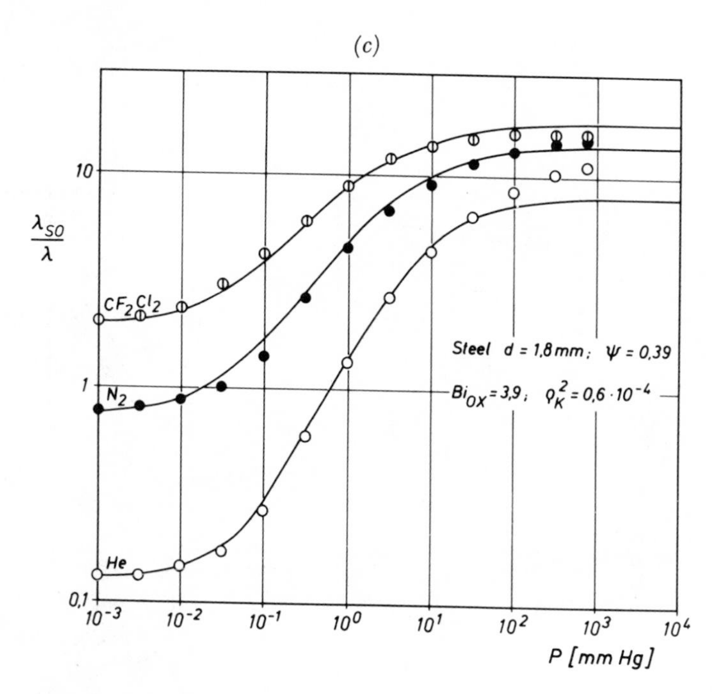

Figure 11. Thermal conductivity λ_{so} of packed beds of equal-sized metallic spheres at various pressures and room temperature according to Ref. 60 and Equation 23. (a) (left) Copper–nitrogen; (b) (left) copper–helium, nitrogen, R 12; (c) (above) steel–helium, nitrogen R 12.

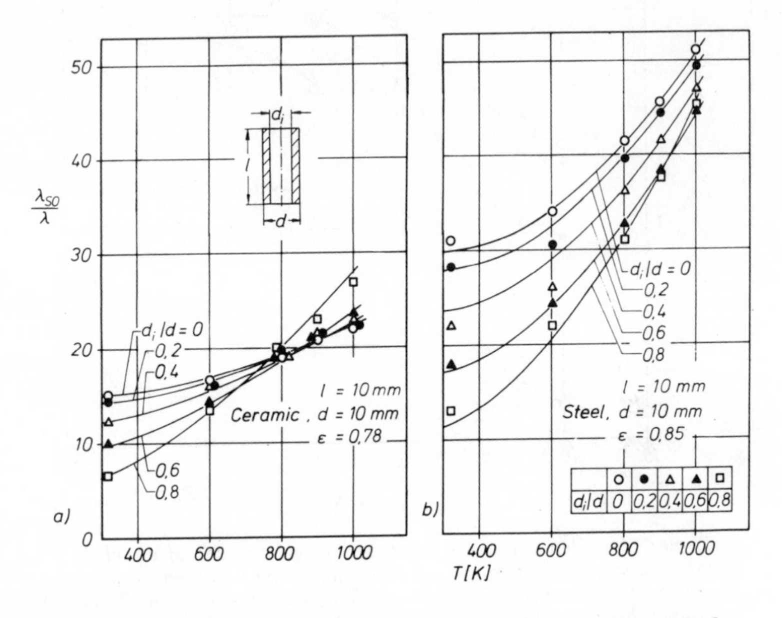

Figure 12. Thermal conductivity λ_{so} of packed beds of hollow cylinders at various temperatures according to Ref. 58 and Equation 23. (a) Ceramic–nitrogen; (b) steel–nitrogen.

packed beds of styrene, ceramic, steel and copper spheres of 10 mm diameter obtained by Bauer (58). The fluid was nitrogen. The full lines represent Eq. 24 including 26. Extrapolation to Pe = 0 gives λ_{SO}/λ according Eq. 23.

Fig. 17 a and b show the influence of the temperature up to 1000 K. The full lines represent Eq. 24, 26. The dotted line is calculated neglecting the heat transfer by radiation for T = 1000 K. In Fig. 18 a and b equivalent data are presented for cylindrical particles both of poor and good conducting material. The shape factor is $F_F = 1,75$ and the equivalent particle diameter $d_F = \sqrt[3]{31/2d}$. The mixing length x_F as a function of the diameter-to-length-ratio d/l is shown in Fig. 19. In Fig. 20 the mixing length x_F for hollow cylinders is plotted against the ratio of the inner to outer diameter d_i/d. Long cylinders with thin walls are most effective with regard to radial mixing of the fluid. The corresponding expression for the prediction of x_F are given in (58). Fig. 21 shows the mixing length x_F for binary mixtures of spheres depending on the mass fraction Δf_1 with the diameter ratio d_1/d_2 as parameter. The shape factor is $F_F = 1,15$ and the mixing length is to be calculated by $1/x_F = \Delta f_1/d_1 + (1-\Delta f_1)/d_2$. Bauer (58) has also given shape factors and mixing lengths for infinitely variable particle-size distributions.

Rewriting Eq. 22 a for practical application gives

$$\dot{q} = -\lambda \left(\frac{\lambda_{SO}}{\lambda} + \frac{\Lambda}{\lambda}\right) \frac{\partial T}{\partial r}$$

or

$$\dot{q} = -\lambda \, (K_1 + K_2 \, Pe) \frac{\partial T}{\partial r} \quad .$$

The parameters K_1 and K_2 are listed up in Table 1 for spheres, cylinders, hollow cylinders, binary mixtures of spheres and continuous granulations of both poor and good conducting material (ceramic, steel).

V.2 Heat transfer between tube walls and packed beds.

The heat transfer between a rigid wall and an adjacent packed bed of spheres is well understood for zero mass flow rate of the fluid. In this case the wall heat transfer coefficient α_W can be predicted by a formula derived by Schlünder (47):

$$\alpha_W = \frac{4\lambda}{d} \left[\left(\frac{\sigma}{d} + 1\right) \ln\left(1 + \frac{d}{\sigma}\right) - 1 \right] + \varepsilon C_S \left(\frac{T_m}{100}\right)^3 \qquad (27)$$

where $\sigma = 4\sigma_o \dfrac{2-\gamma}{\gamma}$.

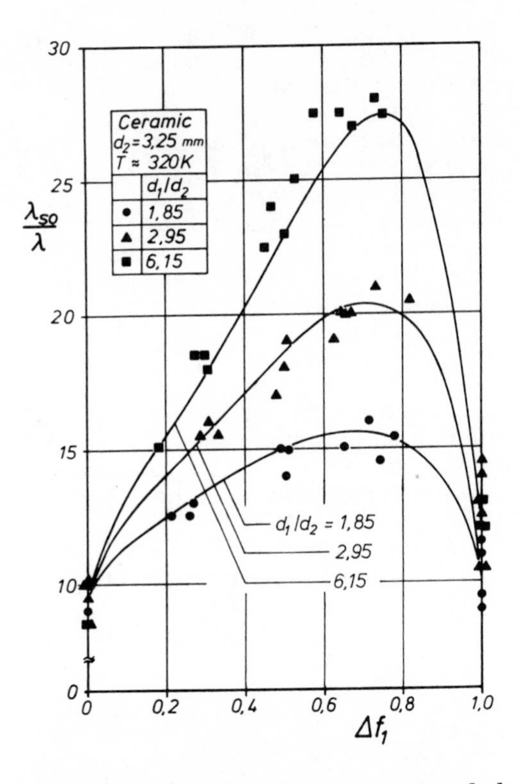

Figure 13a. Thermal conductivity λ_{so} of packed beds of binary mixtures of ceramic spheres as a function of the mass fraction Δf_1 at normal pressure according to Ref. 60 and Equation 23. Various diameter ratios d_1/d_2.

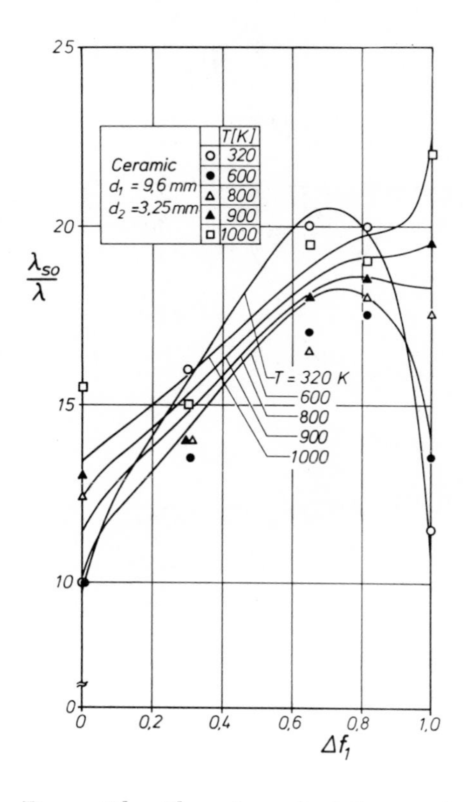

Figure 13b. Thermal conductivity λ_{so} of packed beds of binary mixtures of ceramic spheres as a function of the mass fraction Δf_1 at normal pressure according to Ref. 60 and Equation 23. Various temperatures.

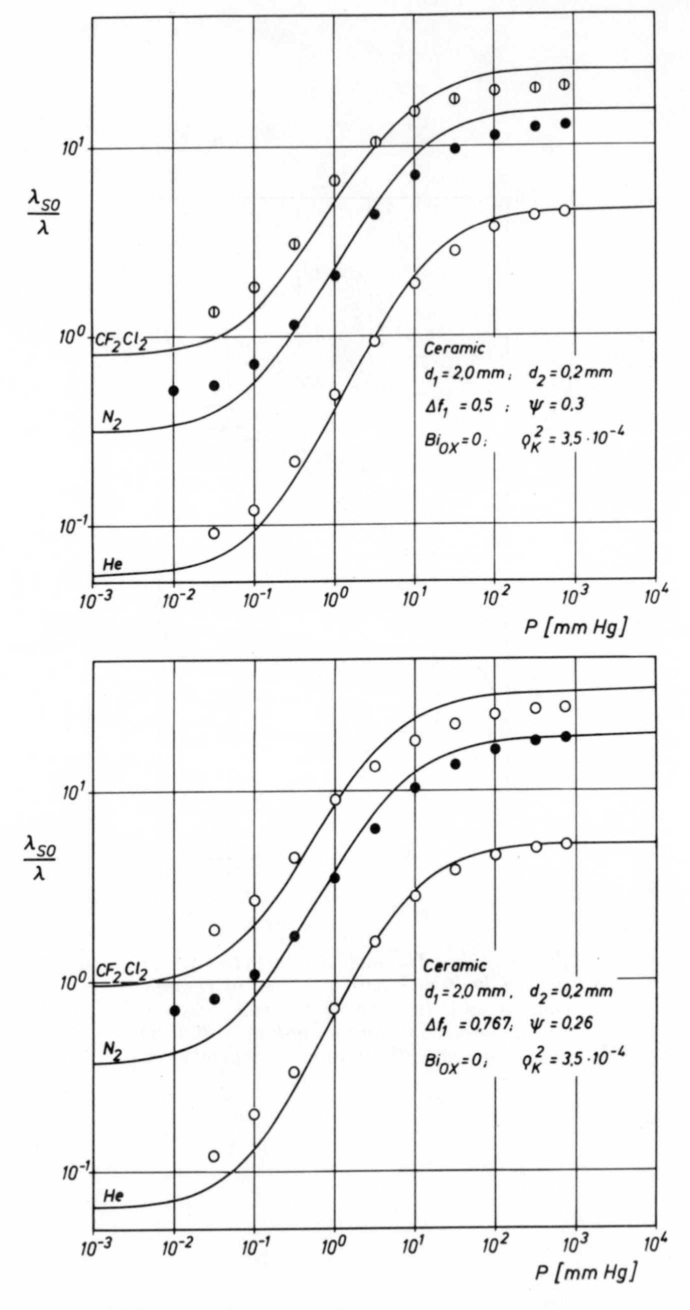

Figure 14a, b. Thermal conductivity of packed beds of binary mixtures of ceramic spheres at various pressures and room temperature according to Ref. 60 and Equation 23

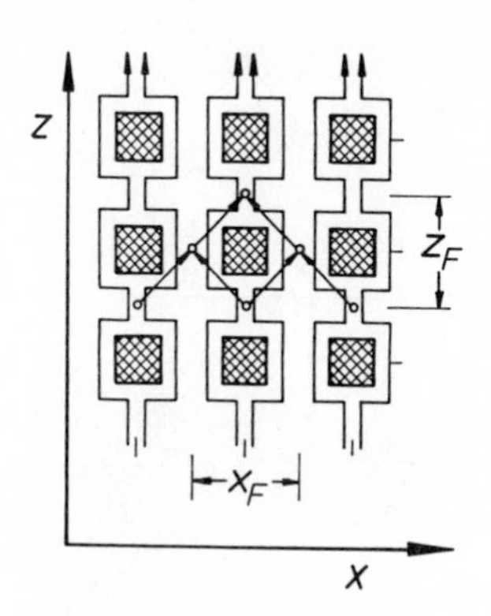

Figure 15. Flow pattern in a packed bed

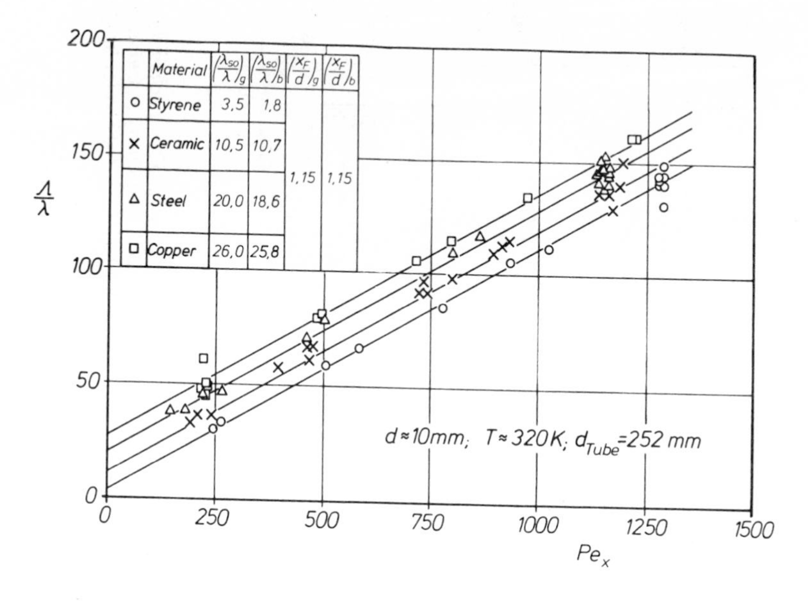

Figure 16. Apparent heat conductivity Λ of packed beds of styrene, ceramic, steel, and copper spheres at room temperatures and normal pressure

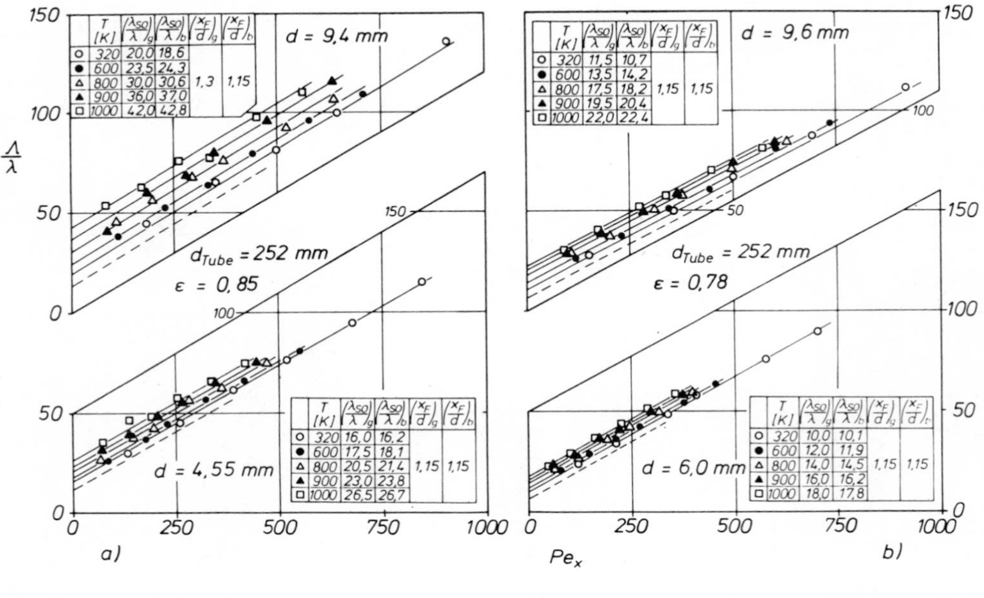

Figure 17. Apparent heat conductivity Λ of packed beds of spheres vs. Pe number at various temperatures and normal pressure. Full lines according to Equations 24 and 26. Pe = 0: λ_{so} according to Equation 23. (a) Steel–nitrogen; (b) ceramic–nitrogen.

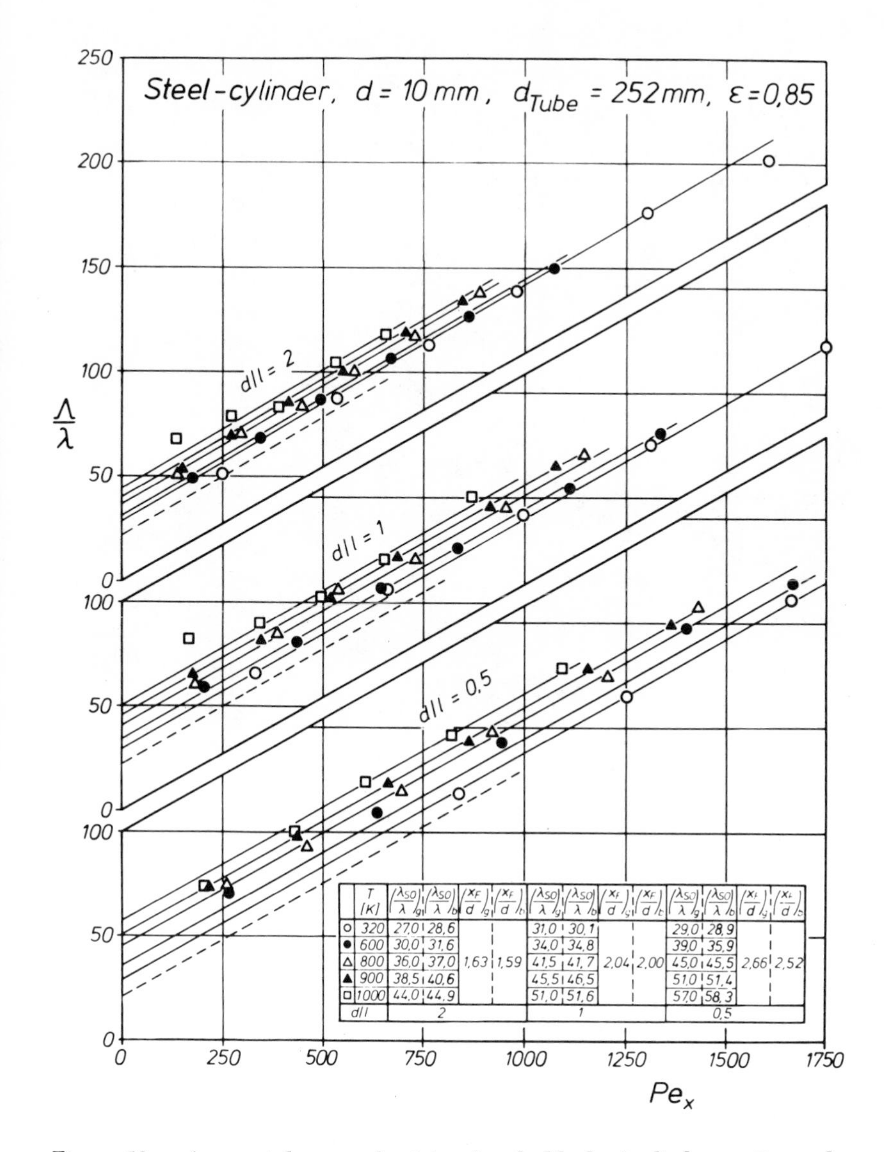

	T [K]	$(\lambda_{so}/\lambda)_g$	$(\lambda_{so}/\lambda)_b$	$(x_F/d)_g$	$(x_F/d)_b$	$(\lambda_{so}/\lambda)_g$	$(\lambda_{so}/\lambda)_b$	$(x_F/d)_g$	$(x_F/d)_b$	$(\lambda_{so}/\lambda)_g$	$(\lambda_{so}/\lambda)_b$	$(x_F/d)_g$	$(x_F/d)_b$
○	320	27,0	28,6	1,63	1,59	31,0	30,1	2,04	2,00	29,0	28,9	2,66	2,52
●	600	30,0	31,6			34,0	34,8			39,0	35,9		
△	800	36,0	37,0			41,5	41,7			45,0	45,5		
▲	900	38,5	40,6			45,5	46,5			51,0	51,4		
□	1000	44,0	44,9			51,0	51,6			57,0	58,3		
d/l		2				1				0,5			

Figure 18a. Apparent heat conductivity of packed beds of cylinders vs. Pe number at various temperatures and normal pressure. Full lines according Equations 24 and 26. Pe = 0: λ_{so} according to Equation 23. Dotted line calculated for 1000°K, however, heat transfer by radiation dropped. Steel–nitrogen.

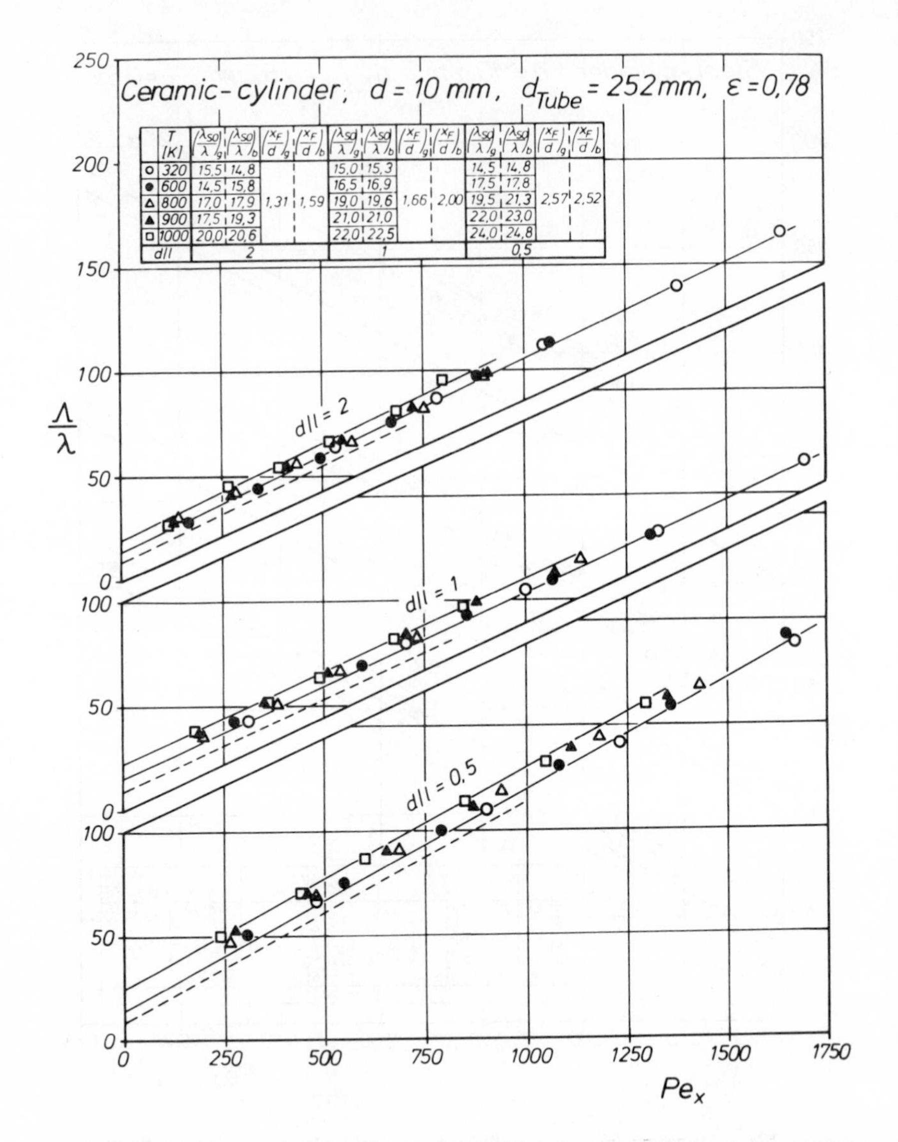

Figure 18b. Apparent heat conductivity of packed beds of cylinders vs. Pe number at various temperatures and normal pressure. Full lines according Equations 24 and 26. Pe = 0: λ_{so} according to Equation 23. Dotted line calculated for 1000°K, however, heat transfer by radiation dropped. Ceramic–nitrogen.

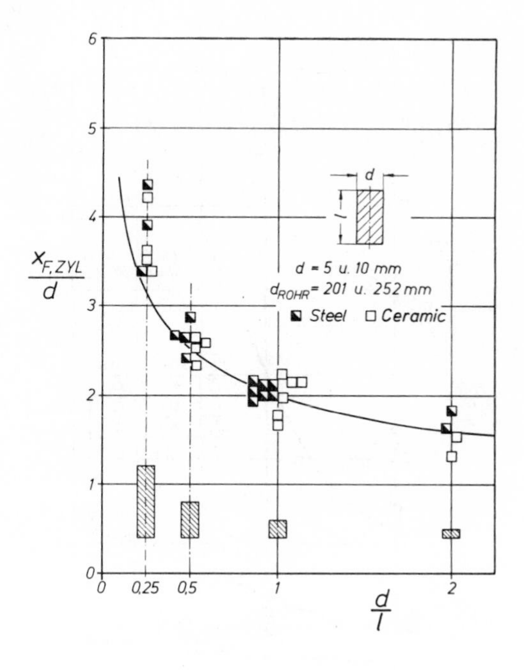

Figure 19. Mixing length x_F for cylindrical particles according to Ref. 58.

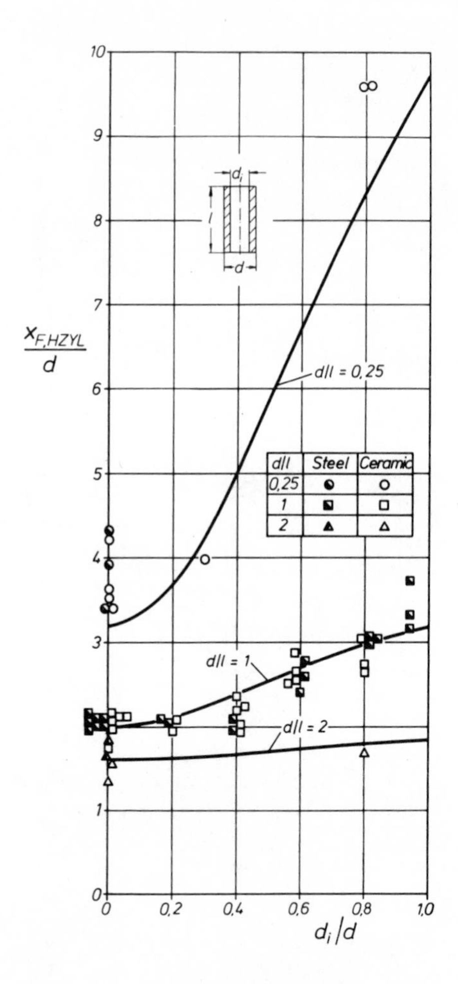

Figure 20. Mixing length for hollow cylinders according to Ref. 58.

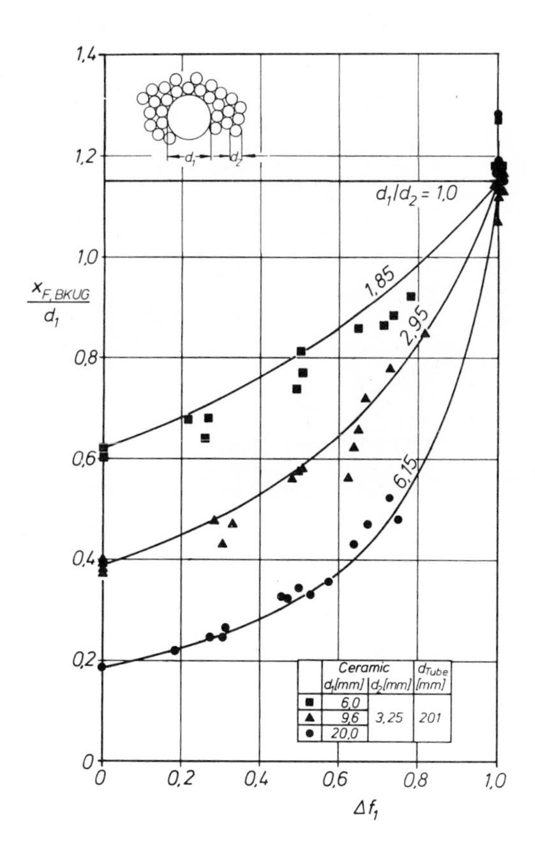

Figure 21. Mixing length for binary mixtures of spheres according to Ref. 58.

Table I.

$$\frac{\Lambda}{\lambda} = K_1 + K_2 \cdot Pe$$

with $Pe = \dot{m}_0 \cdot c_p \cdot d / \lambda$

Binary mixtures of spheres: $\Delta f = \frac{M \text{ (Mass of the large spheres)}}{M_{tot} \text{ (Total mass)}}$

Particle size distribution: $d = \frac{1}{\Sigma \Delta f_i / d_i}$; $\xi = \left[\frac{\Sigma \Delta f_i / d_i^2}{(\Sigma \Delta f_i / d_i)^2} - 1 \right]^{\frac{1}{2}}$; $\Delta f_i = \frac{M_i}{M_{tot}}$, $d_i = \sqrt[3]{\frac{6 M_i}{\rho_s \cdot z \cdot \pi}}$

	Material		d [mm]	T [K]	Sphere	Cylinder			Hollow Cylinder									Binary mixtures of spheres								Particle size distribution				
						l/d			l/d = 0.5, d_i/d			l/d = 1.0, d_i/d			l/d = 4.0, d_i/d			Δf = 0.2, d/d_k		Δf = 0.4, d/d_k		Δf = 0.6, d/d_k		Δf = 0.8, d/d_k		ξ				
						0.5	1.0	4.0	0.2	0.5	0.8	0.2	0.5	0.8	0.2	0.5	0.8	3	6	3	6	3	6	3	6	0	0.25	0.5	0.75	1.0
K_1	poor conducting particles	Example Ceramic-Air	5	300	8.1	10.7	10.9	9.2	10.2	7.8	4.4	10.4	8.1	4.7	6.8	7.4	5.1	10.6	11.4	12.7	14.4	14.4	17.4	14.4	16.5	7.9	10.0	11.5	12.6	13.9
				800	11.1	10.8	11.6	13.6	10.7	10.1	10.3	11.5	11.1	11.9	13.6	14.1	16.9	9.0	8.7	10.1	10.5	11.1	11.9	11.5	12.6	11.0	11.3	11.2	11.5	11.9
				1000	12.9	11.3	12.7	15.3	11.7	11.8	13.3	12.7	13.1	16.0	16.0	17.6	23.2	9.3	8.5	10.2	9.9	11.0	11.2	11.6	11.3	12.8	12.1	12.1	12.2	12.3
			10	300	8.9	11.2	11.0	10.4	10.6	8.5	5.4	11.1	8.9	5.3	10.2	8.6	7.1	10.9	11.5	12.9	14.6	14.7	17.6	14.8	18.5	8.8	10.0	12.1	13.5	14.4
				800	14.9	13.5	14.6	16.4	13.4	13.6	16.1	14.6	15.2	18.7	18.6	20.5	27.2	10.5	9.4	11.5	11.0	12.5	12.6	13.1	13.5	14.6	13.9	13.8	13.9	14.0
				1000	17.1	14.9	16.1	21.5	15.0	16.3	21.7	16.3	16.2	25.3	22.1	25.7	37.5	11.2	9.6	11.3	10.3	12.8	12.2	13.5	13.1	17.1	15.7	15.2	15.1	15.0
	good conducting particles	Example Steel-Air	5	300	17.0	25.6	27.0	21.2	24.0	16.9	7.5	25.3	17.8	3.1	20.1	15.3	6.2	22.1	20.7	23.4	30.5	36.0	45.2	38.6	52.5	16.7	23.5	29.4	34.2	38.6
				800	22.9	27.6	30.0	31.3	26.4	21.1	15.2	23.5	25.6	18.0	31.2	28.0	25.7	20.3	17.2	26.2	24.4	32.6	34.3	35.4	42.7	22.6	26.8	30.5	34.2	38.3
				1000	28.5	31.1	35.2	41.0	30.0	25.5	21.4	34.1	23.4	25.8	40.4	38.3	36.9	21.2	17.0	20.9	23.6	33.2	33.2	36.7	41.8	28.1	31.0	34.4	37.7	40.8
			10	300	13.5	28.8	30.2	24.1	27.1	13.3	9.2	28.4	20.5	10.0	23.0	18.0	10.8	25.7	25.0	34.0	36.7	42.5	51.9	43.7	61.1	13.2	26.4	32.3	38.9	44.4
				800	33.3	37.2	41.8	46.2	35.9	30.3	25.4	40.4	34.3	30.6	47.6	45.2	46.3	26.5	22.5	35.7	31.8	41.5	44.4	45.0	54.1	33.4	36.9	41.1	45.2	49.1
				1000	45.4	45.1	51.3	66.4	44.0	39.6	37.9	50.8	46.5	46.1	66.0	65.2	72.1	23.3	23.2	30.4	32.2	44.5	44.4	43.4	[illegible]	44.6	46.1	49.2	52.5	55.8
	$\frac{d_{Tube}}{d}$																													
K_2	10				0.1	0.15	0.18	0.23	0.15	0.16	0.16	0.19	0.23	0.27	0.34	0.54	0.77	0.041	0.021	0.048	0.026	0.075	0.035	0.076	0.053	0.14				
	25				0.12	0.17	0.22	0.34	0.17	0.18	0.19	0.23	0.28	0.32	0.40	0.63	0.30	0.048	0.024	0.057	0.031	0.070	0.042	0.083	0.062	0.17				
	≥50				0.13	0.19	0.24	0.37	0.19	0.20	0.21	0.25	0.30	0.35	0.43	0.69	1.00	0.052	0.025	0.061	0.034	0.059	0.045	0.036	0.067	0.18				

λ is the heat conductivity of the fluid (gas), σ_o is the mean free path of the gas molecules, γ is the accomodation coefficient, εC_s is the radiation emissivity and T_m is the average temperature between the wall and the first layer of the packed bed. The first term on the right hand side of Eq. 27 accounts for molecular heat conduction through the gaseous gap between the wall and the spheres. The second term accounts for radiation heat transfer. Eq. 27 is based on the assumption that there is only point contact between the spheres and the wall, i.e. $\rho_K = 0$, see Eq. 23. Wunschmann and Schlünder (63), (64) have confirmed the validity of Eq. 27 by unsteady state heat transfer experiments for poor conducting materials like polystyrene and glass spheres. A packed bed resting on an electrically heated copper plate was heated from below for a certain contact time t. Evaluation of these experiments yield the heat transfer coefficient α_W defined by

$$\alpha_W = \frac{\dot{q}}{T_W - \bar{\bar{T}}} \tag{28}$$

where $\dot{q}$ is the heat flux, T_W the wall temperature and $\bar{\bar{T}}$ is the average temperature of the packed bed, determined by a heat balance, as a function of the contact time t. For very short contact times and in particular at low pressure the heat transfer is entirely controlled by Eq. 27. For longer contact times the packed bed may be considered as a quasi-continuum so that Fourier's theory applies giving

$$\alpha_W = \frac{2}{\sqrt{\pi}} \; \frac{\sqrt{\rho c_p \lambda_{SO}}}{\sqrt{t}} \tag{29}$$

where λ_{SO} is to be calculated by Eq. 23. Fig. 22 a, b, c show the experimental data for packed beds of polystyrene, glass and bronze spheres, respectively. The parameter is the gas (air) pressure. On the left α_W according Eq. 27 and on the right according Eq. 29 is indicated. There is good agreement with the theory for the poor conducting materials like polystyrene and glass. The bronze spheres, however, show higher heat transfer coefficients in particular at low pressure and short contact time. In this regime one would expect that the heat transfer is only radiation controlled (α_W = 5 W/m^{2}K at room temperature). Obviously the assumption of point contact must be replaced by another one admitting a certain contact area, i.e. $\rho_K > 0$. A rough estimation gives

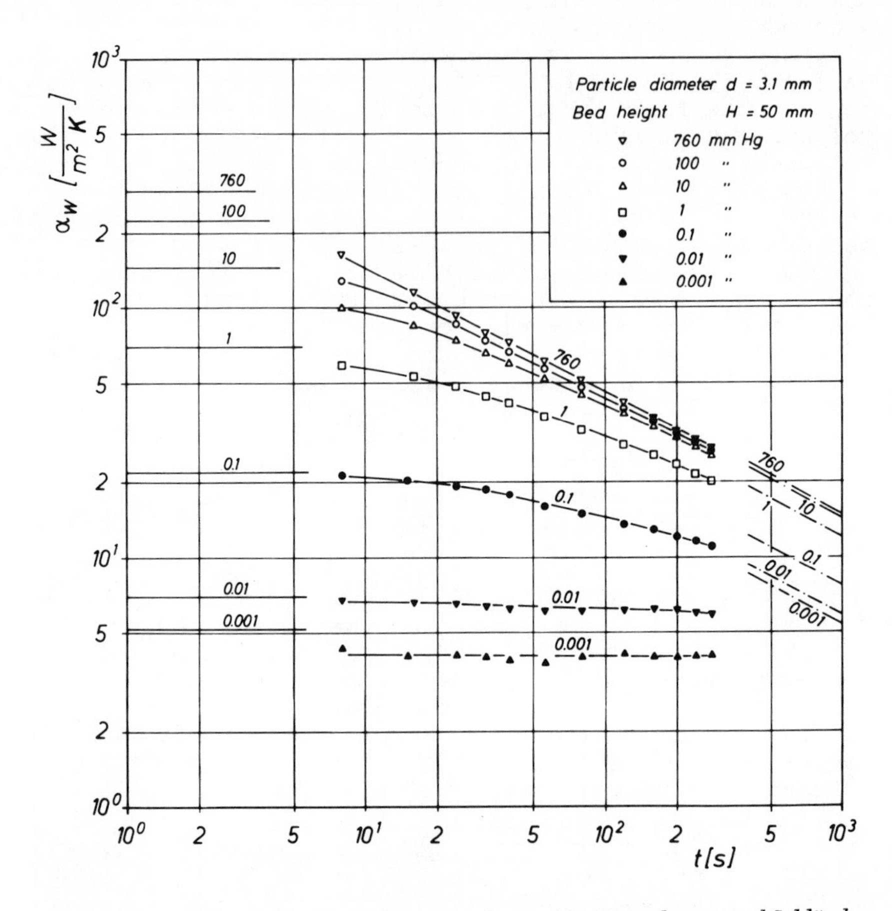

Figure 22a. Wall heat transfer coefficient α_W *obtained by Wunschmann and Schlünder (64) from transient heat transfer experiments at room temperature and various gas (air) pressures. Glass, d = 3,1 mm.*

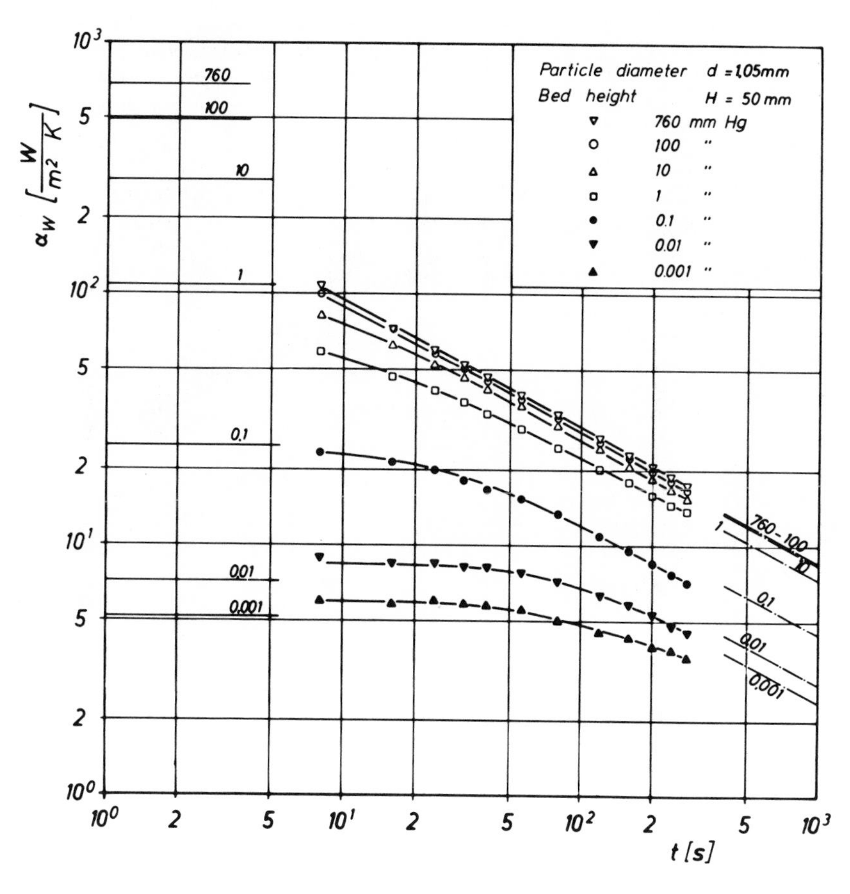

Figure 22b. Wall heat transfer coefficient α_W obtained by Wunschmann and Schlünder (64) from transient heat transfer experiments at room temperature and various gas (air) pressures. Polystyrene, d = 1,05 mm.

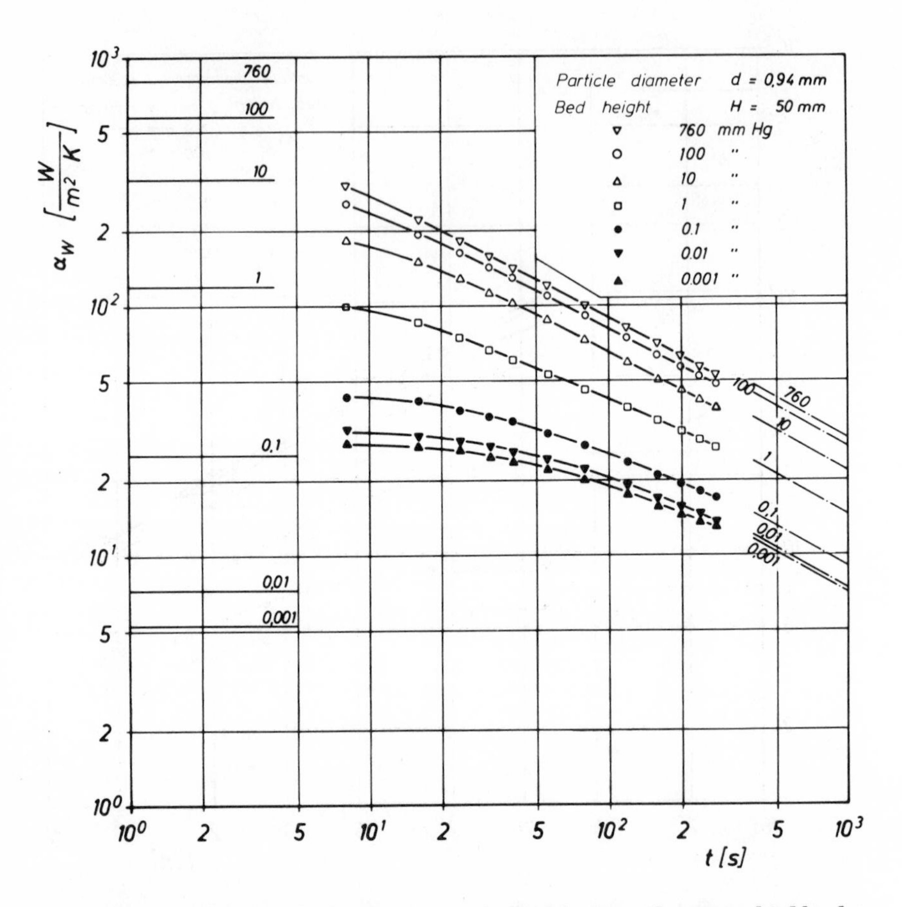

Figure 22c. Wall heat transfer coefficient αw obtained by Wunschmann and Schlünder (64) from transient heat transfer experiments at room temperature and various gas (air) pressures. Bronze, d = 0,94 mm.

$$\alpha_W \cong 2 \frac{\lambda_S}{d} \rho_K + \varepsilon C_S \left(\frac{T_m}{100}\right)^3 \qquad (30)$$

With the data from Fig. 22 c at low pressure and short contact time one obtains $\rho_K \cong 10^{-4}$ which is in the same order of magnitude as has been found from steady state experiments, see e.g. Fig. 12.

Eq. 27 gives average wall heat transfer coefficients obtained by integrating over the local heat transfer rates, which reach a rather high value near the contact point, see Fig. 9. This large variation of the local heat flux with respect to the particle radius may become of practical importance, too. E.g. should a wet porous catalyst be dried in an indirectly heated rotary dryer, one would expect that the drying rate during the constant rate period is controlled by the wall heat transfer coefficient α_W according Eq.27. Experiments, however, have shown that the drying rate is much lower. Fig. 23 shows experimental data (University of Karlsruhe, still unpublished) for magnesium silicate spheres of 6 mm diameter dried in an agitated packed bed contacting a hot plate. The α_W-controlled drying rate can be maintained only during the first few seconds. Later when a constant rate period is observed, the drying rate is much lower. The reason for this considerable rate reduction is, that the particles immidiately dry off near the contact point, because the local heat flux is so high, that the equivalent amount of liquid cannot be brought to the surface of the particles at this point. This means that the drying rate during the constant rate period is no longer heat transfer controlled, but also mass transfer controlled according to the strength of the capillary forces in the porous granules.

Wall heat transfer coefficients in tubular reactors with a fluid passing through can be determined either by direct evaluation of temperature profiles in the packed bed or by analyzing over all heat transfer coefficients, so far the apparent heat conductivity is known. Recently Hennecke and Schlünder (65) have evaluated about 5000 data collected from 14 authors, who have published over all heat transfer coefficients, applying Eq. 22 a and the following ones for the prediction of the apparent heat conductivity of the packed bed Λ . Assuming plug flow they obtained wall heat transfer coefficients depending not only on the Peclet number ($Pe = ud/\kappa$) but also on the ratio of tube diameter D to tube length L as shown in Fig. 24, where the wall Nusselt number $Nu_W = \alpha_W d/\lambda$ is plotted against Pe. u is the superficial velocity of the gas,

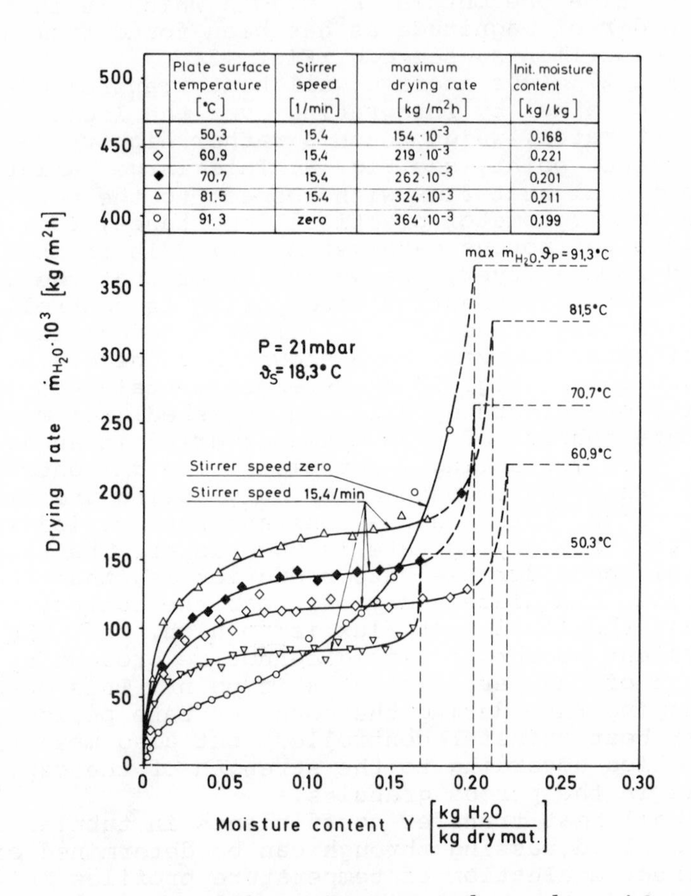

Figure 23. Contact drying of magnesium silicate spheres of 6-mm diameter in an agitated packed bed. Drying rate m *vs. moisture content* Y. *Pressure 21 mbar.*

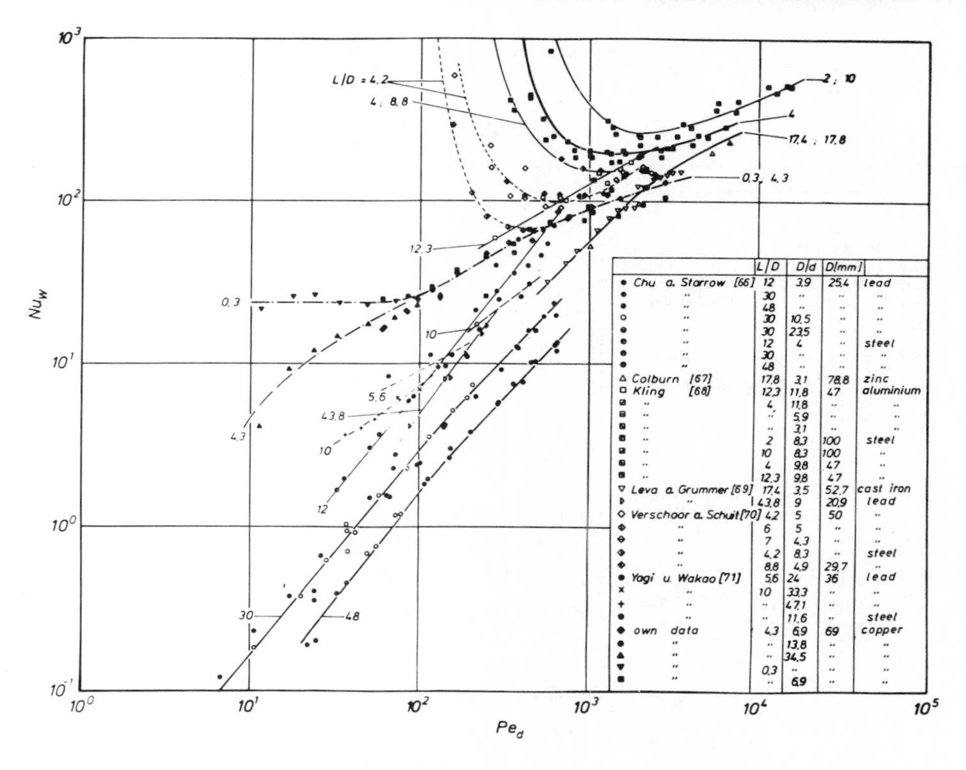

Figure 24. Wall heat transfer coefficients $Nu_w = \alpha_w/d\lambda$ *vs. particle Peclet number* $Pe = ud/\kappa$ *obtained by evaluation of overall heat transfer coefficients in tubular reactors assuming plug flow according to Ref.* 65. *D* = *tube diameter, L* = *tube length.*

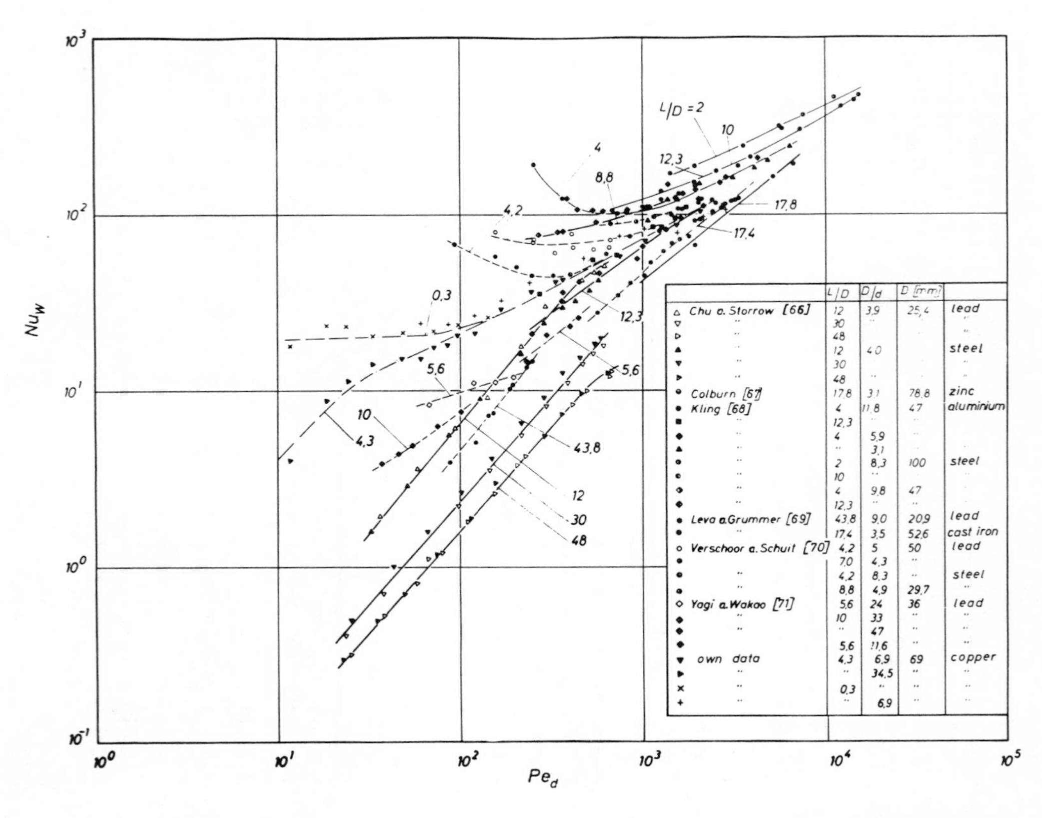

Figure 25. *The same data as in Figure 24. However, evaluated, with a flow rate distribution according to Figure 3.*

d is the particle diameter, κ the thermal diffusivity and λ the thermal conductivity of the gas. Some of the Nu_W-data not only tend towards infinity but also turned out to be negative! Taking into account the radial flow rate distribution as given in Fig. 3 the same evaluation yielded wall heat transfer coefficients Nu_W (Pe, L/D) depending much less on the ratio L/D as shown in Fig. 25. Also no negative values have been obtained. However, for long tubes the wall heat transfer coefficients are still lower by orders of magnitude than those for short tubes. This effect is still not really understood. It may be that this is due to the by pass effect described in section IV. However, this is an assumption and is still subject to further investigation. For practical purposes Hennecke (65) has developed semi-empirical correlations to predict both wall heat transfer coefficients as shwon in Fig. 25 as well as flow rate distributions in packed bed reactors as shown in Fig. 3.

At the present state these correlations together with Eq. 23 for the prediction of the apparent heat conductivity and Eq. 6, 7 for the particle-to-fluid heat transfer may be recommended for application in tubular reactor design. These correlations give fairly reliable parameters for either homogeneous or heterogeneous one or two dimensional models for the mathematical simulation of packed bed reactors.

Symbols

A	surface area		
a_V	surface area per unit volume		
C_S	radiation of the black body		
d	particle diameter		
D, d_{Tube}	bed diameter		
f	cross sectional area		
L	bed height ;	l	particle length
ΔP	pressure drop		
$\dot{q}$	Heat flux		
r, ρ	radial coordinate		
S	gap width		
T	temperature ;	t	time
u	superficial velocity		
$\dot{V}$	flow rate		
y	distance from the tube wall		
α	heat transfer coefficient		
ε	radiation emissivity		
η, ν	dynamic and kinematic viscosity, resp.		
λ	void fraction		
Λ	apparent heat conductivity		
κ	thermal diffusivity		
σ_o	mean free path of the molecules		

Literature Cited

(1) Benenati, R.F. and Brosilow, C.B.
AIChE J. (1962) 3 359

(2) Ridgway, K. and Tarbuck, V.J.
Chem. Engng. Sci. (1968) 23 1147

(3) Martin, H.
Chem. Engng. Sci. (to be published in 1978)

(4) Ergun, S.
Chem. Engng. Progr. (1952) 48 89

(5) Schwartz, C.E. and Smith, J.M.
Ind. Engng. Chem.(1953) 45 1209

(6) Gnielinski, V.
Forsch. Ing. Wesen (1975) 41 145

(7) Ranz, W.E.
Chem. Eng. Progr. (1952) 48 247

(8) Gnielinski, V.
verfahrenstechnik (to be published in 1978)

(9) Schlünder, E.U., Einf. Wärme- u. Stoffübertragung, Vieweg-Verlag Braunschweig, 2.Aufl. (1975)

(10) Hobson, M. and Thodos, G.
Chem. Engng. Progr.(1951) 47 370

(11) Gupta, A.S. and Thodos, G.
Amer.Inst.Chem.Engng.(AIChE) J. (1963) 9 751

(12) Rowe, P.N. and Claxton, K.T.
Trans.Instn.Chem.Engrs. (1965) 43 T321

(13) Glaser, M.B. and Thodos, G.
Amer.Inst.Chem.Engng.(AIChE) J. (1958) 4 63

(14) Malling, G.F. and Thodos, G.
Internat.J.Heat Mass Transfer (1967) 10 489

(15) Gupta, A.S.
Ph.D. Dissertation, Northwestern Univ. Evaston, Ill. (1963)

(16) Thoenes, D. and Kramers, H.
Chem. Engng. Sci. (1958) 8 271

(17) Glaser, H.
Chemie-Ing.-Techn. (1962) 34 468

(18) Jaeschke, L.
Diss. TH Darmstadt (1960)

(19) Bradshaw, R.D. and Myers, J.E.
Amer.Inst.Chem.Engng.(AIChE) J. (1963) 9 590

(20) Gaffney, B.J. and Drew, T.B.
Industr. Engng. Chem. (1950) 42 1120

(21) Wilkins, G.S. and Thodos, G.
Amer.Inst.Chem.Engng.(AIChE) J. (1965) 15 47

(22) v.d. Decken, C.B., Hantke, H.J., Binckebanck, J. and Bachus, K.P.
Chemie-Ing.-Techn. (1960) 32 591

(23) Wilson, E.J. and Geankopolis, C.J.
Industr.Engng.Chem.Fund. (1966) 5 9

(24) Bhattacharya, S.N. and Raja Rao, M.
Indian Chem. Engng. (1967) 9 T65
(25) Evans, G.C. and Gerald, C.F.
Chem. Engng. Progr. (1953) 49 135
(26) Venkateswaran, S.D. and Laddha, G.S.
Indian Chem. Engng. (1966) 8 T33
(27) Williamson, J.E., Bazaire, K.E. and Geankopolis, C.J.
Industr.Engng.Chem.Fund. (1963) 2 126
(28) McCune, L.K. and Wilhelm, R.H.
Industr. Engng. Chem. (1949) 41 1124
(29) Karabelas, A.J., Wegner, T.H. and Hanratty, T.J.
Chem. Engng. Sci. (1971) 26 1581
(30) Jolls, K.R. and Hanratty, T.J.
Amer.Inst.Chem.Engng.(AIChE) J. (1969) 15 199
(31) Soerensen, J.P. and Stewart, W.E.
Chem. Engng. Sci. (1974) 29 827
(32) Kunii, D. and Suzuki, M.
Int. J. Heat and Mass Transfer (1967) 10 845
(33) Schlünder, E.U.
Chem. Engng. Sci. (1977) 32 845
(34) Grootenhuis, P. Mackworth, R.C.A. and Sounders, O.A.
Proc. Instn. Mech. Engrs. (1951) General Discussion on Heat Transfer pp. 363-366
(35) Eichhorn, J. and White, R.R.
Chem. Eng. Progr. Symp. Series (1952) 48 11
(36) Kunii, D. and Smith, J.M.
AIChE J. (1961) 7 29
(37) Löf, G.O.G. and Hawley, R.W.
Ind.Engng.Chem. (1948) 40 1061
(38) Domadien, G.
Revue Inst. Fr. Petrole (1961) 16 1330
(39) Satterfield, C.N. and Resnick, H.
Chem. Engng. Progr. (1954) 50 504
(40) Kunii, D. and Ito, K.; referred to as "to be published" in (32)
(41) Mimura, T. Studies on heat transfer in packed beds, Graduate Thesis, University of Tokio (1963)
(42) Suzuki, K., Studies on heat transfer in packed beds, Graduate Thesis, University of Tokio (1964)
(43) Tokutomi, T., Heat transfer dynamics of packed beds, M.S. thesis, University of Tokio (1966)
(44) Harada, H., Studies on axial heat transfer in packed beds, Graduate Thesis, University of Tokio (1965)
(45) Mitsumori, T., Unsteady-state heat transfer in packed beds, Graduate Thesis, University of Tokio (1966)

(46) Seidel, H.P.
CIT (1965) 37 1125
(47) Schlünder, E.U.
CIT (1971) 43 651
(48) Zehner, P.
VDI-Forschungsheft 558, (1973)
(49) Krischer, O. and Kröll, K. Die wiss. Grundlagen der Trocknungstechnik Bd. 1, Springer-Verlag (1963) Berlin, Göttingen, Heidelberg
(50) Yagi, S. and Kunii, D.
Amer. Inst. Chem. Engrs. (1957) 3 373
(51) Schumann, E.W. and Voss, V.
Fuel (1934) 13 249
(52) Kunii, D. and Smith, J.M.
AIChE J. (1960) 6 71
(53) Schlünder, E.U.
CIT (1966) 38 967
(54) Woodside, W.
Cand. J. Phys. (1958) 36 815
(55) Deissler, R.G. and Boegli, J.S.
Trans. Amer.Soc.Mech.Engrs.(ASME) (1958) 80 141
(56) Wakao, N. and Kato, K.
J. Chem. Engng. Japan (1969) 2 24
(57) Krupitzka, R.
Int. Chem. Engng. (1967) 7 122
(58) Bauer, R.
VDI-Forschungsheft 582 (1977), VDI-Verlag Düsseldorf and Int. Chem. Engng. (1977)
(59) Damköhler, G. (M. Eucken and A. Jakob)
Der Chemieingenieur, Bd. III, 1. Teil, Akad. Verlag GmbH Leipzig (1937)
(60) Imura, S. and Takgegoshi, E.
Nippon Kikai Gakkai Ronbunsku Tokyo (1970) 40 489
(61) Rumpf, H. and Gupte, A.R.
Chem. Ing. Techn. (1971) 6 367
(62) Schlünder, E.U.
Chem. Ing. Techn. (1966) 38 967
(63) Wunschmann, J. and Schlünder, E.U.
5th International Heat Transfer Conference Tokyo (1974), Paper CT2.1
(64) Wunschmann, J. and E.U. Schlünder
verfahrenstechnik (1975) 10
(65) Hennecke, F.W. and Schlünder, E.U.
Chem. Ing. Techn. (1973) 45 277
(66) Leva, M.
Ind. Engng. Chem. (1946) 38 415
(67) Leva, M.
Ind. Engng. Chem. (1947) 39 865
(68) Hanratty, Th.P.
Chem. Engng. Sci. (1954) 230

(69) Zehner, P. and Schlünder, E.U.
Chem. Ing. Techn.(1973) 45 272
(70) Beek, J.
Advances in Chem. Eng. Vol. 3, Academic Press
New York (1962)
(71) Yagi, S. and Kunii, D.
Int. Development in Heat Transfer, Part IV (1961)

RECEIVED March 30, 1978

5

Environmental Reaction Engineering

JOHN H. SEINFELD

Department of Chemical Engineering, California Institute of Technology, Pasadena, CA 91125

Chemical reaction engineering problems associated with environmental systems are numerous. Design of gas cleaning absorption processes, waste water treatment facilities, low-emission combustion processes, and catalytic mufflers are typical problems. A review of the state of environmental reaction engineering cannot be accomplished in a chapter of modest length. Rather than attempt a comprehensive review of environmental reaction engineering, therefore, we have chosen to focus here somewhat more narrowly. Specifically, we will discuss several challenging problems in what might be termed atmospheric reaction engineering.

First, we survey the principal current problems in the chemistry of the polluted atmosphere. In the atmosphere primary gaseous emissions such as oxides of nitrogen, hydrocarbons, and sulfur dioxide undergo a large number of chemical reactions that lead to formation of ozone and oxygenated organic species as well as to formation of submicron particles consisting of, among other constituents, sulfates, nitrates and oxygenated organics. In addition to homogeneous (gas-phase) reactions, heterogeneous (particulate-phase) reactions may also take place in the aerosols generated. Second we discuss the key problems associated with predicting the dynamics of aerosols in the atmosphere. Whereas the physical processes that can affect the dynamics of gaseous pollutants are limited to those that alter concentrations in a unit volume of air through atmospheric transport, those that influence particle behavior are much more diverse. Processes such as nucleation, condensation and coagulation must be considered.

The object of research in atmospheric reaction engineering is to understand as fundamentally as possible the processes that determine the evolution of gaseous and particulate species in the atmosphere, and, from such knowledge, to design source control strategies to meet air quality criteria.

Chemistry of the Urban Atmosphere

Virtually every reaction occurring in the atmosphere is

0-8412-0432-2/78/47-072-162$07.75/0

subject to some degree of uncertainty, whether in the rate constant or in the nature and quantity of the products. In evaluating a mechanism that describes atmospheric chemical dynamics the customary procedure is to compare the results of laboratory experiments, usually in the form of concentration-time profiles, with simulations of the same experiment using the proposed mechanism. A sufficient number of experimental unknowns exists in all such mechanisms that the predicted concentration profiles can be varied somewhat by changing rate constants (and perhaps mechanisms) within accepted bounds. The inherent validity of a mechanism can be judged by evaluating how realistic the parameter values used are and how well the predictions match the data. Since the mechanism will generally not be able to reproduce every set of data to which it is applied, two factors must be considered: (a) identification of the major sources of uncertainty, such as inaccurately known rate constants or mechanisms of individual reactions, and (b) evaluation of so-called "chamber effects," phenomena peculiar to the laboratory system in which the data have been generated. We focus here on the identification of the major sources of uncertainty in urban atmospheric chemistry.

Chemistry of Oxides of Nitrogen and Hydrocarbons. The chemistry of the polluted atmosphere is exceedingly complex. Several hundred chemical reactions are known to occur in a mixture of only a single hydrocarbon, oxides of nitrogen, carbon monoxide, water vapor, and air. The polluted atmosphere contains hundreds of different hydrocarbons, each with its own reactivity and reaction products. The classes of major primary pollutants in the polluted atmosphere are given in Table I. In this section we focus on the chemistry of the oxides of nitrogen and hydrocarbons.

Table I. Classes of Major Primary Pollutants in the Polluted Atmosphere

HYDROCARBONS
Alkanes (e.g. n-butane, isopentane, isooctane)
Cycloalkanes (e.g. cyclohexane, methylcyclopentane)
Olefins (alkenes)(e.g. ethylene, propylene, butene)
Cycloolefins (e.g. cyclohexene)
Alkynes (e.g. acetylene)
Aromatics (e.g. toluene, xylene)
ALDEHYDES, RCHO (e.g. formaldehyde, acetaldehyde)
KETONES, RCOR (e.g. acetone, methylethylketone)
NITRIC OXIDE, NO[a]
CARBON MONOXIDE, CO
SULFUR DIOXIDE, SO_2

[a] "NO_x" is often used to indicate "oxides of nitrogen." In practice NO_x usually refers to the sum, $NO+NO_2$, although it may include such other forms as NO_3 and N_2O_5. Nitrous oxide, N_2O, is relatively inert in the lower atmosphere and is not included in NO_x.

Table II presents a summary of the principal reactions that occur when an atmosphere containing oxides of nitrogen is irradiated with sunlight. Most of the rate constants in Table II are well-established. The most important recent developments in NO_x chemistry are those involving the reactions of OH and HO_2 with NO and NO_2. New values have recently been determined for each of the four rate constants in this set, and the mechanism of the HO_2-NO_2 reaction has been elucidated. Thus, although there are the customary levels of experimental uncertainty associated with each of the rate constants in Table II, the reactions in the NO_x system do not represent serious gaps in our understanding of atmospheric chemistry.

A careful review of the net results of Reactions 1 through 24 in Table II reveals that these reactions alone cannot explain the rapid conversion of nitric oxide to nitrogen dioxide and ozone formation observed in the real atmosphere. In fact, if these reactions alone occurred, the original supply of nitrogen dioxide in the atmosphere would be slightly depleted as irradiation with sunlight occurred, and a small and near constant level of ozone would be created in a few minutes. The key to the observed nitric oxide to nitrogen dioxide conversion lies in a sequence of reactions between free radicals that have been generated and other reactive molecules such as carbon monoxide, hydrocarbons, and aldehydes present in the polluted atmosphere.

A rational sequence of reactions that converts NO to NO_2 involves carbon monoxide. A reaction chain involving the hydroxyl radical and carbon monoxide can in principle drive nitric oxide to nitrogen dioxide in the atmosphere:

$$HO + CO \rightarrow H + CO_2$$

$$H + O_2 + M \rightarrow HO_2 + M$$

$$HO_2 + NO \rightarrow NO_2 + HO$$

Several molecules other than CO present in the polluted atmosphere, such as aldehydes and hydrocarbons, may participate in the sequence of reactions reforming hydroperoxyl radicals from hydroxyl radicals. The reaction path involving formaldehyde is, for example,

$$HCHO + h\nu \rightarrow \begin{cases} H + HCO \\ H_2 + CO \end{cases}$$

$$HCHO + OH \rightarrow HCO + H_2O$$

$$H + O_2 + M \rightarrow HO_2 + M$$

$$HCO + O_2 \rightarrow HO_2 + CO$$

Table II. Principal Inorganic Atmospheric Reactions Involving Oxides of Nitrogen

Reaction	Rate constant @25°C ppm-min units	Reference
1. $NO_2 + h\nu \rightarrow NO + O(^3P)$	variable[a]	1,2
2. $O(^3P) + O_2 + M \rightarrow O_3 + M$	2.0×10^{-5} [b]	3
3. $O_3 + NO \rightarrow NO_2 + O_2$	25.2	3
4. $NO_2 + O(^3P) \rightarrow NO + O_2$	1.34×10^{4}	3
5. $NO_2 + O(^3P) \rightarrow NO_3$	3.4×10^{3} [c]	3
6. $NO + O(^3P) \rightarrow NO_2$	3.6×10^{3} [c]	3
7. $NO_2 + O_3 \rightarrow NO_3 + O_2$	5×10^{-2}	3
8. $NO_3 + NO \rightarrow 2NO_2$	1.3×10^{4}	3
9. $NO_3 + NO_2 \rightarrow N_2O_5$	5.6×10^{3} [c]	3
10. $N_2O_5 \rightarrow NO_2 + NO_3$	22	3
11. $N_2O_5 + H_2O \rightarrow 2HONO_2$	5×10^{-6}	3
12. $NO + NO_2 + H_2O \rightarrow 2HONO$	2.2×10^{-9} [b]	4
13. $HONO + HONO \rightarrow NO + NO_2 + H_2O$	1.4×10^{-3}	4
14. $O_3 + h\nu \rightarrow O_2 + O(^1D)$	variable[a]	3
15. $O_3 + h\nu \rightarrow O_2 + O(^3P)$	variable[a]	3
16. $O(^1D) + M \rightarrow O(^3P) + M$	8.5×10^{4}	3
17. $O(^1D) + H_2O \rightarrow 2OH$	5.1×10^{5}	3
18. $HO_2 + NO_2 \rightarrow HONO + O_2$	< 1.2	5
19. $HO_2 + NO_2 \rightarrow HO_2NO_2$	1.2×10^{3}	5
20. $HO_2NO_2 \rightarrow HO_2 + NO_2$	5.1	6
21. $HO_2 + NO \rightarrow NO_2 + OH$	1.2×10^{4}	7
22. $OH + NO \rightarrow HONO$	1.6×10^{4} [c]	3
23. $OH + NO_2 \rightarrow HONO_2$	1.6×10^{4} [c]	8
24. $HONO + h\nu \rightarrow OH + NO$	variable[a]	9
25. $CO + OH \rightarrow CO_2 + H$	4.4×10^{2}	10-12

[a] The photolysis rate constants can be calculated by (13)

$$k_j = \int_0^\infty \sigma_j(\lambda)\phi_j(\lambda)I(\lambda)\, d\lambda$$

where

Footnotes continued on bottom of next page.

Hydrocarbon oxidation plays a central role in the chemistry of the lower atmosphere. Whereas primary reaction rate constants for olefin, alkane, and aromatic reactions with OH, O_3, and O are, by and large, well-established, certain reaction mechanisms are still uncertain, most notably for olefin-O_3, aromatic-OH, and aromatic-O_3 reactions.

Extensive investigations of the liquid-phase reaction of ozone with olefins have identified many of the reaction intermediates and have established the Criegee zwitterion mechanism as a major reaction pathway. Until recently, the Criegee mechanism has also been widely assumed to apply to the gas-phase ozone-olefin reaction. Although several measurements of gas-phase products are consistent with the Criegee mechanism (14), the mechanism fails to explain the formation of free radical intermediates in low pressure (∿ 2 torr) ozone-olefin reactions and of unusual ozonolysis products both at low and high total pressures (15). Recently, evidence on the nature of certain excited intermediates in gas-phase ozone-olefin reactions has appeared (16). Concurrently, on the basis of thermochemical-kinetic calculations, O'Neal and Blumstein (17) proposed alternatives to the gas-phase Criegee mechanism, involving internal hydrogen abstractions of the initial molozonide in addition to its decomposition to the Criegee fragments. Their mechanism rationalizes most of the unusual products observed in previous studies. The extent to which a certain ozone-olefin reaction will proceed by the Criegee or O'Neal-Blumstein mechanism is still uncertain.

The mechanism of photooxidation of aromatic species in the atmosphere is perhaps the area of greatest uncertainty in atmospheric hydrocarbon chemistry. The principal reaction of aromatics is with the hydroxyl radical. Absolute rate constants have recently been determined at room temperature for the reaction of OH radicals with benzene and toluene (18,19) and with a series of aromatic hydrocarbons (19). Recently, absolute rate constants for the reaction of OH radicals with a series of aromatic hydrocarbons have been determined over the temperature range 296-473K (20). For aromatic-OH reactions, the initial step can be either abstraction or addition to the aromatic ring. For toluene, for example,

Footnotes from previous page

$\sigma_j(\lambda)$ = absorption cross section of species j

$\phi_j(\lambda)$ = quantum yield of the photolysis of species j

$I(\lambda)$ = actinic irradiance of the light.

[b]Units of rate constant are $ppm^{-2} min^{-1}$.

[c]Pseudo second-order rate constant for 1 atm. of air.

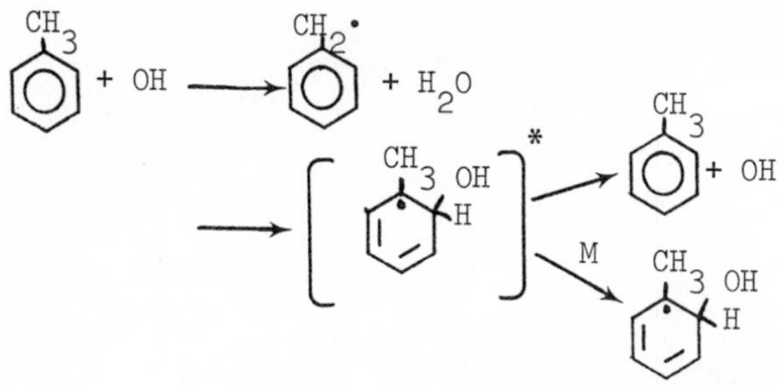

Abstraction occurs mainly from the substituent R-groups rather from the ring. Addition is shown for the ortho position, although addition may occur at any of the carbon atoms of the aromatic ring. The energy-rich OH-adduct may decompose or be stabilized as shown above. The amount of reaction at room temperature proceeding by abstraction is of the order of 2-20% depending on the individual hydrocarbon (20). The OH-aromatic adduct presumably reacts with other atmospheric species such as O_2, NO, or NO_2. A possible mechanism for the O_2 reaction of the toluene adduct, for example, leads to formation of a cresol,

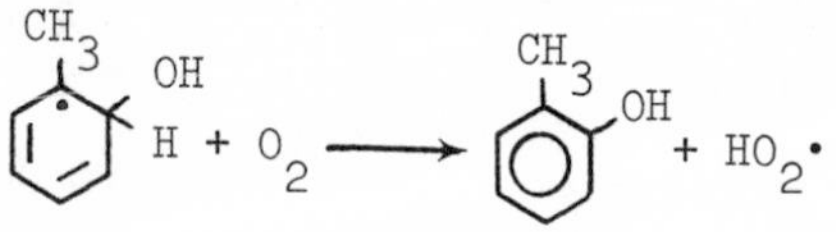

The elucidation of aromatic-OH reaction mechanisms is a key problem in atmospheric chemistry.

Free radical chemistry forms the basis for the conversion of NO to NO_2 and the formation of ozone and organic products. The classes of free radicals important in atmospheric chemistry are, aside from OH and HO_2, alkoxyl radicals (RO), peroxyalkyl radicals (RO_2), and peroxyacyl radicals ($RC(O)O_2$)*. Table III summarizes the reactions of these three radical classes, including OH and HO_2, with NO and NO_2. The set of reactions in Table III are instrumental in determining the rate of conversion of NO and NO_2 and the formation of organic nitrites and nitrates. There exists considerable uncertainty in the rate constants for reactions of RO and RO_2 with NO and NO_2.

---

* Addition to O_2 can be considered as the sole fate of alkyl (R) and acyl (RCO) radicals, leading to peroxyalkyl and peroxyacyl radicals, respectively. The acylate radical (RC(O)O) rapidly dissociates yielding an alkyl radical and CO_2. Hydroxy-peroxyalkyl radicals are formed in olefin-OH reactions. We do not discuss the reactions of hydroxy-peroxyalkyl radicals here.

Table III. Reactions of Alkoxyl, Alkylperoxyl and Acylperoxyl Radicals with NO and NO_2

	NO		NO_2	
Free Radical	Reaction	Rate constant @25°C ppm-min units	Reaction	Rate constant @25°C ppm-min units
OH	$OH + NO \rightarrow HONO$	1.6×10^4	$OH + NO_2 \rightarrow HONO_2$	1.6×10^4
HO_2	$HO_2 + NO \rightarrow NO_2 + OH$	1.2×10^4	$HO_2 + NO_2 \rightarrow HONO + O_2$	<1.2
			$\rightarrow HO_2NO_2$	1.2×10^3
			$(HO_2NO_2 \rightarrow HO_2 + NO_2)$	5.1
RO	$RO + NO \rightarrow RONO$	4.9×10^4 [a]	$RO + NO_2 \rightarrow RONO_2$	1.55×10^4 [b]
	$(RONO + h\nu \rightarrow RO + NO)$		$\rightarrow RCHO + HONO$	1.55×10^3 [b]
RO_2	$RO_2 + NO \rightarrow NO_2 + RO$	c	$RO_2 + NO_2 \rightarrow RO_2NO_2$	5.5×10^3 [d]
	$\rightarrow RONO_2$		$(RO_2NO_2 \rightarrow RO_2 + NO_2)$	
RCO_3	$RCO_3 + NO \rightarrow NO_2 + RCO_2$	4×10^3 [e]	$RCO_3 + NO_2 \rightarrow RCO_3NO_2$	2.07×10^3 [e]
			$(RCO_3NO_2 \rightarrow RCO_3 + NO_2)$	0.0372 [e]

Footnotes for Table III

[a]Rate constants for RO-NO reactions have not been measured directly but have been calculated from measured rates of the reverse reaction and thermodynamic estimates. Batt et al. (21) obtained rate constants for several RO-NO reactions that fall in the range 2.9-5.8x10^4 ppm^{-1} min^{-1}. Mendenhall et al. (22) and Batt et al. (21) determined the rate constant for t-butoxy+NO, obtaining 1.45x10^4 and 6.80x10^4 ppm^{-1} min^{-1}, respectively. We estimate the uncertainty in a given RO-NO rate constant to be a factor of 2-4.

[b]Two reaction paths for RO-NO_2 reactions exist. For methoxy+NO_2, the fraction of reactions proceeding by abstraction have been estimated from 0.08 to 0.23 (23,24). Rate constants for RO-NO_2 have been inferred from measured values of the ratio of the rate constants of RO-NO to RO-NO_2 reactions. For methoxy radicals, this ratio has been estimated from 1.2 to 2.7. (23,25) Thus, the uncertainty in the RO-NO rate constant is compounded by the uncertainty in the RO-NO to RO-NO_2 ratio. Uncertainties in RO-NO_2 rate constants are probably at least a factor of four.

[c]Conversion of NO to NO_2 in the urban atmosphere occurs primarily by reactions of the form RO_2+NO → NO_2+RO. Aside from the HO_2-NO reaction, rate constants have not been measured for RO_2-NO reactions. It has been postulated that longer chain RO_2 radicals ($n \geq 4$) derived from alkanes undergo addition to NO. (26)

[d]Peroxynitrates are formed in the RO_2-NO_2 reaction. By analogy to the HO_2-NO_2 reaction, a small fraction of these reactions may proceed by abstraction. The peroxynitrate may thermally decompose. Rate constants for RO_2-NO_2 reactions and RO_2NO_2 decomposition are not available.

[e]The values shown are from reference 27.

Chemistry of Sulfur Dioxide. Sulfur oxides in the atmosphere can most conveniently be considered as occurring in three forms: sulfur dioxide (SO_2), sulfuric acid (H_2SO_4), and inorganic sulfates. Sulfur dioxide is the anhydrous form of the weak acid, sulfurous acid (H_2SO_3). The salts of this acid are sulfites and bisulfites. Sulfuric acid is the hydrated form of sulfur trioxide (SO_3), which is derived from the oxidation of sulfur dioxide. Sulfur trioxide is intensely hygroscopic, and is immediately converted into sulfuric acid in the atmosphere. Inorganic sulfates are presumably derived from either the reaction of sulfuric acid with cations or the oxidation of sulfites. There is little information available concerning the formation and occurrence of organic sulfates in the atmosphere.

The oxidation of SO_2 to sulfate is an important atmospheric phenomenon. It is now recognized that both homogeneous (gas-phase) and heterogeneous (particulate-phase) processes contribute to SO_2 oxidation in the atmosphere. Possible routes that have been identified are:

1. Homogeneous
   Oxidation of SO_2 to H_2SO_4 by free radicals present in the polluted urban atmosphere (particularly photochemical)
2. Heterogeneous
   a. Liquid-phase oxidation of SO_2 by O_2
   b. Liquid-phase oxidation of SO_2 by O_3
   c. Metal-ion catalyzed, liquid-phase oxidation of SO_2
   d. Catalytic oxidation of SO_2 on particle surfaces.

Sulfur dioxide oxidation rates measured in the laboratory or inferred from atmospheric data display a remarkable variability. The characteristic times of SO_2 oxidation vary from a few minutes to several days. Sulfur dioxide in pure air is very slowly oxidized in the presence of sunlight to sulfuric acid at a rate of about 0.1%/hr. (28). Whereas there is presently inadequate information to characterize fully the chemical processes by which SO_2 is oxidized in polluted urban air, the conversion is much more rapid than in pure air. This accelerated conversion is due to the presence of other air contaminants that generally facilitate the oxidation of SO_2. As noted above, two processes appear to be involved: homogeneous oxidation by components (e.g. free radicals) present in photochemical smog and heterogeneous oxidation predominantly by certain types of aerosols.

Homogeneous (photochemical) oxidation of SO_2 is felt to result from reaction of SO_2 with a variety of free radicals present in photochemical air pollution. Rates of oxidation of SO_2 in Los Angeles have been estimated to range as high as 13%/hr., although these rates cannot necessarily be attributed exclusively to photochemical oxidation.

Heterogeneous oxidation of SO_2 occurs in aerosols in which SO_2 has been absorbed. The oxidation may occur through the action of dissolved oxygen or ozone or may take place catalytically in the presence of metallic compounds, such as manganese, iron, vanadium, aluminum, lead and copper. Prediction of the rate of SO_2 oxidation in a particle has proved to be quite difficult, as one must account for diffusion of gaseous SO_2 to the particle, transfer of SO_2 across the gas-particle interface, and diffusion and reaction of SO_2 within the particle. Relative humidity is a significant factor in the heterogeneous SO_2 oxidation process since the process takes place, in general, in water droplets. In addition, since an acidic pH generally decreases the rate of SO_2 oxidation, the formation of sulfuric acid in an aerosol would tend to be self-limiting unless the acidity is diluted by additional water vapor. In this respect, alkaline metal compounds, such as iron oxide, and ammonia also enhance the oxidation rate by decreasing droplet acidity through their buffering capacity. Extrapolated rates of oxidation by heterogeneous processes in urban air range upwards of 20%/hr.

Meteorology has a substantial effect on the atmospheric oxidation of SO_2. Increased humidity accelerates the heterogeneous oxidation of SO_2, whereas cloud cover might be expected to lower the rate of photochemical processes, and rain will wash out sulfur oxides from the atmosphere. Temperature affects reaction rates and the solubility of gases.

Table IV summarizes a number of SO_2 oxidation rates measured in the laboratory and the atmosphere. The rates vary from a low of 0.1%/hr for photooxidation of SO_2 in clean air to over 2%/min measured in water droplets. Studies reflecting both homogeneous and heterogeneous processes are presented in Table IV. In the next subsections we consider the elements of both homogeneous and heterogeneous processes in an attempt to estimate the contribution of each to the atmospheric oxidation of SO_2.

1. Homogeneous Oxidation of SO_2.

There are a number of homogeneous (gas-phase) reactions for the atmospheric oxidation of SO_2. A thorough review of these reactions has been carried out by Calvert et al. (46). Table V summarizes the rate constant values for the most important of these reactions.

Sulfur dioxide is converted to SO_3 by the reaction (reaction numbering is separate from that in Table I)

$$SO_2 + O(^3P)(+M) \xrightarrow{1} SO_3\ (+M)$$

Calvert et al. (46) recommended the apparent second-order rate constant at 1 atm. in air at 25°C as $k_1 = (5.7 \pm 0.5) \times 10^{-14}\ cm^3\ molec^{-1}\ sec^{-1}$ ($8.3 \times 10^1\ ppm^{-1}\ min^{-1}$). The source of oxygen atoms for reaction 1 is largely from the photolysis of NO_2,

$$NO_2 + h\nu \xrightarrow{2} NO + O(^3P)$$

The primary competition for the oxygen atoms is the reaction

$$O(^3P) + O_2 + M \xrightarrow{3} O_3 + M$$

Table IV. Observed Sulfur Dioxide Oxidation Rates

Experimental conditions	Presumed atmospheric conditions	SO_2 oxidation rate	Reference
Atmospheric study of Canadian smelting area.	150-4200 μg/m^3 SO_2	0.035%/min	29
Sunlamp in smog chamber; high SO_2 concentrations in pure air.	SO_2; sunlight; clear air (reaction unaffected by humidity).	0.1-0.2%/hr	28
Catalyst droplet exposed to high concentrations of SO_2 in humid air.	Natural fog containing 1 μm crystals of $MnSO_4$ in droplets; 2600 μg/m^3 SO_2.	1%/min	30
Artificial fog in smog chamber; very high levels; SO_2 and metal sulfates.	(Levels in smog chamber) 0.6 mg/m^3 SO_2; 2 mg/m^3$MnSO_4$	0.01%/min at 77% RH 2.1%/min at 95% RH	31
Plume of coal-burning power plant.	Found moisture level in plume important.	0.1%/min at 70% RH 0.5%/min at 100% RH	32
$(NH_4)_aSO_4$ formation in water droplets exposed to NH_3 and SO_2.	100 μg/m^3 SO_2; 10 μg/m^3 NH_3; cloud droplet radius of 10 μm	2.5%/min in droplets	33
UV-irradiated gas mixtures; NO_x, hydrocarbons, SO_2; high levels.	Noon sun.	1-3%/hr.	34
Sunlight; 200-2000 μg/m^3 SO_2; trace impurities.	Assuming 300 μg/m^3 SO_2; bright sunlight for 10 hr. would produce 30 μg/m^3 of sulfate.	0.65%/hr. (high rate may be due to trace impurities).	35

Table IV. Observed Sulfur Dioxide Oxidation Rates (Continued)

Experimental conditions	Presumed atmospheric conditions	SO_2 Oxidation rate	Reference
Smog chamber; light; SO_2, NO_x, olefins	SO_2, 260 $\mu g/m^3$; ozone, 100$\mu g/m^3$ olefin, 33 $\mu g/m^3$, bright sunlight.	3%/hr for pentene; 0.4%/hr for propene	36,37
Metallic aerosol particles on Teflon beads in flow reactor; SO_2; water vapor	Natural fog (0.2 g H_2O/m^3) in industrial area; SO_2, 260 $\mu g/m^3$; $MnSO_4$, 50 $\mu g/m^3$	2%/hr.	38
Photochemical reactants; SO_2 in ppm concentrations.	Sunlight; SO_2, 260 $\mu g/m^3$; ozone, 200 $\mu g/m^3$; olefin, 33 $\mu g/m^3$; 40% RH	3%/hr.	39
Atmospheric study of Rouen (industrial city) in winter.	68-242 $\mu g/m^3$ SO_2.	6-25%/hr.	40
Los Angeles air trajectories.	-	1.2-13%/hr.	41
Plume of an oil-fired power plant; airborne sampling.	Catalytic oxidation by vanadium particles. Distance $\leq$25 km	pseudo-second order mechanism; rate constant = 1 $ppm^{-1} hr^{-1}$	42
St. Louis urban plume; airborne sampling.	900-1200 m. altitude 20-25°C; RH=40-60%	10-14%/hr.	43
Plumes of four coal-fired power plants; airborne sampling	RH=32-85%; 10-25°C; Distance $\leq$70 km	∿1%/hr.	44
Smelter plume; airborne sampling.	Catalytic oxidation	pseudo-second order mechanism; rate constant=0.2 $ppm^{-1} hr^{-1}$	45

Table V. Rate Constants and Estimated Contributions to Atmospheric SO_2 Oxidation for Homogeneous Chemical Reactions

Reaction	Rate constant @25°C cm^3 $molec^{-1}$ sec^{-1}	Estimated contribution to SO_2 oxidation rate, % hr^{-1}
$SO_2+O(^3P)(+M) \rightarrow SO_3(+M)$	$(5.7\pm0.5)x10^{-14}$	$4.6x10^{-3}$
$SO_2+HO_2 \rightarrow OH+SO_3$	$(8.7\pm1.3)x10^{-16}$ [a]	0.75
$SO_2+CH_3O_2 \rightarrow CH_3O+SO_3$ $\rightarrow CH_3O_2SO_2$	$(5.3\pm2.5)x10^{-15}$ [b]	0.46
$SO_2+OH(+M) \rightarrow HOSO_2(+M)$	$(1.1\pm0.3)x10^{-2}$ [c]	1.0

[a] The study of Payne et al. (47) provides the only experimental estimate of this reaction. Measurements of the rate of the SO_2-HO_2 reaction were made relative to those of $2HO_2 \rightarrow H_2O_2+O_2$. They derived the estimate $k_{SO_2-HO_2}/k^{1/2}_{HO_2-HO_2} = (4.8\pm0.7x10^{-16}(cm^3\ molec^{-1}\ sec^{-1})^{1/2}$. Calvert et al. (46) discuss the available values for $k_{HO_2-HO_2}$, and based on this discussion recommended the lower limit for $k_{SO_2-HO_2}$ given in the table.

[b] It is not possible to determine the extent to which each of the reactions occur from the existing data. Thermodynamic arguments favor the formation of CH_3O and SO_3 rather than $CH_3O_2SO_2$.

[c] The value shown was recommended by Calvert et al. (46) based on an average of the results of the most extensive studies at high pressure: Atkinson et al. (48), Cox (49), and Castleman and Tang (50).

Oxygen atoms can be considered to be in a steady state as a result of reactions 2 and 3 (reaction 1 has a negligible effect on the concentration of oxygen atoms), $[O]_{ss} = k_2[NO_2]/k_3[O_2][M]$. The rate of reaction 1 is estimated from $k_1[O]^2_{ss}[SO_2]$, and thus the characteristic time for SO_2 oxidation by reaction 1 is $\tau_1 = k_3[O_2][M]/k_1k_2[NO_2]$. Assuming $[NO_2] = 0.1$ ppm, $[O_2] = 2.1x10^5$ ppm, $[M] = 10^6$ ppm, and $k_2 = 0.4\ min^{-1}$, a value typical of Los Angeles noonday intensities, we obtain $\tau_1 \cong 1.3x10^6$ min. The corresponding oxidation rate in % hr^{-1} is given in Table V.

The characteristic time for the reaction

$$SO_2 + HO_2 \xrightarrow{4} OH + SO_3$$

is given by $\tau_4 = \{k_4[HO_2]\}^{-1}$. Hydroperoxyl radical concentrations in ambient air have not been measured. Simulations of smog photochemistry (51) yield approximate HO_2 concentrations of 10^{-4} ppm. On this basis, using the value of k_4 given in Table V, we estimate $\tau_4 \cong 0.8x10^4$ min.

The characteristic time for the reaction

$$SO_2 + CH_3O_2 \xrightarrow{5a} CH_3O + SO_3$$
$$\xrightarrow{5b} CH_3O_2SO_2$$

is given by $\tau_5 = \{k_5[CH_3O_2]\}^{-1}$. The methylperoxyl radical is one of the most abundant of the organic free radicals in the polluted atmosphere. Simulations of smog photochemistry (51) yield approximate CH_3O_2 concentrations of 10^{-5} ppm. Thus, we estimate $\tau_6 \cong 1.3x10^4$ min.

The characteristic time for the reaction

$$SO_2 + OH\ (+M) \xrightarrow{6} HOSO_2\ (+M)$$

is given by $\tau_6 = \{k_6[OH]\}^{-1}$. Hydroxyl radical concentration measurements in ambient air were reported by Wang eg al. (52). Peak OH concentrations in urban air were found to exceed 10^7 molecules/cm^3 ($\sim 10^{-7}$ ppm). Based on this value of [OH], we obtain $\tau_d \cong 0.6x10^4$ min. The fate of the $HOSO_2$ product has not been established with certainty; it is usually assumed that is hydrates in some manner to form sulfuric acid.

Table V summarizes the estimated contributions of the homogeneous reactions discussed in this section to the overall rate of SO_2 oxidation in the atmosphere. The total estimated SO_2 oxidation rate from these processes in a smoggy atmosphere is 2.2%/hr., a value comparable to those inferred from ambient measurements of SO_2 to sulfate conversion rates.

2. Heterogeneous Oxidation of SO_2

As noted above, the heterogeneous oxidation of SO_2 may take place by the following mechanisms:

a. Liquid-phase oxidation of SO_2 by O_2
b. Liquid-phase oxidation of SO_2 by O_3
c. Metal-ion catalyzed, liquid-phase oxidation of SO_2
d. Catalytic oxidation of SO_2 on particle surfaces.

In spite of the fact that the liquid-phase (uncatalyzed) oxidation of SO_2 by O_2 has been studied for many years, there does not exist a clear understanding of the primary reaction mechanism. The rate of sulfate formation is usually expressed as first-order in the concentration of sulfite ion,

$$\frac{d[SO_4^{=}]}{dt} = k_s[SO_3^{=}]$$

A recent value of k_s determined from an extensive set of experiments is (53)

$$k_s = k_1 + k_2[H^+]^{1/2} + k_3 p_{O_2}[H^+]^{-1}$$

where, at 298°K, $k_1 = 4.8 \times 10^{-3}$ sec^{-1}, $k_2 = 4.9$ sec^{-1} $(mole/\ell)^{-1/2}$, $k_3 = 3.9 \times 10^{-12}$ sec^{-1} $(mole/\ell)$ atm^{-1}, and p_{O_2} is the partial pressure of O_2 in the gas phase.

Sulfur dioxide is oxidized in aqueous solution by ozone. A recently determined rate expression for sulfate formation in aqueous solution by this route is (53)

$$\frac{d[SO_4^{=}]}{dt} = k_4 K_{HO_3} p_{O_3}[HSO_3^-][H^+]^{-0.1}$$

where, at 298°K $k_4 = 4.4 \times 10^4$ $(mole/\ell)^{-0.9}$ sec^{-1}, $K_{HO_3} = 0.0123$ atm, and p_{O_3} is the partial pressure of ozone in the gas phase.

Larson et al. (53) conclude that, owing to the relatively small amount of liquid water involved, neither the O_2 nor the O_3 oxidation is fast enough to produce significant quantities of sulfate in the liquid phase at humidities less than saturation. These reactions could only occur at a significant rate under saturated conditions, i.e. in fogs or clouds where the liquid water content may exceed 0.1 g/m^3. For cloud conditions of $[O_3]$ = 0.05 ppm, $[H_2O]$ = 0.6 g/m^3, $[SO_2]$ = $[NH_3]$ = 0.01 ppm, the rate for SO_2 oxidation by O_3 is in the range of 1-4%/hr.

The metal-ion catalyzed, liquid-phase oxidation of SO_2 has received considerable attention as a mechanism for SO_2 conversion in plumes and contaminated droplets. In general, the mechanisms proposed are lengthy, and the derived rate expressions are largely empirical. Table VI summarizes a variety of studies on this process. Observed rates vary substantially depending on the particular catalyst, relative humidity, and other conditions.

Novakov et al. (54) have suggested that the surface of soot particles serves as a catalyst for the oxidation of SO_2. Such a process might be of importance in a plume containing significant quantities of carbonaceous particles or in an atmosphere where motor vehicle soot aerosol is present. Very little is presently known about the rates or mechanisms of this process.

Aerosol Chemistry. The general relationship between gaseous and particulate pollutants in the urban atmosphere is depicted in Figure 1. Aerosols may be emitted directly from sources or be formed in the atmosphere as the result of condensation of secondary vapors formed in gas-phase reactions.

The essential elements of urban aerosol chemistry are shown in Figure 2, in which we have represented the chemistry in terms of the conversion of SO_2, NO_x and hydrocarbons to particulate sulfate, nitrate, and organics, respectively. Table VII summarizes the key unknown aspects of the processes depicted in Figure 2. There are many features of atmospheric aerosol chemistry that must be elucidated before we understand fully the formation and growth of atmospheric particles.

Dynamics of Urban Aerosols

Table VIII summarizes the physical processes that affect the evolution of aerosol in a unit volume of atmosphere. To develop the general dynamic equation governing aerosol behavior let us assume that the aerosol is composed of liquid droplets of M chemical species. We let c_i denote the concentration of species i in a droplet, i = 1,2,..., M, and D_p denote the diameter of the particle. We then define $n(D_p, c_1, \ldots, c_M, \tilde{r}, t)$ as the size-composition distribution function, such that $n\, dD_p\, dc_1 \ldots dc_M$ is the number of particles per unit volume of atmosphere at location $\tilde{r}$ at time t of diameter D_p to $D_p + dD_p$ and of composition c_i to c_i+dc_i (moles l^{-1}) of species i, i = 1,2,..., M.* The total particle number density (cm^{-3}) at location $\tilde{r}$ at time t is

$$N(\tilde{r},t) = \int_0^\infty \cdots \int_0^\infty n(D_p, c_1, \ldots, c_M, \tilde{r}, t)\, dD_p\, dc_1 \ldots dc_M. \quad (1)$$

The distribution of particles by particle diameter ($\mu m^{-1}\ cm^{-3}$) is defined by $n_0(D_p,\tilde{r},t)$ and given by

$$n_0(D_p,\tilde{r},t) = \int_0^\infty \cdots \int_0^\infty n(D_p, c_1, \ldots, c_M, \tilde{r}, t)\, dc_1 \ldots dc_M. \quad (2)$$

The general equation governing the size composition distribution function was derived by Chu and Seinfeld (67). The equation

*Throughout we give representative units for various quantities. Numerical conversion factors associated with these units are not explicitly indicated in the equations.

Table VI. Metal-Ion Catalyzed, Liquid-Phase Oxidation of SO_2

Author	Type of mechanism	Rate coefficient and/ or expression	Comments
Fuller and Crist (55)	Cu^{2+} catalyst; mannitol inhibitor	$k_s=0.013+2.5[Cu^{2+}]$	$25^{o}C$
Basset and Parker (56)	Metal salts		Formation of complexes such as $[O_2\ Mn(SO_3)_2]^{2-}$ and rapid oxidation.
Junge and Ryan (57)	Fe^{2+} catalyst with and without NH_3		Conversion rate = $1.8x10^{-4}$ % min^{-1}
Foster (58)	Metal salts $2SO_2+2H_2O+O_2 \rightarrow 2H_2SO_4$	SO_2 conversion rate = 0.09% min^{-1} for Mn, 0.15-1.5% min^{-1} for Fe.	Theoretical study; rates for Mn and Fe depend on many factors; rate for Fe catalyzed oxidation is pH dependent.
Matteson, Stober, and Luther (59)	SO_2 oxidation catalyzed by metal salts; $Mn^{2+}+SO_2 \rightleftarrows Mn \cdot SO_2^{2+}$ $2Mn \cdot SO_2^{2+}+O_2$ $\rightleftarrows [(Mn \cdot SO_2^{2+})_2 \cdot O_2]$ $\rightleftarrows 2Mn \cdot SO_3^{2+}$ $Mn \cdot SO_3^{2+}+H_2O \rightleftarrows Mn^{2+}+HSO_4^-+H^+$ $HSO_4^-+H^+ \rightleftarrows H_2SO_4$	$-\frac{d[SO_2]_g}{dt} = k_1[Mn^{2+}]_0^2$ $k_1=2.4x10^5\ M^{-1}s^{-1}$	Negligible $SO_4^{=}$ formation for RH<95%; similar mechanism may be responsible for catalysis by other metal salts.

Table VI. Metal-Ion Catalyzed, Liquid-Phase Oxidation of SO_2 (Continued)

Author	Type of mechanism	Rate coefficient and/ or expression	Comments
Cheng, Corn, and Frohliger (38)	SO_2 oxidation catalyzed by NH_3; $2SO_2 + 2H_2O + O_2 \xrightarrow{catalyst} 2H_2SO_4$	SO_2 conversion rate ∿.03%/min with Mn^{2+} levels typical of urban industrial atmosphere; ∿.33%/min with levels typical of plume from coal powered plant.	Oxidation rate estimated by extrapolation to atmospheric conditions.
Chen and Barron (60)	Sulfite oxidation catalyzed by cobalt ions; free radical mechanism; CO(III) reduced.	$-\frac{d[SO_2]_g}{dt} = k[Co(H_2O)_6^{3+}]^{1/2} x [SO_{3=}]^{3/2}$	Could not determine specific value for k.
Brimblecombe and Spedding (61)	SO_2 oxidation by O_2 with trace Fe catalyst	$-\frac{d[S(IV)]}{dt} = k[Fe(III)][S(IV)]$ $k = 100\ M^{-1}\ s^{-1}$; SO_2 conversion rate ∿3.2%/day in fog assuming 28 $\mu g/m^3$ SO_2 and 10^{-6}M Fe(III)	Possibility of Fe(III) contamination discussed.
Freiberg (62)	SO_2 oxidation catalyzed by Fe^2	$\frac{d[SO_4^=]}{dt} = K_0 K_s^2 [H_2SO_3]^2 x [Fe^{3+}]/[H^+]^3$ K_s = 1st dissociation constant of H_2SO_3	Rate increases rapidly with RH and decreases by about one order of magnitude with 5^oC increase in temperature.

Table VI. Metal-Ion Catalyzed, Liquid-Phase Oxidation of SO_2 (Continued)

Author	Type of mechanism	Rate coefficient and/ or expression	Comments
Freiberg (63)	SO_2 oxidation catalyzed by Fe	Same as above, except K_0 a complex function of $[Fe^{3+}]$	Rate dependence changes from $[SO_2]^2/[H^+]^3$ to $[SO_2]/[H^+]$ as pH or $[SO_2]$ increases.
Barrie and Georgii (64)	Mn and Fe catalysts.	$\frac{d[SO_4^{=}]}{dt} = k[SO_2]_g$	8^oC and 25^oC; 2.1 mm diam. droplets; 10^{-6} to 10^{-4} M for Mn and Fe; SO_2 concentrations 0.01-1.0 ppm. In pH range 2-4.5 the catalytic effectiveness was $Mn^{++} > Fe^{++} > Fe^{3+}$. Increase in T from 8° to 25^oC caused an increase in Mn^{++} catalyzed oxidation rate of 5-10 in pH range 2-4.5.

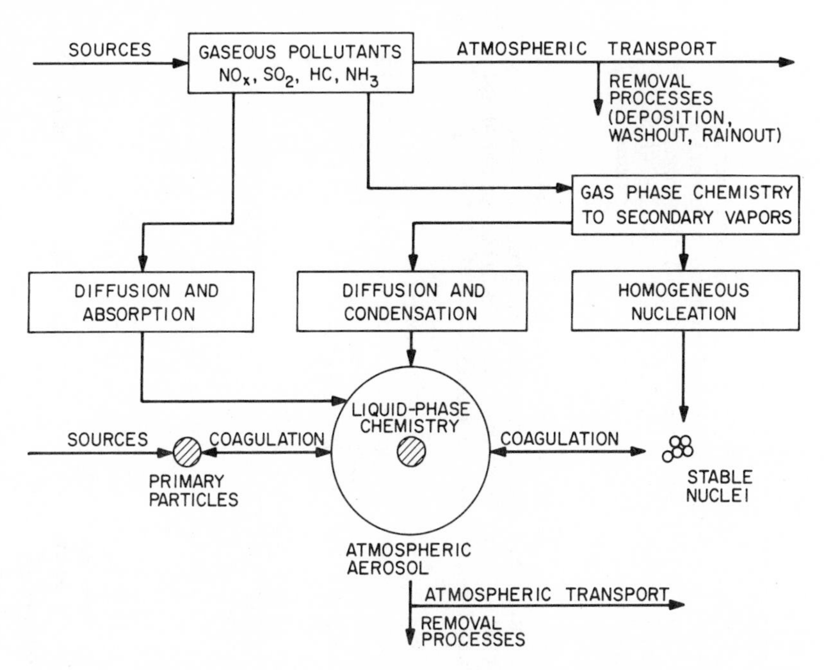

Figure 1. Relationships among primary and secondary gaseous and particulate air pollutants

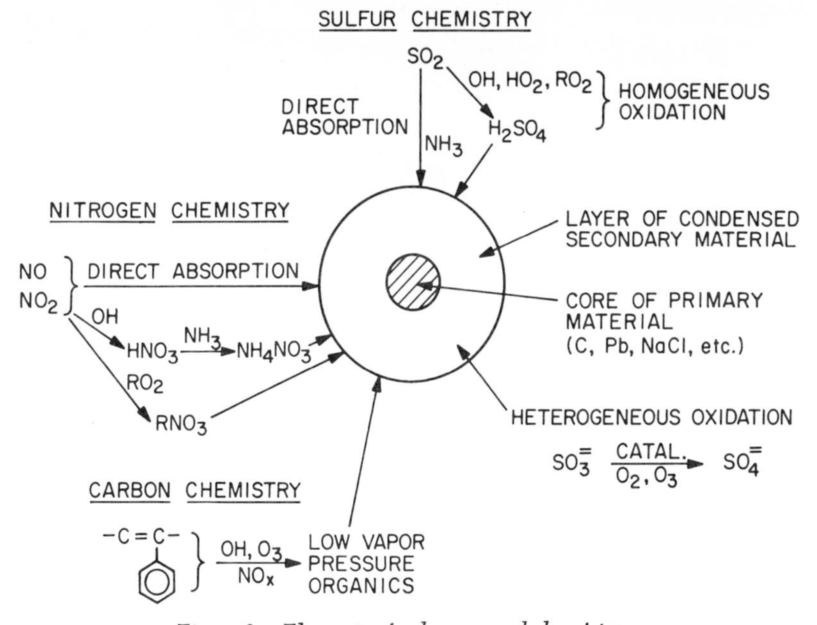

Figure 2. Elements of urban aerosol chemistry

Table VII. Problems in Atmospheric Aerosol Chemistry

Process	Chemistry	Discussion
Oxidation of SO_2 to sulfate	Homogeneous, gas-phase reactions.	Free radical reactions (see Table V). Rate constants for SO_2-OH and SO_2-HO_2 measured but still somewhat uncertain. Rate constants for SO_2-RO and SO_2-RO_2 reactions unknown. Mechanism for the SO_2-OH reaction unknown; product $HOSO_2$ assumed to be hydrated to H_2SO_4.
	Heterogeneous	
	a. SO_2 oxidation in aqueous solution by O_2 and O_3	Mechanisms of O_2 and O_3 oxidation of SO_2 in solution still unknown. Empirical rate equations are available. Both processes are rather slow compared to catalytic processes, but may be important in fogs or clouds.
	b. Metal ion catalyzed SO_2 oxidation in aqueous solution (Fe,Mn).	Mechanisms of oxidation are generally unknown. Although many empirical rate expressions are available (see Table VI) substantial uncertainty exists in prediction of rates. Processes can be important in plumes.
	c. Catalytic oxidation of SO_2 on particle surfaces.	Virtually nothing known about rates or mechanisms. Probably not too important in atmosphere as a whole but possibly substantial in plumes.
Conversion of NO_x to Nitrate (65)	Homogeneous	
	a. $OH+NO_2 \rightarrow HNO_3$ $HNO_3+NH_3 \rightarrow NH_4NO_3$	OH-NO_2 rate constant well established but OH levels are uncertain. HNO_3 vapor pressure too high for direct condensation, but HNO_3 may react with NH_3. Rate constant for HNO_3-NH_3 reaction unknown. This reaction could be a substantial source of NH_4NO_3 in aerosols.

Table VII. Problems in Atmospheric Aerosol Chemistry (Continued)

Process	Chemistry	Discussion
	b. Organic nitrate formation $RO + NO_2 \rightarrow RONO_2$ $RO_2 + NO_2 \rightarrow RO_2NO_2$	Rate constants for $RO-NO_2$ reactions not well established, and those for RO_2-NO_2 reactions are unknown. Subsequent chemistry of $RONO_2$ and RO_2NO_2 uncertain.
	Heterogeneous a. Liquid-phase	Process involves absorption of NO and NO_2 in presence of NH_3. Nitrate concentrations in solution from this process can be substantial. Extent of importance of this process in the atmosphere is unknown, as is the existence of any liquid-phase catalytic mechanisms.
	b. Catalytic oxidation of NO_x on particle surfaces.	Virtually nothing known about rates or mechanisms.
Conversion of Hydrocarbons to Particulate Organics (66)	Homogeneous oxidation of hydrocarbons by OH and O_3 to low vapor pressure organic species.	Rate constants for reaction of hydrocarbons with OH and O_3 are reasonably well known. Mechanisms of formation of the low vapor pressure organic species are still speculative, particularly for aromatics, although hydrocarbons that lead to aerosol precursors have been identified, e.g. cyclic and di-olefins.

Table VIII. Physical Processes Affecting the Evolution of Aercsol Number Density Distributions

Process	Definition	Important Features
Homogeneous nucleation of a gaseous species	Agglomeration of molecules to form a stable particle.	Particles formed in this way are very small. High concentrations and low vapor pressures are required to initiate homogeneous nucleation.
Heterogeneous condensation.	Growth of particles by diffusion of gaseous species to the particle surface followed by absorption or adsorption.	The driving force for growth is the difference between the ambient partial pressure and the vapor pressure just above the surface of the particle. The vapor pressure above the surface of a particle depends on the size of the particle (the Kelvin effect) and the composition of the particle.
Coagulation	The process of collision of two particles to form a single particle.	Since it is a second-order process (i.e., its rate is proportional to the square of the local number density), high concentrations are required for appreciable rates. Because large particles (over 1 μm in diameter) generally do not exist in concentrations high enough to produce appreciable rates, coagulation is most important for small particles. Different mechanisms can bring two particles together; the efficiencies of these mechanisms depend on the sizes of the two particles.

Table VIII. Physical Processes Affecting the Evolution of Aerosol Number Density Distributions (Continued)

Process	Definition	Important Features
Brownian diffusion	Diffusion of particles as a result of collisions with gas molecules.	Brownian diffusivity depends on particle size; it is most important for small particles.
Turbulent diffusion	Motion of particles with the turbulent airflow.	Turbulent diffusion of aerosols can be treated analogously to that of gaseous species.
Gravitational settling	Settling of particles due to gravity.	This is important only for large particles (over 10 μm in diameter); it is the mechanism primarily responsible for the cutoff of the aerosol size spectrum at the upper end of the particle size range.
Deposition	Loss occuring when particles cross streamlines and deposit on surfaces, such as grass.	Deposition is primarily important for particles larger than 1 μm in diameter.
Rainout in clouds	Loss of particles through their role as condensation nuclei in clouds.	
Washout under clouds	Scavenging of particles by falling raindrops.	

includes the effects of coagulation, heteromolecular nucleation, heterogeneous condensation, and aerosol phase chemical reactions, in addition to advection and turbulent diffusion. The full general dynamic equation governing the mean size-composition distribution function $\bar{n}(D_p, c_1, \ldots, c_M, \underset{\sim}{r}, t)$ is

$$\frac{\partial \bar{n}}{\partial t} + \sum_{i=1}^{3} \frac{\partial}{\partial r_i} (\bar{u}_i \bar{n}) + \frac{\partial}{\partial D_p} \left(\bar{F}_{D_p} n \right) + \sum_{i=1}^{M} \frac{\partial}{\partial c_i} (\bar{F}_i \bar{n}) - u_s \frac{\partial \bar{n}}{\partial r_3}$$

$$= \sum_{i=1}^{3} \frac{\partial}{\partial r_i} \left[(K_{ii} + \mathcal{D}) \frac{\partial \bar{n}}{\partial r_i} \right]$$

$$+ \frac{1}{2} \int_0^{D_p/2^{1/3}} \int_0^{c_1} \cdots \int_0^{c_M} \beta \left[\left(D_p^3 - \tilde{D}_p^3 \right)^{1/3}, \tilde{D}_p \right]$$

$$\cdot \bar{n} \left[\left(D_p^3 - \tilde{D}_p^3 \right)^{1/3}, c_1 - \tilde{c}_1, \ldots, c_M - \tilde{c}_M, \underset{\sim}{r}, t) \right]$$

$$\cdot n(\tilde{D}_p, \tilde{c}_1, \ldots, c_M, \underset{\sim}{r}, t) \frac{D_p^2}{(D_p^3 - \tilde{D}_p^3)^{2/3}} d\tilde{D}_p \, d\tilde{c}_1 \ldots d\tilde{c}_M$$

$$- \int_0^{\infty} \cdots \int_0^{\infty} \beta(D_p, \tilde{D}_p) \bar{n}(D_p, c_1, \ldots, c_M, \underset{\sim}{r}, t)$$

$$\cdot \bar{n}(\tilde{D}_p, \tilde{c}_1, \ldots, \tilde{c}_M, \underset{\sim}{r}, t) \, d\tilde{c}_1 \ldots d\tilde{c}_M d\tilde{D}_p$$

$$+ S_0(D_p, c_1, \ldots, c_M, t) + S_1(D_p, c_1, \ldots, c_M, \underset{\sim}{r}, t) \quad , \tag{3}$$

where $\bar{F}_{D_p} = dD_p/dt$, $\bar{F}_i = dc_i/dt$, u_s is the size-dependent sedimentation velocity, S_o is the rate of formation of particles by heteromolecular nucleation, and S_1 is the rate of introduction of particles from sources.

In the last several years a great deal has been learned about air pollution aerosols (68-71). Although our knowledge is still far from complete, there is general agreement on the nature of urban aerosol size distributions and their interpretation (70). It has been established that the principal growth mechanism for urban atmospheric aerosols in the 0.1-1.0 µm diameter size range (the so-called Accumulation Mode) is gas-to-particle conversion (68,71). The major secondary components in atmospheric aerosols have been identified as sulfates, nitrates and particulate organic species (68,71,72,73). The qualitative picture of polluted tropospheric aerosols that has evolved is that primary aerosol provides the surface upon which secondary species condense, so that eventually the volume of secondary aerosol substantially exceeds that of the original primary aerosol.

Assessment of the effect of gaseous and particulate primary emission controls on atmospheric particulate concentrations and properties necessitates the development of a mathematical model capable of relating primary emissions of gaseous and particulate pollutants to the size and chemical composition distribution of atmospheric aerosols.

We have developed the following aspects of a general aerosol model: coagulation and heterogeneous condensation flux expressions, a numerical method for solution of the general dynamic equation, SO_2 homogeneous chemistry, and NO_x homogeneous and heterogeneous chemistry. Also, we have obtained solutions to special cases of equation (3) in order to elucidate the features of size distribution dynamics with simultaneous coagulation and condensation. For the urban-scale aerosol model we developed a linear gas-particle material balance model capable of predicting steady-state levels of gaseous precursors and secondary aerosol constituents (74). We then studied the gas-to-particle conversion process through the dynamics of the size distribution of photochemical aerosols (75).

The area of greatest uncertainty concerns the chemical nature of the aerosol and the influence of homogeneous and heterogeneous chemistry on the size and composition distribution of the aerosol. Thus, it is now necessary to synthesize our knowledge of both the homogeneous and heterogeneous chemistry of SO_2, NO_x, and organic species to describe the size and composition dynamics of the aerosol. The basic mathematical aspects of this task have been completed; the key issue now is the aerosol chemistry. Future work will concern the elucidation and synthesis of current knowledge on aerosol chemistry with the object of developing a mathematical model for the general urban aerosol.

LITERATURE CITED

1. Bass, A. M., Ledford, A. E., Jr., Laufer, A. H., "Extinction Coefficients of NO_2 and N_2O_4," Jour. of Research Natl. Bureau of Standards - A. Physics and Chemistry (1976) 80A,143
2. Hall, T. C., Jr., Blacet, F. E., "Separation of the Absorption of Spectra of NO_2 and N_2O_4 in the Range 2400-5000Å," J. Chem. Phys. (1952) 20, 1745.
3. Hampson, R. F., Jr., Garvin, D., "Chemical Kinetic and Photochemical Data for Modeling Atmospheric Chemistry," NBS Technical Note 866 (1975).
4. Chan, W. H., Nordstrom, R. J., Calvert, J. G., Shaw, J. H., "Kinetic Study of HONO Formation and Decay Reactions in Gaseous Mixtures of HONO, NO, NO_2, H_2O, and N_2," Environ. Sci. Technol. (1976) 10, 674.
5. Howard, C. J., "Kinetics of the Reaction of HO_2 with NO_2," J. Chem. Phys. (1978) in press.
6. Graham, R. A., Winer, A. M., Pitts, J. N., Jr., "Temperature Dependence of the Uni-molecular Decomposition of Pernitric Acid and Its Atmospheric Implications," Chem. Phys. Lett. (1978) in press.
7. Howard, C. J., Evenson, K. M., "Kinetics of the Reaction of HO_2 with NO," Geophys. Res.Lett. (1977) 4, 437.
8. Tsang, W., Garvin, D., Brown, R. L., "NBS Chemical Kinetics Data Survey - The Formation of Nitric Acid from Hydroxyl and Nitrogen Dioxide," Natl. Bureau of Standards (1977).
9. Cox, R. A., Derwent, R. G., "The Ultraviolet Absorption Spectrum of Gaseous Nitrous Acid," J. Photochem. (1977) 6, 23.
10. Sie, B. K. T., Simonaitis, R., Heicklen, J., "The Reaction of OH with CO," Int. J. Chem. Kinetics (1976) 8, 85.
11. Chan, W. H. Uselman, W. N., Calvert, J. G., Shaw, J. H., "The Pressure Dependence of the Rate Constant for the Reaction: $HO + CO \rightarrow H + CO_2$," Chem. Phys. Lett. (1977) 45, 240.
12. Cox, R. A., Derwent, R. G., Holt, P. M., "Relative Rate Constants for the Reactions of OH Radicals with H_2, CH_4, CO, NO, and HONO at Atmospheric Pressure at 296°K," J. Chem. Soc. Faraday Trans. I (1976) 72, 2031.
13. Schere, K. L., Demerjian, K. L., "Calculation of Selected Photolytic Rate Constants Over a Diurnal Range," Environmental Protection Agency Report EPA-600/4-77-015 (1977).
14. Walter, T. A., Bufalini, J. J., Gay, B. W., Jr., "Mechanism for Olefin-Ozone Reactions," Environ. Sci. Technol. (1977) 11, 382.
15. Atkinson, R., Finlayson, B. J., Pitts, J. N., Jr., "Photoionization Mass Spectrometer Studies of Gas Phase Ozone-Olefin Reactions," J. Amer. Chem. Soc. (1973) 95, 7592.
16. Pitts, J. N., Jr., Kummer, W. A., Steer, R. P., Finlayson, B. J., "The Chemiluminescent Reactions of Ozone with Olefins and Organic Sulfides," Adv. Chem. (1972) 113, 246.

17. O'Neal, H. E., Blumstein, C., "A New Mechanism for Gas Phase Ozone-Olefin Reactions," Int. J. Chem. Kinetics (1973) 5, 397.
18. Davis, D. D., Bollinger, W., Fischer, S., "A Kinetics Study of the Reaction of the OH Free Radical with Aromatic Compounds. I. Absolute Rate Constants for Reaction with Benzene and Toluene at 300°K," J. Phys. Chem. (1975) 79, 293.
19. Hansen, D. A., Atkinson, R., Pitts, J. N., Jr., "Rate Constants for the Reaction of OH Radicals with a Series of Aromatic Hydrocarbons," J. Phys. Chem. (1975) 79, 1763.
20. Perry, R. A., Atkinson, R., Pitts, J. N., Jr., "Kinetics and Mechanism of the Gas Phase Reaction of OH Radicals with Aromatic Hydrocarbons Over the Temperature Range 296-473K," J. Phys. Chem. (1977) 81, 296.
21. Batt, L., McCulloch, R. D., Milne, R. T., "Thermochemical and Kinetic Studies of Alkyl Nitrites (RONO) - D(RO-NO), The Reactions between RO and NO, and the Decomposition of RO," Int. J. Chem. Kinetics Symposium (1975) 1, 441.
22. Mendenhall, G., Golden, D. M., Benson, S. W., "The Very-Low-Pressure Pyrolysis (VLPP) of n-Propyl Nitrate, ter-Butyl Nitrite, and Methyl Nitrite, Rate Constants for Some Alkoxy Radical Reactions," Int. J. Chem. Kinetics (1975) 7, 725.
23. Weibe, H. A., Villa, A., Hellman, T. M., Heicklen, J., "Photolysis of Methyl Nitrite in the Presence of Nitric Oxide, Nitrogen Dioxide, and Oxygen," J. Amer. Chem. Soc. (1973) 95, 7.
24. Barker, J. R., Benson, S. W., Golden, D. M., "Re Decomposition of Dimethyl Peroxide and the Rate Constant for CH_3O + $O_2 \rightarrow CH_2O + HO_2$," Int. J. Chem. Kinetics (1977) 9, 31.
25. Baker, G., Shaw, R., "Reactions of Methye, Ethoxye, and t-Butoxye with Nitric Oxide and with Nitrogen Dioxide," J. Chem. Soc., (London) (1965) 6965.
26. Darnall, K. R., Carter, W. P. L., Winer, A. M., Lloyd, A. C., Pitts, J. N., Jr. "Importance of RO_2 + NO in Alkyl Nitrate Formation from C_4-C_6 Alkane Photooxidations under Simulated Atmospheric Conditions," J. Phys. Chem. (1976) 80, 1948.
27. Cox, R. A., Roffey, M. J., "Thermal Decomposition of Peroxyacetylnitrate in the Presence of Nitric Oxide," Environ. Sci. Technol. (1977) 11, 900.
28. Gerhard, E. R., Johnstone, H. F., "Photochemical Oxidation of Sulfur Dioxide in Air," Ind. Eng. Chem. (1955) 47, 972.
29. Katz, M., "Photoelectric Determination of Atmospheric SO_2 Employing Dilute Starch-Iodine Solutions," Anal. Chem. (1950) 22, 1040.
30. Johnstone, H. F., Coughanowr, D. R., "Absorption of Sulfur Dioxide from Air: Oxidation in Drops Containing Dissolved Catalysts," Ind. Eng. Chem. (1958) 50, 1169.
31. Johnstone, H. F., Moll, A. J., "Formation of Sulfuric Acid in Fogs," Ind. Eng. Chem. (1960) 52, 861

32. Gartrell, F. E., Thomas, F. W., Carpenter, S. B., "Atmospheric Oxidation of SO_2 in Coal-Burning Power Plant Plumes," Am. Ind. Hyg. J. (1963) 24, 113.
33. Van Den Heuvel, A. P., Mason, B. J., "The Formation of Ammonium Sulfate in Water Droplets Exposed to Gaseous SO_2 and NH_3," Q.J.R. Met. Soc. (1963) 89, 271.
34. Urone, P., Lutsep, H., Noyes, C. M., Parcher, J. F., "Static Studies of Sulfur Dioxide Reactions in Air," Environ. Sci. Technol. (1968) 2, 611.
35. Cox, R. A., Penkett, S. A., "The Photooxidation of Sulphur Dioxide in Sunlight," Atmospheric Environment (1970) 4, 425
36. Cox, R. A., Penkett, S. A. "Photooxidation of Atmospheric SO_2," Nature (1971) 229, 486.
37. Cox, R. A., Penkett, S. A., "Oxidation of Atmospheric SO_2 by Products of the Ozone-Olefin Reaction," Nature (1971) 230, 321.
38. Cheng, R. T., Corn, M., Frohliger, J. I., "Contribution to the Reaction Kinetics of Water Soluble Aerosols and SO_2 in Air at ppm Concentrations," Atmospheric Environment (1971) 5, 987.
39. Cox, R. A., Penkett, S. A., "Aerosol Formation from Sulphur Dioxide in the Presence of Ozone and Olefinic Hydrocarbons," J. Chem. Soc. (1972) 68, 1735.
40. Benarie, M., Nonat, A., Menard, T., "Etude de la Transformation de l'anhydride Sulfureux en Acide Sulfurique en Relation avec les Donnees Climatologiques, dans un Ensembly Urbain in a Caractere Industriel, Rouen," Atmospheric Environment (1973) 7, 403.
41. Roberts, P. T., Friedlander, S. K., "Conversion of SO_2 to Sulfur Particulate in the Los Angeles Atmosphere," Env. Health Perspectives (1975) 10, 103.
42. Newman, L., Forrest, J., Manowitz, B. "The Application of an Isotopic Ratio Technique to a Study of the Atmospheric Oxidation of Sulfur Dioxide in the Plume from an Oil-Fired Power Plant," Atmospheric Environment (1975) 9, 959.
43. Alkezweeny, A. J., Powell, D. C., "Estimation of Transformation Rate of SO_2 to SO_4 from Atmospheric Concentration Data," Atmospheric Environment (1977) 11, 179.
44. Forrest, J., Newman, L., "Further Studies on the Oxidation of Sulfur Dioxide in Coal-Fired Power Plant Plumes," Atmospheric Environment (1977) 11, 465.
45. Forrest, J., Newman, L. "Oxidation of Sulfur Dioxide in the Sudbury Smelter Plume," Atmospheric Environment (1977) 11, 517.
46. Calvert, J. G., Su, F., Bottenheim, J. W., Strausz, O. P., "Mechanism of the Homogeneous Oxidation of Sulfur Dioxide in the Troposphere," Atmos. Environ. (1978) in press.
47. Payne, W. A., Stief, L. J., Davis, D. D., "A Kinetics Study of the Reaction of HO_2 with SO_2 and NO," J. Amer. Chem. Soc. (1973) 95, 7614.

48. Atkinson, R., Perry, R. A., Pitts, J. N., Jr., "Rate Constants for Reactions of the OH Radical with NO_2 (M=Ar and N_2) and SO_2 (M=Ar)," J. Chem. Phys. (1976) 65, 306.
49. Cox, R. A., "The Photolysis of Gaseous Nitrous Acid - A Technique for Obtaining Kinetic Data in Atmospheric Photooxidation Reactions," Int. J. Chem. Kinetics Symp. (1975) 1, 379.
50. Castleman, A. W., Jr., Tang, I. N., "Kinetics of the Association Reaction of SO_2 with the Hydroxyl Radical," J. Photochem. (1976/77) 6, 349.
51. Sander, S. P., Seinfeld, J. H., "Chemical Kinetics of Homogeneous Atmospheric Oxidation of Sulfur Dioxide," Environ. Sci. Technol. (1976) 10, 1114.
52. Wang, C. C., Davis, L. I., Jr., Wu, C. H., Japar, S., Niki,H., Weinstock, B., "Hydroxyl Radical Concentrations Measured in Ambient Air," Science (1975) 189, 797.
53. Larson, T. V., Horike, N. R., Harrison, H., "Oxidation of Sulfur Dioxide by Oxygen and Ozone in Aqueous Solution: A Kinetic Study with Significance to Atmospheric Rate Processes," Atmospheric Environment (1978) in press.
54. Novakov, T., Chang, S. G., Harker, A. B., "Sulfates in Pollution Particulates: Catalytic Oxidation of SO_2 on Carbon Particles," Science (1974) 186, 259.
55. Fuller, E. C., Crist, R. H., "The Rate of Oxidation of Sulfite Ions by Oxygen," J. Am. Chem. Soc. (1941) 63, 1644.
56. Bassett, H., Parker, W. G., "The Oxidation of Sulfurous Acid," J. Chem. Soc. Pt. 2. (1951) 1540
57. Junge, C. E., Ryan, T. G., "Study of the SO_2 Oxidation in Solution and its Role in Atmospheric Chemistry," Q.J.R. Met. Soc., (1958) 84, 46.
58. Foster, P. M., "The Oxidation of SO_2 in Power Station Plumes," Atmospheric Environment (1969) 3, 157.
59. Matteson, J. J., Stober, W., Luther, H., "Kinetics of the Oxidation of SO_2 by Aerosols of Manganese Sulfate," Ind. Eng. Chem. Fund. (1969) 8, 677.
60. Chen, T., Barron, C. H., "Some Aspects of the Homogeneous Kinetics of Sulfite Oxidation," Ind. Eng. Chem. Fund., (1972) 11, 446.
61. Brimblecomb, P., Spedding, D. J., "The Catalytic Oxidation of Micromolar Aqueous SO_2 -- I. Oxidation in Dilute Solutions Containing Iron (III). Atmospheric Environment, (1974) 8, 937.
62. Freiberg, J., "Effects of Relative Humidity and Temperature on Iron-Catalyzed Oxidation of SO_2 in Atmospheric Aerosols," Env. Sci. Tech. (1974) 8, 731.
63. Freiberg, J., "The Mechanism of Iron Catalyzed Oxidation of SO_2 in Oxygenated Solutions," Atmos. Env. (1975) 9, 661

64. Barrie, L., Georgii, H. W., "An Experimental Investigation of the Absorption of Sulfur Dioxide by Water Drops Containing Heavy Metal Ions," Atmospheric Environment (1976) 10, 743.
65. Orel, A. E., Seinfeld, J. H., "Nitrate Formation in Atmospheric Aerosols," Environ. Sci. Technol. (1977) 11, 1000.
66. "Aerosols," Chapt. 3 in Ozone and Other Photochemical Oxidants, National Academy of Sciences (1977).
67. Chu, K. J., Seinfeld, J. H., "Formulation and Initial Application of a Dynamic Model for Urban Aerosols," Atmospheric Environment (1975) 9, 375.
68. Hidy, G.M., "Summary of the California Aerosol Characterization Experiment," J. Air Poll. Control Ass. (1975) 25, 1106.
69. Willeke, K., Whitby, K. T., Clark, W. E., Marple, V. A., "Size Distributions of Denver Aerosols - A Comparison of Two Sites," Atmospheric Environment (1974) 8, 609.
70. Willeke, K., Whitby, K. T., "Atmospheric Aerosols: Size Distribution Interpretation," J. Air Poll. Control Ass. (1975) 25, 529.
71. Whitby, K. T., Liu, B. Y. H., Kittleson, D. B., Progress Report on Sulfur Aerosol Research of the Particle Technology Laboratory of the U. of Minnesota (1977).
72. Grosjean, D., Friedlander, S. K., "Gas-Particle Distribution Factors for Organic and Other Pollutants in the Los Angeles Atmosphere," J. Air Poll. Control Ass. (1975) 25, 1038.
73. National Academy of Sciences, "Aerosols," Chpt. 3 of Ozone and Other Photochemical Oxidants (1977).
74. Peterson, T. W., Seinfeld, J. H., "Mathematical Model for Transport, Interconversion and Removal of Gaseous and Particulate Air Pollutants - Application to the Urban Plume," Atmospheric Environment (1978) 12, xxx
75. Jerskey, T. N., Seinfeld, J. H., "Aerosol Formation and Growth in the Urban Atmosphere," AIChE J., submitted for publication (1978).

RECEIVED February 17, 1978

6

The Design of Gas-Solids Fluid Bed and Related Reactors

W. P. M. VAN SWAAIJ

Twente University of Technology, Enschede, Netherlands

In fluidization it is attempted to overcome the problems in handling granulair solids or powder and to improve its heat transport properties. This is done by passing a fluid through a dense swarm of particles, thus reducing internal friction and cohesion between the particles. The particles then become in a dynamic state of equilibrium, on the average their weight is just balanced by the drag force exerted by the fluid on the particles.

The gas solids suspension, which may contain larger empty spaces called bubbles, can to a certain extent be considered and handled like a liquid which is of great advantage in process operations.

Due to the extensive mixing caused by the fast flowing bubbles, heat transfer and heat transport rates are very high (see eg. (1)).

Fluidized beds have been studied during the last 35 years with an effort that is almost unique for a single type of process operation. Many thousands of articles, patents, several textbooks (1-6) and many review articles on special subjects appeared, while a considerable number of symposia have been devoted to this subject. This reflects also a wide spread use of fluid beds in physical operations and as a chemical reactor in chemical, petroleum, environmental, metallurgical and energy industries. Old applications like gasification of coal (Winkler generator, 1926) and combustion of coal are reviving.

Furthermore, fluidization is a fascinating subject, an "El Dorado" for model builders and a rich source of Ph-D. programs. It is impossible to give a short review of the problem "fluidized bed reactors", as this would fill a complete series of textbooks. We shall only consider relatively recent developments in a few areas from the point of view of design/process development:

0-8412-0432-2/78/47-072-193$07.50/0

modelling bubbling bed reactors, co-current and counter current reactors.
Fig. 1 shows the different types of gas-solid fluid bed and related reactors. The bubbling bed reactors (a) are the oldest type and most fluid bed studies refer to this regime. In modern fluid bed cat cracking regenerators gas velocities are so high that individual bubbles become vague and so much of the solids is entrained to the cyclones (not shown) that also the bed level is not clearly defined anymore (b). (This regime is presently called turbulent beds). Especially with the introduction of zeolite catalyst it became advantageous to carry out cat cracking in a riser reactor where catalyst is transported with the gas phase at a high velocity thus realising short contact times and less backmixing.

At gasvelocities between (b) and (c) there is a transport regime with a higher solidsconcentration called "fast" bed. Counter current contactors are used as strippers (e.g. in cat crackers) or as chemical reactors if counter current is required (e.g. for heat exchange, high solid phase conversion, adsorption, etc).

Bubbling Bed Reactors, Particle Selection

One of the first problems in design/process development of a fluid bed reactor is the prediction of the type of fluidization that will be obtained in the large scale reactor unit. Many process variables such as temperature, pressure, etc. will be chosen on other arguments, but in the selection of the particle size and the fluidization velocity, considerations on the state of fluidization often play a major role. Some rules of the tumb were already available (7), but during the past few years a more systematic approach to this problem has been discussed in open literature.

Geldart et al. (8,9) represented the different types of fluidization as a function of d_p and $\rho_p-\rho_g$ (Fig. 2). Many more characteristics on the different fluidization states than indicated in Fig. 2 are given in their papers. The difference in fluidization between A and B powders is easily demonstrated in a bed expansion graph (*see* Fig. 3).

The characteristics of B powders are relatively simple. Directly beyond the point of minimum fluidization bubbles are formed and the average dense phase porosity doesn't change. In large scale units bubbles grow rapidly to large sizes by coalescense, dense phase mixing is moderate and the apparent "viscosity" is high.

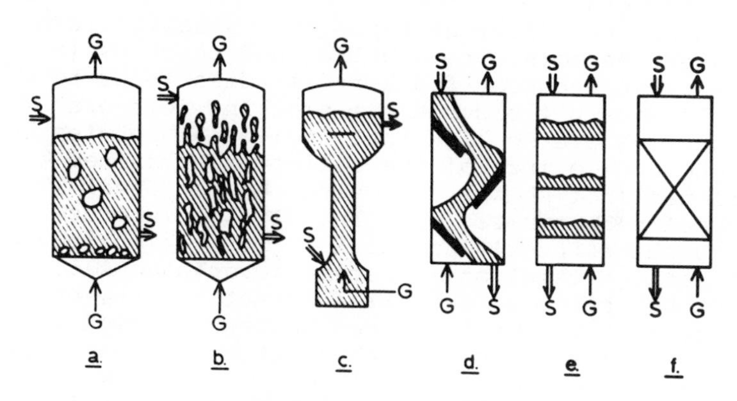

Figure 1. Fluid bed and related reactors. (a) Bubbling bed (with or without internals); (b) turbulent bed; (c) pneumatic transport (riser and fast bed); (d) countercurrent baffle column; (e) plate column (sieve trays, bubble caps, etc.); (f) packed column (bubble flow or trickle flow).

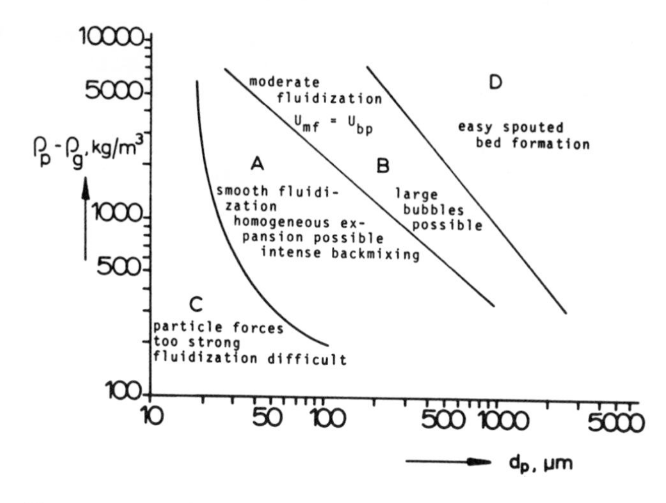

Figure 2. Powder classification according to Geldart (8, 9) (ambient conditions)

A-powders behave quite differently. There is a homogeneous (bubble free) expansion at low gas velocities and if bubble formation starts the total bed expansion first decreases to a minimum value and then increases again. In large units bubbles grow in size but may reach a state of dynamic equilibrium between coalescence and splitting and therefore a maximum size (10,11). Even at low gas velocities strong convection currents and intense backmixing exist in the dense phase. The dense phase expansion of A-powders is not only of academic interest, but is essential to obtain smooth fluidization on a large scale (at large bed heights). This can be illustrated with the experiments of de Groot (7) (see Fig. 4). The rapid decrease in bed expansion with bed diameter for B-powders reflects the formation of large bubbles which causes intense shacking of the apparatus and formation of dead zones (7). It is surprising that not much more attention has been paid to these facts.

The essence of the A-type expansion phenomena has been described by Rietema (12), Morooka et al. (13) and van Swaaij and Zuiderweg (14), de Jong and Nomden (15) and Bayens and Geldart (9), but the importance for large scale fluidization has not been generally recognized and is not mentioned in the recent handbooks.

From the point of view of design, the dense phase expansion under full fluidization conditions is most important. This expansion can be measured by the bed colapse technique (Rietema (12) , de Vries et al.(16)) in which dense phase expansion can be separated from expansion due to bubbles and the permeability of the dense phase be estimated. The high bed expansion at U_{bp} can be easily broken up by stirring with a rod and then the dense phase expansion under full fluidization conditions is roughly obtained; the original expansion is restored, however, if the bed is left undisturbed again. The author observed in two dimensional homogeneous bed experiments a tendency for injected single large bubbles to decrease in size if the bed expansion was below the dense phase expansion under full fluidization conditions and to increase in size if the bed expansion was higher. This indicates that a kind of dynamic equilibrium between bubble gas and dense phase exists under full fluidization conditions, which is of course very important for mass transfer (to be discussed later). Little information about this equilibrium dense phase expansion is available in the open literature. Rowe (17) gives some data on dense phase through flow which seems to be very high. The measuring

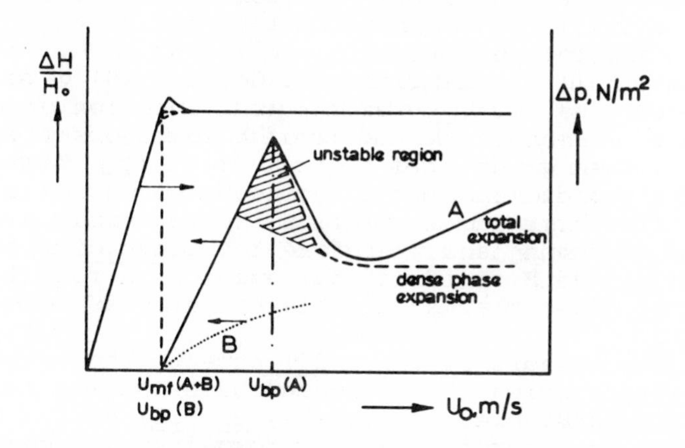

Figure 3. Bed expansion graph for A- and B-type fluidization

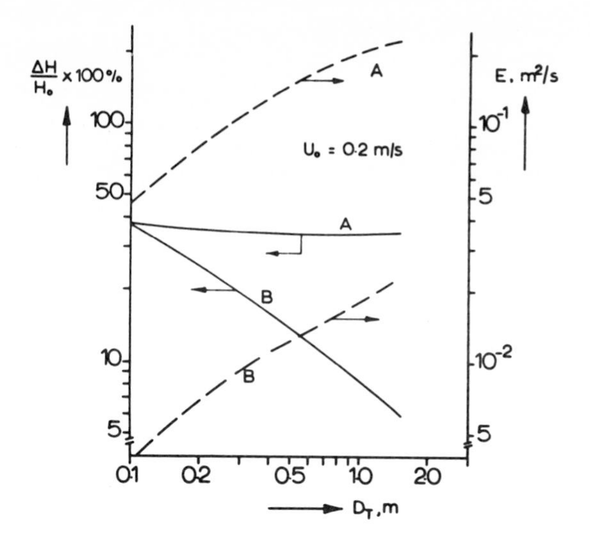

Figure 4. Bed expansion and mixing coefficient of A and B powders. Data from de Groot (7) (ambient conditions, air–silica).

technique has not yet been indicated. As the range of homogeneous fluidization is important for the type of fluidization, the empirical relations and theories about this regime will be shortly discussed. De Jong and Nomden (15) and Bayers and Geldart (9) gave emperical relations for U_{bp} and U_{mf} as a function of the particle diameter. The range of homogeneous fluidization is then given by $U_{mf} < U_o < U_{bp}$ (see Fig. 5). The whole phenomenon of homogeneous expansion is difficult to understand. A complete theory should describe this phenomenon (and therefore the state of fluidization, A and B fluidization, etc.) as a fluidization property and not as a particle property as done by Geldart.

Some authors (18, 19, 20) come to the conclusion in their theories that homogeneous expansion should always be unstable. Others (21, 22, 23) derive different functions for the maximum porosity ε_{bp} above which no stable homogeneous operation is possible.

$$Ga_p = f\{\varepsilon_{bp}\} \quad (1)$$

$$Ga_p = \frac{\rho_p{}^2 d_p{}^3 g}{\eta^2}$$

Oltrogge (23) and Verloop (22) introduce in their derivations of equation (1) a stabilizing elastic behaviour of the bed which is ascribed to the hydrodynamics of the flow of a fluid through particle layers. However, this explanation cannot be valid for the low Reynolds numbers encountered in homogeneous gas-solids fluidization as was recently shown by Mutsers and Rietema (21). Rietema (22) showed the essential role of interpartical contact in fluidization. This contact can be demonstrated e.g. by electrical conductivity measurements. Due to this contact, particles exert cohesion forces on each other. For particles with $d_p < 100\ \mu m$ the cohesion number

$$Co = \frac{\text{cohesion force between particles in contact}}{\text{gravity force on particle}}$$

becomes much larger than one (22).

The elasticity E_e of the particle structure within the dense phase, a consequence of the particle forces, was considered in the stability criterium for homogeneous fluidization of Mutsers and Rietema:

$$N_F = \frac{\rho_p^3 d_p^4 g^2}{\eta^2 E_e} < \frac{150(1-\varepsilon_{bp})}{\varepsilon_{bp}^2(3-2\varepsilon_{bp})} \quad (2)$$

ε should be smaller than ε_{bp}.

In combination with the minimum fluidization criterium equation (2) can be used to predict the range of homogeneous fluidization at different conditions. A problem however, is to find the value of E_e on the forehand and generally measurements will be required. E_e was found to depend on particle size and size distribution, gas properties and degree of expansion, etc. (24). An indication of the correctness of the theory of Rietema is given by the tilted bed experiment (*see* Fig. 6). The bubble point is reached at the same velocity where no tilting without yielding of the powder structure is possible anymore. In experiments with a centrifugal field applied to a fluid bed, it was shown by Mutsers and Rietema (26) that ε_{bp} was a function of g^2/η^2 as indicated by equation (2) and not of g/η^2 as would be indicated by equation (1). Agbin et al. (27) showed that application of magnetic forces can extend the range of homogeneous fluidization. So it can be concluded that although inter particle forces are weak, they play an important role in the bubble formation (*see* also Donsi and Massimilla (28)).

The theory of Rietema must still be extended to indicate the boundary between A and C fluidization, describe stability of bubbles formed, etc. It will be clear that B type fluidization, which shows no dense phase expansion, cannot be compared with A type fluidization. This has led to considerable misunderstanding because in the early university investigations B type fluidization was often studied while process developers tried to avoid this regime because of problems with large scale fluid beds, specially at large bed heights (29).

Reactor Models of the Bubbling Fluid Bed

Much work has been done on modelling bubbling fluid bed reactors and many reviews are available. Recently Horio and Wen (30) gave an excellent review of the different models. At least 17 models have been discussed and devided into three catagories (*see* Table 1). They distinguished 6 parameters about which different assumptions were made (in fact there are more): method of dividing the phase (3 different assumptions); method of flow assignment (5); cloud volume (3); gas

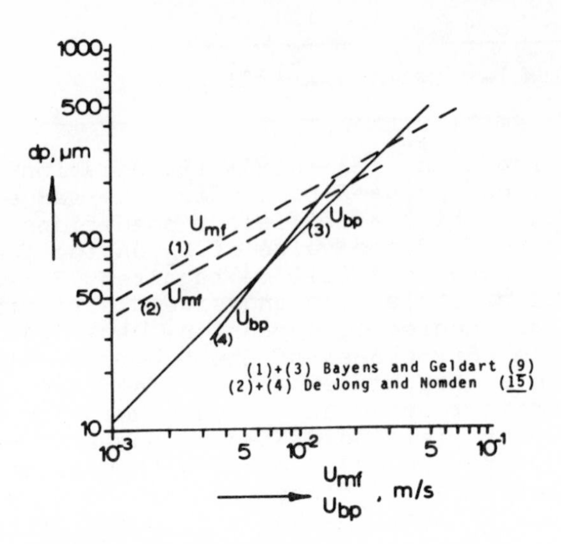

Figure 5. Minimum bubbling velocity and minimum fluidization velocity of cat-cracking catalyst at ambient conditions

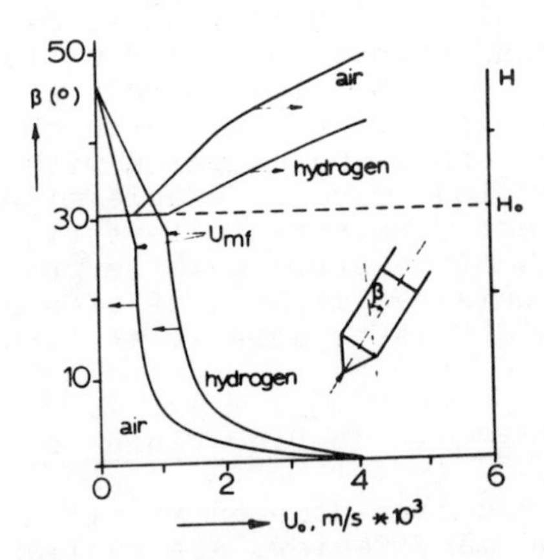

Figure 6. Maximum angle in tilted bed experiments of Mutsers and Rietema (24) with homogeneously expanded cracking catalyst at ambient conditions

exchange coefficients (4); bubble diameter (4); effect of jets (2).

Table 1. Fluid bed reactor models

Level I (31, 32, 33, 34, 35, 36)	Parameters constant along the bed Parameters not related to bubble size, but have to be measured separately
Level II (37, 1, 38)	Adjustable bubble size relates parameters, which are assumed constant along the bed
Level III (39, 40, 41, 42, 43, 44, 38, 45)	Parameters related to bubble size, which is varied according to empirial relations

It is clear that by permutation many more models can be constructed. In fact the situation is still worse. Fig. 7 presents a possible model. The parameters to be filled in have not been invented by the author, but most of them are under discussion in literature and different proposals, backed by experimental evidence, are given for them. Even this model is of course not complete; radial distribution of parameters, adsorption, scaling up factors for parameters, etc, have not been specified.

To test different models for their suitability, one has to formulate the objectives that are aimed for in the application of the model because there does not exist an "ideal" fluid bed model. Most models have been tested in rather small scale units on their overall conversion prediction, sometimes inserting emperical relations for parameters measured under similar conditions. Chavarie and Grace (46) & Fryer and Potter (47) also tested profiles. In this work we shall consider the usefulness of different models in process development / scaling-up activities. A model is considered to be useful if it can reduce the numbers and the complexity of the experiments necessary for a safe scaling-up or design procedure, and if it adequately describes the operation performance of the full scale reactor.

To place different models into perspective let us assume a simple case of a first order heterogenously catalysed chemical reaction to be carried out in a bubbling bed. We will consider a slow reaction rate (0.1 -

$0.5\ s^{-1}$) because for higher rates other types of reactor may be more appropriate and we will aim for a high relative conversion (90-95%). The last assumption is important because it will amplify the differences in the models and it refers to practically important situations. (Regenerators for catalytic cracking units, oxidation of HCl, oxichlorination and many other organic reactions). We shall assume a bubbling bed regime of A-type fluidization with a superficial velocity of 0.1 - 0.2 m/s, and a large diameter full scale bed ($D_T \approx$ 3-4 m). We will not consider internals or multitubular beds. For the last category different models should be applied (see e.g. Raghuraman and Potter (48)).

Now generally the problem is to find the conversion from c/c_o = f (distributor, H, D_T, mass flow, type of gas, solids properties, P, T, disengaging zone design, etc.), but here we shall only treat the problem: conversion = f(H,D_T). The other variables are taken constant (nearly atmospheric pressure, moderate temperature, etc.).

First we shall consider the possibilities of the level I models. As an example we shall take the model of van Deemter (Fig. 8). It can easily be demonstrated (16, 49) that in most cases the mass transfer is the limiting factor for conversion. Therefore the conversion equation simplifies to:

$$\frac{c}{c_o} = \exp - \left[\frac{N_\alpha N_r}{N_\alpha + N_r}\right] \quad (3) \text{ or for large } N_r \tag{3}$$

$$\frac{c}{c_o} = \exp - N_\alpha \quad (4) \quad N_\alpha = \frac{\alpha H}{U_o} \quad N_r = \frac{kH}{U_o} \tag{4}$$

The simplest (but expensive) way of obtaining the relevant parameter N_α is to measure the conversion of a first order reaction at different scales and conditions. This has been carried out on larger scale fluid beds by Botton (50) , van Swaaij and Zuiderweg (49) and de Vries et al. (16) (including D_T = 3m). It should be realised that by this technique the product

$$\alpha = \overline{k_m S} \quad (5) \qquad k_m = \text{mass transfer coefficient}, \quad S = \text{interfacial area}$$

is measured. This is a common practice in gas/liquid contactors where $k_m S$ is measured with a slow reaction or with transient absorption experiments. It should be proven, however, that it is also possible to separate mass transfer to dense phase and reaction in the dense

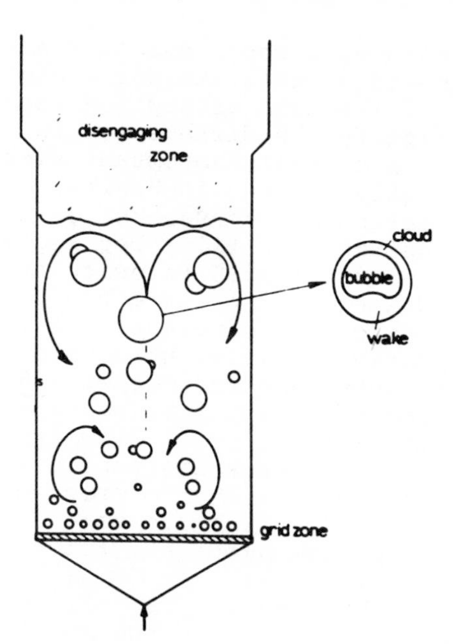

Figure 7. "Complex" fluid bed model

Parameters: (1) jets on distributor or initial bubble size; (2) division of flow through bubble and dense phase (magnitude and direction); (3) axial profile of 2; (4) bubble size profile (and distribution in bubble velocity); (6) axial profile of 5; (7) volume of cloud (relation to size and rising velocity); (8) solids within the bubbles; (9) volume of wake; (10) mass transfer rates (bc-ce-cw-we-ce) (relation to size and rising velocities); (11) intensity of gas mixing within dense phase (circulation patterns or diffusion); (12) expansion of dense phase; (13) influence of disengaging zone.

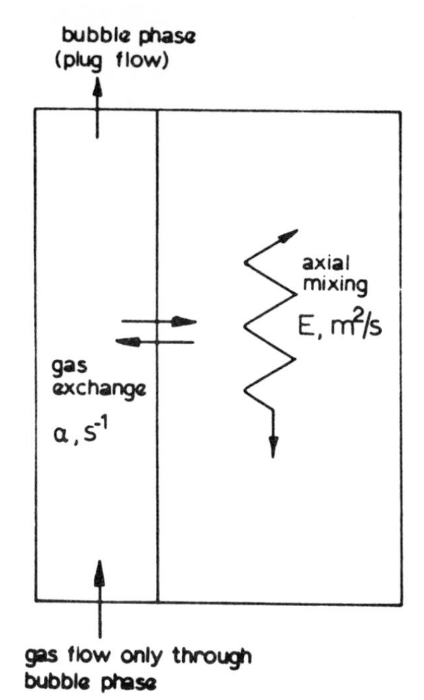

Figure 8. Simplified two-phase fluid bed model (van Deemter)

phase for fluid beds in the same way. Van Swaaij and Zuiderweg (49) carried out these tests for both silica and sand (Fig. 9). The observed α should be independent of the reaction rate applied. It follows from Fig. 9 that this is true for low values of the reaction rate constant. At higher reaction rates the resistance for mass transfer, H_α, decreases (α increases). Such a behaviour was also observed by Furisaki (51), Gilliland and Knudsen (52) and Myauchi and Morooka (53). Possible explanations are:

1) mass transfer enhanced by reaction (see gas/liquid concept)
2) axial mass transfer profile (see bubbling assemblage model (43) and van Swaaij and Zuiderweg (49))
3) reaction takes place within bubble phase wake, etc. see Kunii and Levenspiel (1)
4) reaction in disengaging zone where contacting is better.

The last explanation has been given by Myauchi and Furusaki (54). Their model calculation is also shown in Fig. 9. Other ways to measure N_α by non-reacting tracer tests are summerized in Table II.

Table II Different methods for the determination of the number of transfer units.

Method	Measurements required	Parameters	Ref.
Conversion	- conversion - kinetics - solids-mixing (or dense phase mixing) - (bed and dense phase expansion)	N_E of N_B N_α, N_r ϕ	(49, 50) (16)
Back-mixing	- axial backmixing profiles - solids mixing - bed and dense phase expansion	N_B, N_E, N_α	(50, 35)
Residence Time Distribution	- RTD curves - bed and dense phase expansion - solids mixing or dense phase mixing	N_B or N_E, N_α σ, ϕ	(7, 49, 16, 50)

They were proven to give the same results as the conversion tests (49, 50, 16). Results from these conversion and tracer tests have been condensed into a design method (14). For large diameter beds and A-type fluidization it was found that:

$$H_\alpha = (1.8 - \frac{1.06}{D_T^{0.25}})\ (3.5 - \frac{2.5}{H^{0.25}}) \qquad (6)$$

(D_T, H_α and H in meters; see Fig. 10).

H_α is also influenced by gas velocity (slightly) particle size distribution (e.g. fines (particles with $d_p < 44\mu$) percentage), temperature and pressure. Recommended values have been given elsewhere (14), but more data are needed. Models of level I are flexible and can cope with the complex phenomena occuring in the fluid bed. The parameters should be measured separately which is costly and time consuming but reliable (This was called "Cautious empirsm" by Davidson (55)).

We shall now consider the possibilities of the level II models and take the Kunii and Levenspiel model (1) as an example. The main features of the model are given in Fig. 11. All parameters are related to the effective bubble size which is here the fitting parameter. Although the concepts are not completely compatible, it is possible to separate here $\overline{K_m}$ and $\overline{S}$ in the product $\overline{K_m S}$ of the level I models especially for type A fluidization where clouds are extremely thin. Applying this model outside the region where it has been tested (mainly small scale units) will give the following problems:

- $U_{df} \neq U_{mf}$ (see Fig. 3). This affects the mass transfer relations
- Bubble rising velocity, at least for A-type fluidization differs from the equation:

$$V_b = U_o - U_{mf} + 0.711\ \sqrt{gd_b} \qquad (7)$$

This affects both the mass transfer coefficient and S. Real bubbles in swarms in A-type fluidization may move much faster (Drinkenburg, (56), Oltrogge (23), Morooka et al. (13)). Bubble rising velocity for a given bubble size seems to depend somewhat on the distance from the distribution plate (Calderbank (57)) and strongly on the bed dimension (11) (see Fig. 12). This originates from the strong interaction between solids flow (or mixing) and bubble flow patterns.

- Bubble diameter cannot be easily predicted. Fig. 13 gives results based on different predictions (reviews

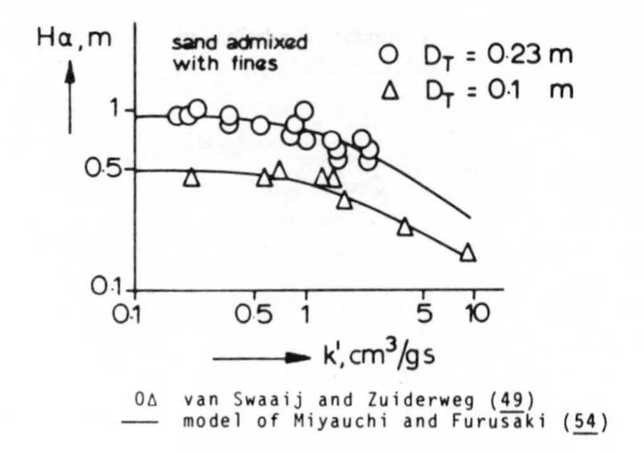

Figure 9. Observed height of a transfer unit as a function of the reaction rate constant

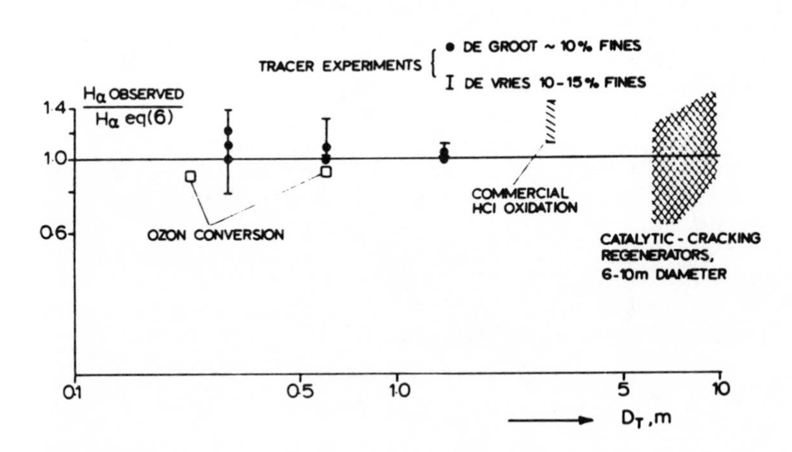

Fluidization and its Application

Figure 10. Comparison of experimental values of Hα for large scale units with (van Swaaij and Zuiderweg (14))

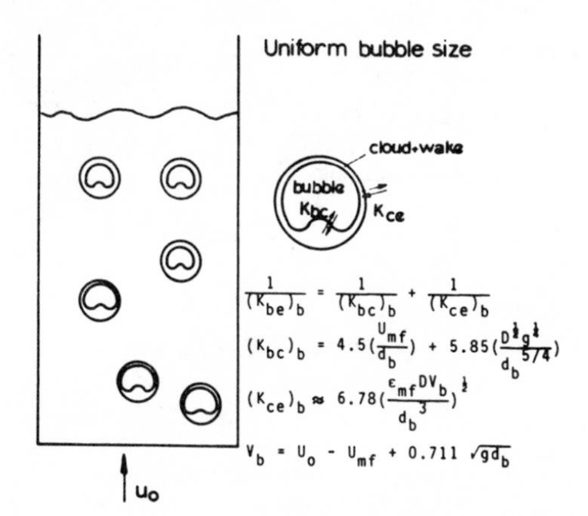

Fluidization Engineering

Figure 11. Model of Kunii and Levenspiel (1) (level II model)

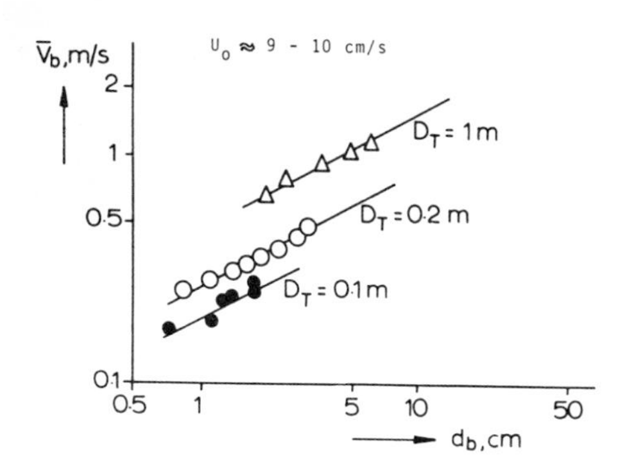

Chemie-Ingenieur-Technik

Figure 12. Average bubble velocity as a function of bubble sizes at different scales (Werther (11))

and recent work, see Mori and Wen (58), Rowe (59) and Danton et al.(60)). Some authors have shown evidence for maximum stable bubble size (Botton (50), Matsen(61) Werther (10)) for A-type fluidization.
Based on a statistical coalescence model and with the concept of the maximum bubble size Werther (11) found:

$$\frac{d_b}{cm} = 0.83 \sqrt[3]{1+0.272\left(\frac{U_o-U_{mf}}{cm/s}\right)} \left\{1+0.0684 \frac{h}{cm}\right\}^{1.21} \quad (8)$$

for $h < h^*$

h^* is the distance from the distributor where the bubble size reaches its maximum. If $h > h^*$, h^* should replace h. Typical values are $h^* = 0.7 - 2m$ (11), which means that most small scale experimental units are in the coalescense regime or at slugflow while the large scale units may easily reach a stable bubble size. The mass transfer performance is very sensitive to bubble size.

- The mass transfer coefficient K_{be} is based on a single bubble. Apart from the problem of the many different available models for the latter process, it has been found (Chiba et al.(62), Drinkenburg (56, 63), Kato and Wen (43)) that bubbles in swarms have much higher exchange coefficients (up to four times) than single ones, possibly due to interaction between bubbles, splitting, coalescense, gas leaking to dense phase. The physical picture suggested by the model may not be relevant and therefore misleading. No influence of the molecular diffusion coefficient on the mass transfer (under full fluidizing conditions) could be found (Drinkenburg (56, 63), Fontaine et al. (64) (except close to U_{mf}), De Vries et al. (16)) in contrast with theory. Because in many investigations it was found that the mass transfer coefficient decreases with increasing bubble size, the following empirical relations have been suggested (43),(65),(66).

$$(K_{be})_b = \frac{11}{d_b} \qquad \text{bubble swarms} \quad (9)$$

$$(K_{be})_b = \frac{3}{d_b} \qquad \text{single bubbles} \quad (10)$$

However, Hoebink and Rietema (67) found a reverse trend (for bubbles with $d_b > 5cm$) which they ascribed to zigzag movements. The large spread in bubble sizes within a swarm can make the overal average behaviour different from that of a uniform swarm (see Schlünder (68)).

The conclusion is that with level II models, instead of the product kmS, an effective bubble size will have to be determined. In small scale units this bubble size was often close to observed average bubble sizes (Kunii and Levenspiel (1)). However, sometimes the model led to inconsistent results (49) and the assumptions made are too far from reality (U_{df}, ε_{df}, V_b, mechanisms of mass transfer) to make extrapolation to other conditions possible. No essential reduction in experimental effort needed for scaling-up can be obtained with these models.

Possibilities of the Level III Models. For these models we will take the model of Kato and Wen (43) (Fig. 14) as an example and a recent one of Werther (11). The bubble size is varied with distance from the bottom according to an empirical relation. In the case of Kato and Wen the mass transfer is related to bubble size via equation (9) and the rising velocity to bubble size. A property of a model such as that of Kato and Wen is that due to the emperical bubble size profile the model predicts relatively higher conversion if higher reaction rates are applied (*see* Fig. 9). However, other explanations also have been suggested for this phenomenon. The model does not take the observed different rising velocities for bubbles at different scales into account and seems to be more suitable for Type-B fluidization (where bubbles rapidly grow to large sizes) and higher reaction rates. Werther (11) calculated, from the local bubble size and rising velocity at different scales, average values for kmS in which an emperical value for km was used, as a first approximation assumed to be independent of S. He found very good agreement with conversion and mass transfer results of Avedesian and Davidson (69), van Swaaij and Zuiderweg (49), de Groot (7) and de Vries et al. (16) (*see* Fig. 15). From these results the following maximum stable bubble sizes can be deduced (Table III).

Table III: Maximum Stable Bubble Sizes.

% fines ($dp<44\mu m$)	estimated $\overline{dp}$	H_α, cm	d_b max, cm
7	74	386	38
12	66	326	25
17	61	297	20
20	57	250	18
30	48	169	12

D_T = 3m $H \approx 10$m U_o = 15 cm/s
porous silica catalyst.
Data calculated from de Vries et al. (16) using theory of (11).

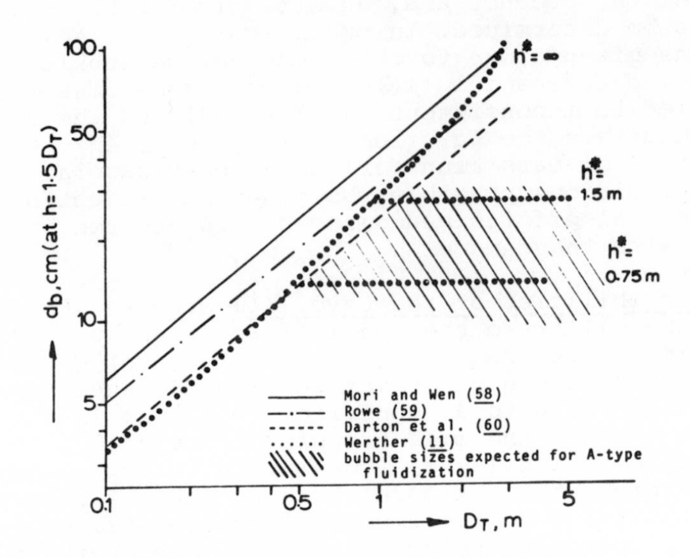

Figure 13. Bubble sizes at h = *1.5*D_T *as a function of bed size* U = *.15m/sec,* $U_{mf} << U_o$*, porous plate distributor.*

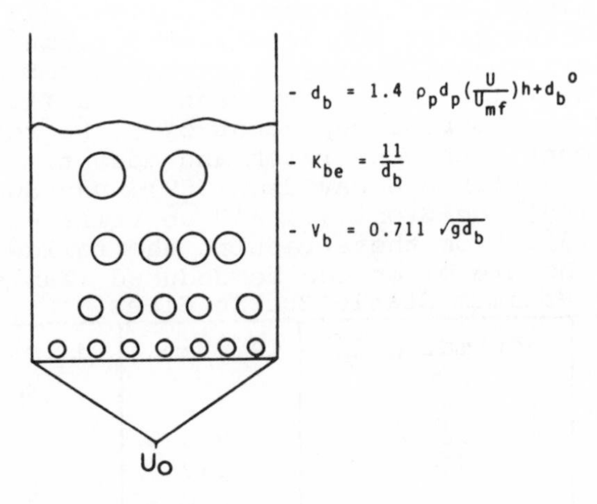

Figure 14. Level III model (Kato and Wen (43))

These maximum stable bubble sizes can be compared with the theory of Davidson and Harrison (2) on stable bubble sizes. This theory, however, predicts bubble sizes of less than 1 cm for the conditions of Table III. Furthermore, like the mixing coefficient E, the maximum bubble sizes may be a function of D_T. No data are available to test this influence.

According to the bubble size measurements of Werther if $h < h^*$ bubble sizes are only little influenced by particle properties while for larger units and higher bed heights the different values of $d_b max$ with different particle sizes (and size distributions) makes the performance of the fluid bed reactor strongly dependent on particle properties. These facts were indeed observed in conversion experiments (49, 16) and this can now be related to bubble growth. The model of Werther can be brought into the form of the Kato and Wen model for conversion prediction at higher reaction rates. However, for these conditions probably other types of reactors (to be discussed later) are better suited than dense beds. It should be realised that profiles of kmS can also be introduced in simple two phase models (see 49).

In the development of bubble fluid bed models, first theoretical relations for bubble properties based on single bubbles were introduced which led to elegant models with only the effective bubble diameter as fitting parameter (level II models). The assumptions were too far from reality, however, and more and more empirical relations have been introduced in later models. To a certain extent level III models can be considered as modified level I models with empirical relations for $d_b = f(h)$, $V_b = f(h)$ and $km = f(d_b)$ together leading to $\alpha_{loc} = f(h)$. These empirical relations are still rather incomplete and uncertain, however (maximum bubble size, mass transfer coefficient, etc.).

This means that in scaling-up activities of fluidized beds, level I or modified level I models in combination with experiments for each system are still required, especially if high temperatures and pressures are applied. Little data are available for these conditions. Also influence of adsorption needs more study.

Co-Current "Fluid Bed" Reactors

In Fig. 16 different regimes are indicated that may be obtained by increasing the gasvelocity of a fluid bed of fine particles (70). There is much confusion about these regimes because they depend on particle size and reactor diameter applied. This has been

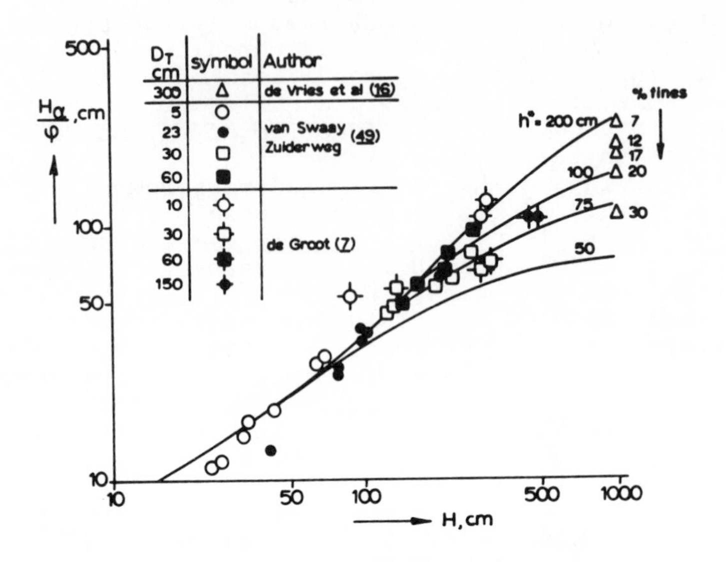

Figure 15. Height of a transfer unit as a function of the bed height for silica (adopted from Werther (11)) $\varphi = V_b/\sqrt{gd_b}$

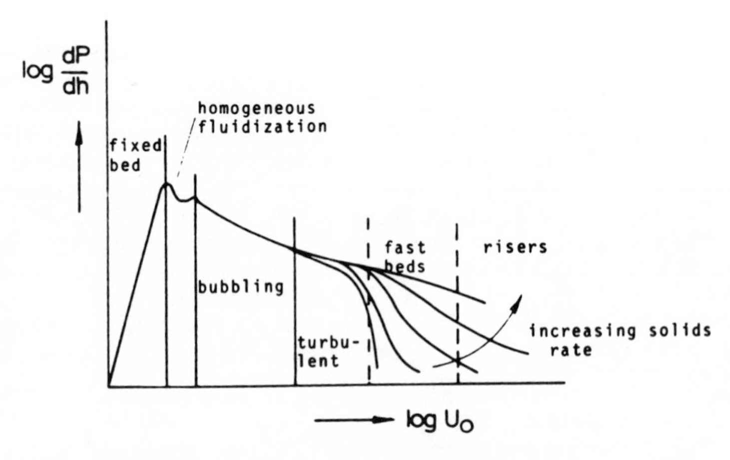

Figure 16. Fluidization and pneumatic conveying of small particles (e.g. FCC catalyst) (See also *Yerushalmi*) (70)

indicated in Table IV.

Table IV. Different regimes in co-current Fluidization

→ increasing U_o

↓ D_T and d_p

$d_p<100\mu$ large bed diameter	A_1 dense fl.bed	A_2 turbulent bed	A_3 fast bed	A_4 riser	
$d_p<100\mu$ small bed diameter	B_1 dense fl.bed	B_2 slugging bed	B_3 turb. bed	B_4 fast bed	B_5 riser
$d_p>150\mu$ relatively small bed diam.	C_1 dense fl.bed	C_2 slugging bed	C_3 choking bed	C_4 fast bed	C_5 riser

Table V. Transitions in Co-Current Regimes (Tentative)

Transition	Criteria
C_1, C_2, A/B, B_1/B_2	slugging criteria (see e.g. (71)
A_1/A_2, B_2/B_3	$U_o \approx 0.5$ m/s (ambient conditions)
A_2/A_3, B_3/B_4	$U_o \approx 2$ m/s (ambient conditions)
A_3/A_4, B_4/B_5	arbitrary say $U_o \approx 10$ m/s (ambient conditions)
C_3/C_4	$U_t/\sqrt{gD_T} < 0.35$ (73), see also (72), (74).

Quantitative information about the different regimes is far from complete, Table V is only a first attempt and much more work has to be done in this area.

The transition to turbulent beds with cat cracking particles can be observed between 0.4 and 0.8 m/s at ambient conditions and modern regenerators of cat cracking units operate close to or within the turbulent regime.

In slugging beds the transition to turbulent beds can be easily identified but in larger beds a more gradual change may occur. Little is known about mass transfer in this area. Analysis of regenerators (14) showed that the height of a mass transfer unit for this regime is not much different from beds at lower velocities. The high load to the cyclones increases the heat capacity of the disengaging zone thus preventing strong increases in temperature due to "afterburning" of CO by unconverted oxygen. Therefore this high velocity regime may be interesting from an operational point of view and not necessarily because of higher mass transfer rate. At still higher velocities the bed density becomes much dependent on the solids rate and in fact the "fast bed" regime (see Yerushalmi (70)) is the lower gas velocity part of the pneumatic transport (riser) regime. Data on this regime have been given by Yousfi et al. (72), (75), Yerushalmi (70), van Swaaij et al. (76).

The following advantages are claimed: good contact between gas and solids (no pertinent data are available however), in comparison with risers high solids concentration (up to 25%) and low gas velocity 1-15 m/s so without excessive reactor length reactions with intermediate rates can be processed (see Table VI), high heat backmixing rates, and high heat wall transfer coefficients (75, 77) (about equal to those of a dense bed) and high heat transfer rates between gas and particles. Furthermore, it is claimed that the gas phase is close to plugflow (no data yet), cohesive solids can easily be handled and scaling-up is relatively simple (by Lurgi cited in (70)).

Disadvantages, however, can be erosion and attrition and problems in gas-solids separation. Furthermore, there may be doubt about some of the claims. Fig. 17 shows a fast bed or riser set-up. Next to it a typical pressure drop graph is given schematically. The accelleration zone may extend to 4 or more meters length even in small scale units (78).

Data of Yerushalmi et al. (70) are taken from the accelleration zone. This also follows from their pressure drop profiles. For scaling-up this is a factor to be taken into account.

In the developed regime, especially in fast bed operation, the observed pressure drop was found to be smaller than that corresponding to the solids hold-up (see van Swaaij et al. (76)). Van Swaaij et al. demonstrated that this could be related to a shear stress on the wall opposite in direction to the flow. From their experiments it followed that this was caused by

Table VI. Length required for fast bed reactors.

k packed bed (sec^{-1})	H (95% conversion), m; no mass transfer limitation	H (95% conversion), m with mass transfer limitation: $H\alpha$ = 2m
0.1	500	500
1	50	56
3	17	23
10	5	11
30	1.7	9.8

First order reaction, U_o = 5 m/s, solids hold-up 18% vol., plugflow.

downflow of solids along the wall. Typical values for a 18 cm diameter fast bed with cracking catalyst were: U_o = 4.7 m/s, average solids flux 133 kg/m^2s, downflow flux at the wall 430 kg/m^2s. Nakamura and Capes (80,81) made a model for this type of flow. It is clear that downflow of solids along the wall complicates the contacting/mixing pattern of fast beds. Critical conversion, mass transfer and mixing tests at different scales will be necessary to evaluate the fast bed regime.

Riser reactors are operating at higher gasvelocities 10-50 m/s. This means that they are suitable for reactions requiring only a very short contact time. Here the solids flow can be expected to be more close to plugflow (no downflow along the walls). Much information on this type of reactors is available within oil and chemical companies. Little data appeared in the open literature on mass transfer, conversion and mixing.

Counter Current Gas-Solids Reactors

Solids staging in fluid bed processes can be obtained in tray columns so that counter current operation is possible (see Fig. 1). These types of reactor have been reviewed by Varma (82) and will not be discussed here. A special variant is the baffle plate (see Fig. 1), which can handle large solids flow (e.g. in strippers of cat crackers).

New counter-current operations have been introduced by Claus et al. (83) involving bubble flow and trickle flow through packed beds (Fig. 18). In particular trickle flow of A-type particles through "open"

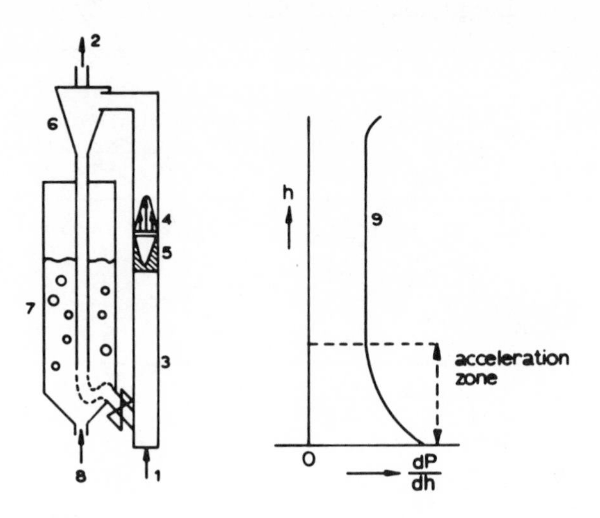

Figure 17. Fast bed or riser

(1) Gas inlet; (2) gas outlet; (3) fast bed (or riser); (4) gas velocity profile; (5) solids density profile; (6) cyclone; (7) hopper; (8) additional gas inlet for fluid bed; (9) pressure drop profile.

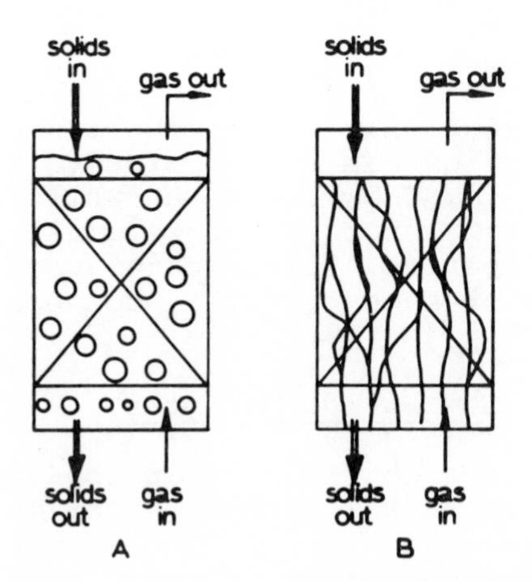

Figure 18. Packed bed gas–solids countercurrent contractors. (A) Bubble flow; (B) trickle flow (gas phase continuous).

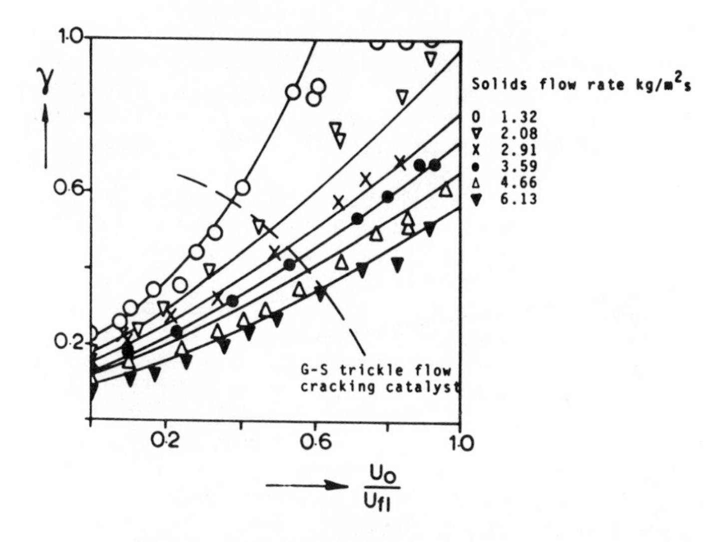

Figure 19. Fraction of solids supported by the gas phase as a function of the gas velocity relative to flooding (Roes and van Swaaij (84))

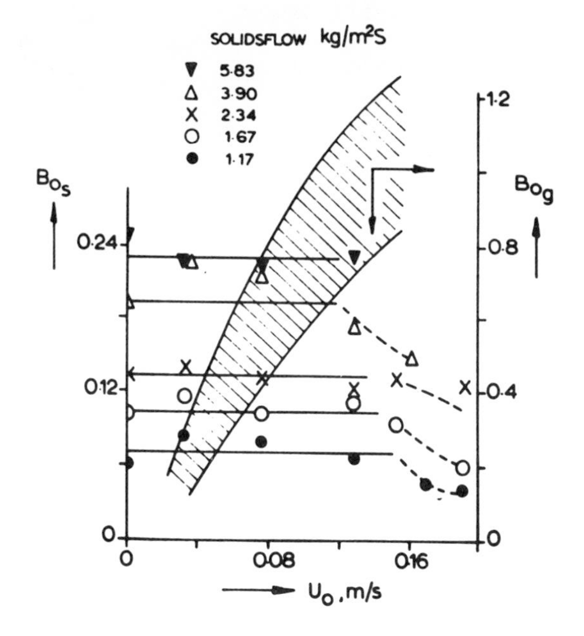

Figure 20. Axial mixing of gas phase and solids phase in gas–solids trickle flow as a function of the gas velocity. Cracking catalyst through 1,5 cm PALL ring packing. Shaded area: gas phase results.

packings (e.g. pall rings) was found to be interesting (see Roes and van Swaaij (84)). The pressure drop was found to be lower than that corresponding to the solids hold-up in the column (Fig. 19), which means that only part of the solid is supported by the gas phase, the rest by the packing. In most cases of trickle flow of liquid, all the liquid is supported by the packing.

Also other properties are roughly similar to gas liquid trickle flow (loading flooding, etc.). Fig. 20 gives typical data for axial dispersion in gas and solids phases. For higher gas and solid rates both phases show a reasonable approach to plugflow (85). Mass transfer was found to be high and axial mixing will still be a limiting factor in many circumstances (85).

The advantages of trickle flow reactors are: low pressure drop, good contacting properties and simple construction in comparison with tray columns. These properties are combined with general advantages of counter current operation: extraction of reactants or products out of the reaction zone possible (e.g. for equilibrium reactions), efficient heat exchange between solids and gas phase, etc. The radial heat transport, wall heat transfer coefficients and scaling-up factors are not yet known.

Application of this new type of operation can be expected in adsorption, flue gas cleaning, equilibrium reactions, etc.

List of Symbols

Bo_g Gas phase Bodenstein number $\frac{U_o d(\text{packing})}{\varepsilon D_g}$

Bo_s Solids phase Bodenstein number $\frac{U_s d(\text{packing})}{\delta D_s}$

C reactant concentration kmol/m^3

C_o reactant concentration at reactor inlet kmol/m^3

d_b bubble diameter m

$d_b max$ maximum stable bubble diameter m

d_p particle diameter m

d(packing) diameter of packing element m

D_T bed diameter m

D molecular diffusion coefficient m^2/s

D_g axial dispersion coefficient gas phase m^2/s

D_s axial dispersion coefficient solids phase m^2/s

E solids mixing coefficient m^2/s

E_e elasticity of dense phase N/m^2

f volume fraction of reactor occupied by the gas in dense phase

Ga_p Gallilei number $\frac{\rho_p{}^2 dp^3 g}{\eta^2}$

g accelleration due to gravity m/s^2

H expanded bed height m

H_o settled bed height m

ΔH $= H-H_o$ m

$H_\alpha = \frac{H}{N\alpha} = \frac{U_o}{\alpha}$ height of a transfer unit m

k_m mass transfer coefficient m/s

k first order reaction rate constant (based on bed volume) s^{-1}

k' $= \frac{k}{\rho_b}$ first order reaction rate constant cm^3/gs

$(K_{be})_b$ mass transfer coefficient bubble/emulsion phase (pro unit bubble volume) s^{-1}

$(K_{bc})_b$ mass transfer coefficient bubble/cloud (pro unit bubble volume) s^{-1}

$(K_{ce})_b$ mass transfer coefficient cloud/emulsion (pro unit bubble volume) s^{-1}

$N_\alpha = \frac{\alpha H}{U_o}$ number of mass transfer units

$N_r = \frac{kH}{U_o}$ number of reaction units

$N_E = \frac{UoH}{fE}$ number of dense phase mixing units

N_B number of backmixing units (see (32)).

P pressure N/m^2

Δp pressure drop N/m^2

S specific bubble area m^2/m^3 bed

U_o superficial fluidization velocity m/s

U_{fl} superficial gas velocity at flooding m/s

U_{mf} superficial minimum fluidization velocity m/s

U_{bp} superficial minimum bubbling velocity m/s

U_t terminal particle falling velocity m/s

U_s superficial solids velocity m/s

V_b bubble rising velocity m/s

α mass transfer coefficient (pro unit bed volume) s^{-1}

α_{loc} local value of α

β angle of tilted bed (Fig. 6)

γ fraction of solids supported by the gas phase

δ solids hold-up m^3 particles/m^3 column

ε_{bp} bed porosity at bubble point

ε_{mf} bed porosity at minimum fluidization

ϕ fractional bubble hold-up

ρ_g gas density kg/m^3

ρ_p particle density kg/m^3

ρ_b bed density g/cm^3

η dynamic fluid viscosity Ns/m^2
σ relative standard deviation of residence time distribution

A bar over a symbol means: average value over the bed volume.

Summary

Literature on fluid bed and related reactors is reviewed with special attention to reactor design and scaling-up problems. Advances have been made in predicting and understanding different fluidization regimes, which show quite different properties with regard to application in chemical reactors. Many dense fluid bed reactor models have been proposed, based on bubble properties.

Early bubble models were too much idealized and therefore in conflict with observed facts, especially for large scale reactors. Also the predictive power of these models is limited. In recent models empirical relations are introduced for the three main factors determining the exchange between bubbles and dense phase: bubble size, rising velocity and mass transfer coefficients. As the empirical relations have a limited range of validity, these models can often not replace general two phase models applied in combination with experimentally observed parameters.

New developments in related reactors are fast beds and trickle flow of gas solids suspensions.

Literature cited

(1) Kunii, D. and Levenspiel, O. "Fluidization Engineering", Wiley & Sons, New York(1969).
(2) Davidson, J.F. and Harrison, D. "Fluidized Particles" Cambridge Univ.Press (1963).
(3) Leva, M. "Fluidization" McGraw Hill, New York (1959)
(4) Othmer, D.F. "Fluidization" Reinhold, New York (1956).
(5) Zenz, F.A. and Othmer, D.F. "Fluidization and Fluid Particle Systems", Reinhold New York (1960).
(6) Davidson, J.F. and Harrison, D. "Fluidization" Academic Press, New York (1971).
(7) De Groot, J.H. in Drinkenburg, A.A.H., Proc.Int.Symp. on Fluidization, 348, Neth. Univ.Press (1967).
(8) Geldart, D. Powder Technol. (1973), 7, 285.
(9) Bayers, J. and Geldart, D. in Angelino, H. et al. "Proc. of the Int.Symp. Fluidization and its Appl." 263, Cepadues-editions, Toulouse (1973).
(10) Werther, J. in D.L. Kearns, "Fluidization Technol. I", 215 Hemisphere Publ.Co., Washington (1976).
(11) Werther, J. Chem.Ing.Tek. (1977) 49 (no.11), 777.
(12) Rietema, K. in Drinkenburg, A.A.H., Proc.Int.Symp. on Fluidization, 159, Neth. Univ.Press, Amsterdam (1967).
(13) Morooka, S., Kato, Y. and Miyauchi, T., J.Chem.Eng. Japan (1972), 5 (161).
(14) Van Swaaij, W.P.M., Zuiderweg, F.J. in Angelino H. et al., Proc. of the Internat. Symp. "Fluidization and its Appl." 454, Cepadues-edit. Toulouse (1973).

(15) De Jong, J.A.H., and Nomden, J.F., Powder Technol, (1974) 9, 91.
(16) De Vries, R.J., van Swaaij, W.P.M., Mantovani, C. and Heykoop, A., Proc. 5th Eur. Symp. on Chem.React.Engg. B9-59, Elsevier Publ.Co., Amsterdam, (1972).
(17) Rowe, P.N. 5th Internat. Symp. on Chem.React.Engg., Houston 1978 (to be publ.)
(18) Pigford, R.H. and Baron, T. Ind.Eng.Chem.Fund, (1965) 4, 81.
(19) Jackson, R. Trans.Inst.Chem. Engg. (1963) 41, 13.
(20) Murray, J.D., J.Fluid.Mech. (1965) 21, 465.
(21) Molerus, O., Chem.Ing.Tek. (1967) 39, 341.
(22) Verloop, J. and Heertjes, P.M., Chem.Engg.Sci. (1970) 25, 825.
(23) Oltrogge, R.D., Thesis Univ. of Michigan, 1972.
(24) Mutsers, S.M.P. and Rietema, K., Powder Technology (1977) 18, 239.
(25) Rietema, K. in Brauer, H. et al., Preprints European Congress, "Transfer Processes in Particle Systems", Nuremberg E12, Dessertationsdruck GmbH & Co. K.G. Schadel, Bamberg FRG (1977).
(26) Mutsers, S.M.P. and Rietema K., Powder Technology (1977) 18, 249.
(27) Agbin, J.A., Nienow, A. and Rowe, P.N., Chem.Engg.Sci. (1971) 26, 1293.
(28) Donsi, G. and Massimila, L. in Angelino, H. et al., Proc. of the Internat.Symp. "Fluidization and its Appl." 41, Cepadues-editions, Toulouse, 1973.
(29) Van Swaaij, W.P.M. and Zuiderweg, F.J., Proc.Int. Symp. on Chem.React.Engg. C1-3, Elsevier Publ.Co., Amsterdam (1972).
(30) Horio, M. and Wen, C.Y., "Fluidization Theories and Applications" A.I.Ch.E. Symp.series (1977) 73 (161), 9.
(31) Shen, C.Y. and Johnstone, H.F., A.I.Ch.E.J. (1955) 1, 349.
(32) Van Deemter, J.J., Chem. Engg.Sci. (1961) 13, 143.
(33) Van Deemter, J.J. in Drinkenburg, A.A.H., Proc.Int. Symp. on Fluid.,334, Neth. Univ.Press, Amsterdam (1967).
(34) Johnstone, H.F., Batchelor, J.D. and Shen, C.Y. A.I.Ch.E.J. (1955) 1, 318.
(35) May, W.G., Chem.Eng.Prog. (1959) 55 (12) 49.
(36) Kobayashi, H., Arai, F., Chem.Eng. Tokyo (1967) 31 239.
(37) Orcutt, J.C., Davidson, J.F. and Pigford, R.L., Chem.Engg.Progr.Symp.Series no.38 (1962) 58, 1
(38) Fryer, C. and Potter, O.E. Ind.Eng.Chem.Fund. (1972) 11, 338.
(39) Mannuro, T. and Munchi, J. J.Ind.Chem. Tokyo (1965) 68, 126.
(40) Toor, F.D. and Calderbank, P.H. in Drinkenburg, A.A. H., Proc.Internat.Symp. on Fluidization, 373, Neth. Univ.Press, Amsterdam(1967)
(41) Patridge, B.A. and Rowe, P.N., Trans.Inst.Chem. Engrs. (1966) 44 T349.
(42) Kobayashi, H., Arai, F., Chiba, T. and Tamaka, Y. Chem.Eng.Tokyo (1969) 33 274.
(43) Kato, K. and Wen, C.Y., Chem.Eng.Sci.(1969) 24, 1351.
(44) Mori, S. and Munchi, J., J.Chem.Eng.Japan (1972) 5 251.
(45) Mori, S. and Wen, C.Y. in Keairns, D.L. "Fluidization Technology I", 179 Hemisphere Publ. Co. Washington (1976).
(46) Chavarie, C. and Grace, J. R., Ind.Eng.Chem.Fund. (1975) 14 75,79 and 86.
(47) Fryer, C. and Potter, O.E. in Keairns, D.L. "Fluidization Technology I", 171, Hemisphere Publ.Co., Washington (1976).
(48) Raghuraman, J. and Potter, O.E., 5th Internat.Symp. on Chem.React.Engg., Houston (1978), to be published.
(49) Van Swaaij, W.P.M. and Zuiderweg, F.J., Proc. 5th Europ.Symp. on Chem.React. Engg, B9-25, Elsevier Publ. Co., Amsterdam (1972).
(50) Botton, R.J. Chem.Eng.Progr. Symp. series (1968) 66,101.
(51) Furusaki, S., A.I.Ch.E.J. (1973) 19, 1009.
(52) Gilleland, E.R. and Knudsen, C.W., Chem.Eng.Progr.Symp. Ser. (1971) 67, 116, 168.

(53) Miyauchi, T. and Morooka, S. Kagaku Kogaku (1969) 33,880.
(54) Miyauchi, T. and Furusaki, S. A.I.Ch.E.J.(1974) 20,1087.
(55) Davidson, J.F., Paper presented for Chisa, Prague (1975).
(56) Drinkenburg, A.H.H., Thesis Eindhoven (1970).
(57) Calderbank, P.H., Pereisa, J. and Burgers, J.M. in Keairns, D.L. "Fluidization Technology", 115 Hemisphere Publ.Co. Washington (1976).
(58) Mori, S. and Wen, C.Y., A.I.Ch.E.J., (1975) 21, 109.
(59) Rowe, P.N.,Chem.Eng.Sci., (1976) 31, 285.
(60) Darton, R.C., Lallauze, R.D. Davidson, J.F., Harrison, D., Trans.Inst.Chem.Engrs. (1977) 55 (4) 274.
(61) Matsen, J.M., A.I.Ch.E.Symp. Ser. (1973)69(128), 30.
(62) Chiba, T. and Kobayashi, H. Chem.Eng.Sci.(1970)25,1375.
(63) Drinkenburg, A.A.H., and Rietema, K., Chem.Eng.Sci. (1973) 28, 259.
(64) Fontaine, R.W. and Harriot, P., Chem.Eng.Sci. (1972) 27 2189.
(65) Kobayashi, H., Arai, F. and Sumagawa, T., Chem.Engg. Tokyo (1967) 31, 239.
(66) Toei, R., Matsuno, R., Mijakana, H., Nishiya, K. and Komagawa, K., Chem.Engg. Tokyo, (1968) 32, 565.
(67) Hoebink, J. and Rietema, K. in Brauer, H.L. et al. "Preprints Europ.Congr.Transfer Processes in Particle Systems" Nuremburg, K51, K.G. Schadel Bamberg, F.R.G. 1977.
(68) Schlünder, E.U., Chem.Eng. Sci. (1977) 32, 845.
(69) Avedesian, M.M., Davidson, J.F., Trans.Inst.Chem.Eng. (1973) 51, 121.
(70) Yerushalmi, J., Turner, D.H. and Squires, A.M., Ind.Eng. Chem.Process.Des.Dev.(1976) 15, 47.
(71) Stewart, P.S.B. and Davidson, J.F., Powder Technol. (1967) 1, 61.
(72) Yousfi, Y. and Gau, G., Chem.Eng.Sci.(1974) 29, 1939.
(73) Yang, W.C. "A criterion for Fast Fluidization", Proc. of Pneumotransport 3, BHRA Fluid Engg., 1976.
(74) Leung, L.S. and Wiles, R.J. Ind.Eng.Chem.Process Des. Dev.(1976) 15, 552.
(75) Yousfi, Y., Gau, G., and Le Goff, P. in Angelino, H. Proc.Int.Symp."Fluidisation and its Applic." 571 Capadues edit.Toulouse(1973).
(76) Van Swaaij, W.P.M., Buurman, C. and van Breugel, J.W., Chem.Eng.Sci.(1970) 25,1818.
(77) Kiang, K.D., Nack, K.T., Oseley, J.H. in Keairns, D. L., "Fluidization Technology II", 471 Hemisphere Publ.Co. Washington, 1976.
(78) Stemerding, S. Chem.Eng.Sci. (1962) 17, 559.
(79) Yousfi, Y., Thesis University of Nancy (1973).
(80) Capes, C.E. and Nakamura, K., Can.J. of Chem.Eng. (1973) 51, 31.
(81) Nakamura, K. and Capes, C.E. Can.J. of Chem.Eng. (1973) 51, 39.
(82) Varma, Y.B.G., Powder Techn. (1975) 12, 167.
(83) Claus, G., Vergnes, F. and Le Goff, P. Can.J.Chem.Eng. (1976) 54, 143.
(84) Roes, A.W.M. and Van Swaaij, W.P.M., Chem.E.J. (to be published).
(85) Roes, A.W.M. and Van Swaaij, W.P.M., Chem.E.J. (to be published).

RECEIVED March 30, 1978

7

Gas-Liquid Reactors

J. C. CHARPENTIER

Laboratoire des Sciences du Génie Chimique CNRS-ENSIC 1, rue Grandville—54042 Nancy Cedex—France

1. Introduction

Gas-liquid reactions and absorptions widely used in four main fields of the chemical and parachemical industry are

. liquid-phase processes : oxidation, hydrogenation, sulfonation, nitration, halogenation, alkylation, sulfation, polycondensation, polymerization.

. gas scrubbing : CO_2, H_2S, CO, SO_2, NO, NO_2, N_xOy, HF, SiF_4, HCl, CL_2, P_2O_5, hydrocarbons etc (very often to combat air pollution).

. manufacture of pure products : H_2SO_4, HNO_3, $BaCO_3$, $BaCl_2$, adipic acid, nitrates, phosphates, etc.

. biological systems : fermentation, oxidation of sludges, production of proteins from hydrocarbons, biological oxidations.

At the heart of these processes is the absorber or the reactor, of a particular configuration best suited to the chemical absorption or reaction being carried out. Its selection, design, sizing, and performance depend on the hydrodynamics and axial dispersion, mass and heat transfer, and reaction kinetics.

Two books deal almost exclusively with the subject of mass transfer with chemical reaction, the admirably clear expositions of Astarita (1) and Danckwerts (2). Since then a flood of theoretical and experimental work has been reported on gas absorption and related separations. The principal object of this review is to present some techniques, results, and opinions published mainly during the three or four last years on mass transfer and hydrodynamics that I consider as the more salient concerning as well the background as the experiments.

This necessitates some theoretical presentation of mass transfer with and without chemical reaction that will be strictly focused in the first part of this review to the case of desorption or when a rise in temperature is accompanying and is great enough to affect the rate of gas absorption. Indeed many general papers and also many papers confined to specific reaction scheme (simultane-

0-8412-0432-2/78/47-072-223$09.75/0

ous absorption of several gases, consecutive or parallel reactions in the liquid phases... with the simultaneous and sophisticated mathematical techniques (very often reduced to a mathematical problem rather than a gas-liquid problem) have been published for the isothermal absorption and it seems now possible to tackle such cases with the help of the last decade literature. The design of gas-liquid equipment is determined by two major considerations : the distribution of components between phases in a state of thermodynamic equilibrium (solubility of gases in liquid) and the rate at which mass transfer occurs under prevailing conditions (liquid diffusivity and chemical reaction). Therefore physico-chemical parameters as well as interfacial parameters for the specified hydrodynamics working regime are practically always required. Obtaining such data is often a challenging problem, so wide is the range of solutes and solvent the chemical engineer or researcher may encounter. Although theory on solubility and diffusivity has not yet progressed to the point where purely theoretical predictions are possible, many strides have been made in that direction quite recently and will be commented in the second part of this review. And naturally the third part will be devoted to the interfacial area and mass transfer coefficient for the accompanying hydrodynamics. The techniques of measurement will be commented and some data and correlations will be presented for the case of trickle-bed reactors, and well stirred tank. Finally one must always keep in mind that the relationships of the physico-chemical or interfacial parameters must be only used in the absence of the experimental data, and after careful study of the conditions of validity. So the last part of the review will be devoted to the simulation of the process of absorption in industrial absorbers by laboratory apparatus with a well defined and known interfacial parameters in which experiments are carried out to obtain information that could be directly applied to design.

2. Mass transfer with and without chemical reaction : desorption and exothermic absorption

Although the situation in gas-agitated liquid reactor is a rather complicated one as often diffusion, convection and reaction proceed simultaneously and the nature of the convective movements of both gas and liquid are ill-defined, however several useful predictions have been performed to describe the behavior of such high complicated system in using highly-simplified models which simulate the situation sufficiently well for practical purposes without introducing a large number of parameters : the film model and the surface-renewal models where the hydrodynamic properties are accounted for by a single parameter either the film thickness δ, or the contact time θ or the surface renewal frequency S. The question of which of these three models is more true to life is less important than the question of which one is found in practice to lead to accurate predictions as it appears that in many circumstances the difference between predictions mode on the basis of the three models will be less than the uncertainties about the values of the physical quantities, such as diffusivities or solubilities, used in the calculation : in fact the three models can be regarded as interchangeable for many purposes and it is then merely a question of convenience which of the three is used.

21 . Desorption with or without chemical reaction

Though the subject matter of absorption of gases in liquid, without or with chemical reaction, has received considerable attention, however there were little available literature and general presentation on desorption of volatile substances until the welcome and excellent review on the mechanism of the various types of desorption processes published in 1976 by Shah and Sharma (3). The authors reviewed, analysed and illustrated the process of desorption, without or with chemical reaction, with the help of film and penetration theories in the following cases (a) desorption without reaction (b) desorption preceded or accompanied by a liquid phase reaction (c) desorption accompanied by a pseudo first order gas-liquid reaction (d) desorption accompanied by a zero order gas-liquid reaction (e) desorption accompanied by a pseudo first order consecutive gas-liquid reaction (f) desorption of a volatile reactant and (g) some aspects of process design of desorption columns with or without chemical reactions.

Just as in the case of absorption accompanied by a reaction of a gas in liquid, desorption accompanied by a gas-liquid reaction |A + zC (non volatile) → B (volatile)| may conform to various mechanisms. Results have been summarized and the concentration profiles have been schematically depicted on each side of the interfacial area (x = 0) by Shah and Sharma (3) in Figure 1 in the case in which a gas A dissolves into the liquid phase and then reacts irreversibly with a non-volatile species C that is present in the liquid phase to give the volatile species B according to the reaction

$$A + Z_1C \xrightarrow{k_A} Z_2B$$

The volatile species B is also capable of reacting with reactant C according to the reaction

$$B + Z_3C \xrightarrow{k_B} \text{Products}$$

In Figure 1 the second reaction is considered to be slow and hence is unlikely to occur in the film (i.e. the diffusion-reac-

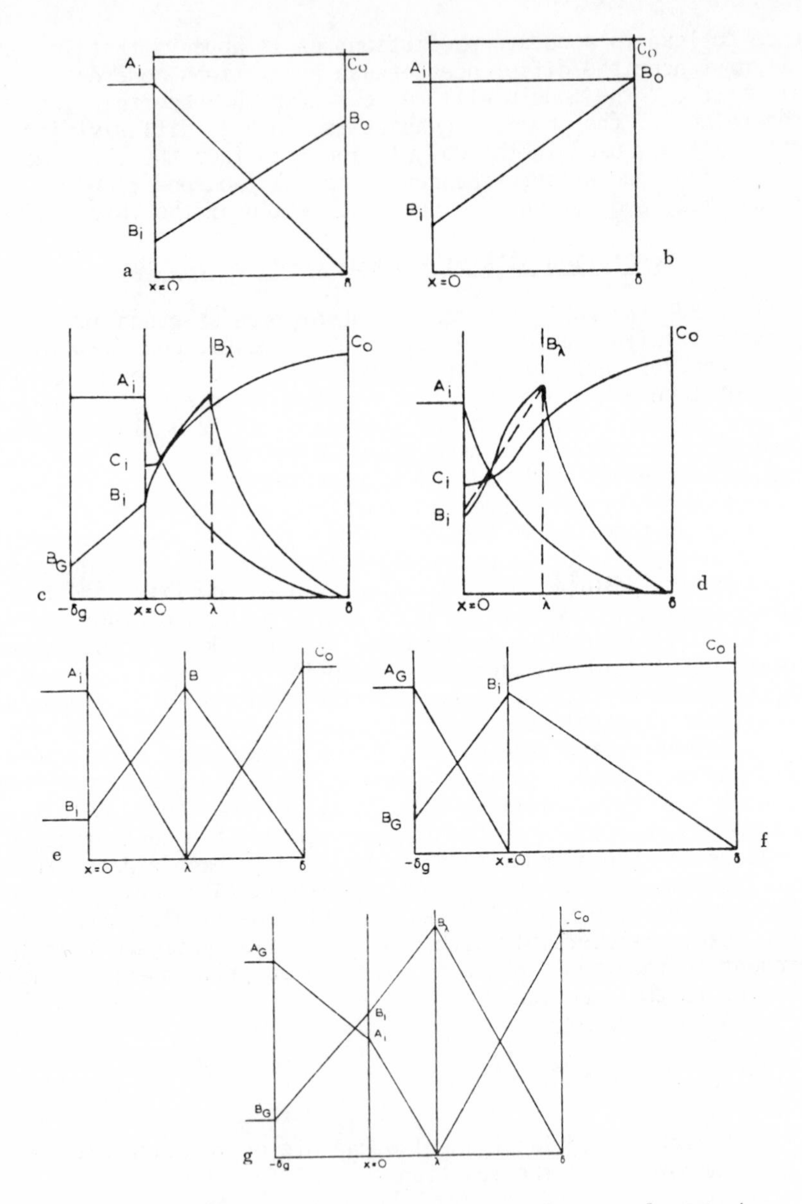

Figure 1. Typical concentration profiles for simultaneous absorption/desorption. Typical concentration profiles for instantaneous reaction between the gas A and reactant C—film theory.

(a) Diffusion controlled–slow reaction; (b) kinetically controlled–slow reaction; (c) gas film controlled desorption–fast reaction; (d) liquid film controlled desorption–fast reaction; (e) liquid film controlled absorption–instantaneous reaction between A and C; (f) gas film controlled absorption–instantaneous reaction between A and C; (g) concentration profiles for A, B, and C: instantaneous reaction between A and C—both gas and liquid phase resistances are comparable.

tion parameter or Hatta number $Ha_B = \sqrt{D_B k_C C_o}/k_L << 1$).

If now both reactions are important (liquid phase oxidation of hydrocarbons by O_2, chlorination of liquid isobutylene, liquid phase conversion of absorbed O_2 and C_2H_4 to acetaldehyde...) the reaction between B and C is almost complete in the film ($Ha_B > 3$) and desorption of species B will occur only if the reaction between B and C is not instantaneous. For simplicity Shah and Sharma (3) have analyzed the case where $Z_1 = Z_2 = Z_3 = 1$ and both of the reactions are pseudo first order with respect to species A and B ($k_A' = k_A C_o$ and $k_B' = k_B C_o$ being the pseudo first order constants). When the gas phase resistance is not important and depending on the values of the ratio $K = k_A/k_B$, species B will not desorb for $K \leq 1$ or will eventually desorb, for $K > 1$ if

$$\frac{K^{1.5}}{K-1} \frac{1}{\tanh(\sqrt{K Ha_B})} > \left(\frac{B_i}{A_i} + \frac{K}{K-1}\right) \frac{1}{\tanh(Ha_B)}$$

22. Supersaturation in gas-liquid reactions involving a volatile product

In many practical cases physical desorption and absorption processes occur simultaneously and then the diffusion film in the liquid phase may supersaturate with gases resulting in the bubble formation and subsequent interfacial turbulence (simultaneous absorption of O_2 and N_2 into water and desorption of C_3H_6). Also, as seen above, absorption accompanied by chemical reaction often results in the formation of a volatile product that must diffuse back into the gas phase to avoid supersaturation of the bulk liquid during the absorption process. This results in a counter diffusion phenomenon which can cause supersaturation in the diffusional boundary layer. If the supersaturation were followed by bubble nucleation, absorption-desorption rates are enhanced due to a reduction in diffusion film and occurence of turbulence in the liquid. The film and penetration theory equations are not applicable to calculation of mass transfer rates under such supersaturation conditions. However a knowledge of the conditions causing the supersaturation may be useful whatever supersaturation may be desirable or not. In 1974 and 1976 Shah, Sharma and coworkers (5, 6) extended the analysis of Brian et al. (4) for the simple physical absorption-desorption process to gas-liquid reactions involving a volatile product and have evaluated the criteria for supersaturation in the following cases of gas-liquid reactions, (a) absorption of a gas A accompanied by an irreversible non-zeroth order liquid phase reaction generating a volatile product, (b) absorption of a gas A accompanied by a zeroth order liquid phase reaction, (c) simultaneous absorption of two gases A and B accompanied by the liquid phase reaction of $A_{(g)} \xrightarrow{k} B_{(g)}$, (d) simultaneous absorption of two gases A and B accompanied by the liquid phase reactions $A_{(g)} \xrightarrow{k_1} B_{(g)}$ and $B_{(g)} \xrightarrow{k_1'}$ non-volatile product, (e) simultaneous absorption of two gases A and E, $A_{(g)} \xrightarrow{k_1} C_{(g)}$ and $E \xrightarrow{k'}$ non-volatile product, (f) absorption of a gas A accompanied by an instantaneous reaction generating a volatile product that reacts also to generate a non-volatile product, (g) simultaneous absorption-desorption accompanied by a second-order irreversible reaction, (h) simultaneous absorption of two reacting gases generating a volatile product, (i) absorption of a gas A accompanied by a zeroth, first or second order liquid phase reaction generating a volatile product B in the presence of gas phase resistance for both A and B. The authors presented diagrams for the critical supersaturation in the plane $(p_{Bo}-p_{BG})/p_{AG}$ versus Ha taking into account the gas phase resistance of gas A and/or the volatile product P (p_{AG} and p_{BG} are the partial pressure of A and B in gas and p_{Bo} is the partial pressure of B in equilibrium with concentration of B in bulk liquid). In these diagrams, regions were located where no supersaturation is possible or supersaturation occurs either in liquid film or in bulk liquid.

Shah et al. (6) also evaluated the situations where the temperature effects at the gas-liquid interface occurs and obtained the supersaturation criteria for non-isothermal physical absorption-desorption processes when the temperature effects are small but non negligible. The results indicate that the necessary condition is, in general, different from that for the isothermal case and involves the ratio of the solubility coefficients of the diffusing species.

It is clear from these studies that supersaturation during simultaneous absorption-desorption is possible in any physical or chemical system as long as the diffusivity of the desorbing species is less than the one for the absorbing species and the concentration distribution of at least one of the diffusion species is non linear which is usually found in practice. Experimental investigations such as those presented in ref. (7, 24) are necessitated now to test and to complete the theoretical work of these authors especially when gas phase resistance and thermal effects are important. Let us see now how this has been recently undertaken for the case of non-isothermal gas absorption with chemical reaction.

23. Steady state multiplicity of adiabatic gas-liquid reactors

Many important gas-liquid reactions such as oxidation, hydrogenation, nitration, sulfonation and chlorination are carried out in adiabatic continuous stirred-tank reactor over a wide range of temperature in which a continuous shift from chemical to mass transfer control can happen. The interactions between the solubility, the diffusional resistances and the chemical reaction may cause the occurence of sustained periodic oscillations and steady state multiplicity as observed experimentally by Hancock and Kenney (8) and Ding et al. (9). This problem of steady state multiplicity of a gas-liquid CSTR has only received a significant attention these 3 or 4 last years with the remarkable theoretical approaches of Hoffman, Sharma and D. Luss (10, 12) continued and completed by Raghuram and Shah (11).

The first development presented by Hofman et al. (10) was concerning a single irreversible second order gas-liquid reaction, A(gas) + B(liq) → R(liq) + L(gas) with the following assumptions : (a) the liquid phase consists of non-volatile components and no evaporation into the gas phase is occuring, i.e. the temperature is well below the boiling point of the liquid, (b) the physical properties of the gas and liquid, the interfacial parameters k_L and a, and the volumetric liquid flowrate q_l leaving the reactor and diffusivities of all reacting species are independant of temperature and conversion and also the diffusivities are equal, (c) the heat of solution $(-\Delta H_S)$ and reaction $(-\Delta H_R)$ are independent of temperature, (d) the solubility of the species A follows the Henry's law, is independent of conversion and its temperature dependance is expressed by $\log(x_A/y_A P) = (D_1/T) - D_2$, (d) the gas and the liquid are at the same temperature within the reactor, (e) the total pressure of the gas bubbles and the gas holdup are constant and independent of temperature, conversion and position within the reactor, (e) the liquid feed contains no dissolved gaseous reactant and the gas phase resistance is negligible.

Thus the material and energy balances are,

$$q_{gf}y_{Af} - q_{go}y_A = F_1 C_{Al} + r_A V_R = E_A k_L a(C_{Ai} - C_{Al})V_R = E_A^* k_L a C_{Ai} V_R$$

$$q_{lf}x_{Bf} - q_{lo}x_B = r_B V_R$$

$$r_A V_R(-\Delta H_R) + F_1 C_{Al}(-\Delta H_S) + (q_{lf}C_{pl} + q_{gf}C_{pg})T_f - (q_{lo}C_{pl} + q_{go}C_{pg})T_o = 0$$

where q_g and q_l are the liquid and gas molar flowrates, F_l is the constant liquid volumetric flowrate, C_{Al} is the bulk concentration of the dissolved gas, C_{pl} and C_{pg} are the heat capacities of the pure liquid and gas and the subscripts f and o refer to the feed and effluent streams. E_A is the enhancement factor and E_A^* is the reaction factor ($E_A^* = E_A$ for $C_{Al} = 0$ i.e. the reaction occurs in the liquid film and not in the bulk). The reaction factor was evaluated in using the Van Krevelen-Hoftijzer approximation (E^* versus Ha) modified by Teramoto (13) and computed by an iterative solution of three equations involving $\theta = k_L a\tau/\varepsilon$, $\alpha = \varepsilon/a\delta$ (τ and ε are the liquid residence time and holdup), and the instantaneous enhancement factor E_i. It was found by Hoffman et al. (10) that the deviations between the exact values of E_A^* and those computed by this approximation technique were always within 4 % whatever E_A^* is smaller or larger than unity.

The steady state solutions are classically determined from the intersection of the heat removal Q_1 and heat generation Q_{II} curves as a function of the temperature T (Figure 2) that may be presented from the mass and energy balances in the shape

$$Q_I \equiv T - \frac{q_{1f}C_{p1}T_{1f}+q_{gf}C_{pg}T_{gf}}{q_{1f}C_{p1}+q_{gf}C_{pg}} = T-T^* = \frac{(-\Delta H_R)r_AV_R+F_1C_{A1}(-\Delta H_S)}{q_{1f}C_{p1}+q_{gf}C_{pg}} \equiv Q_{II}$$

and the necessary condition for the stability of any steady state is $dQ_I/dT > dQ_{II}/dT$. A numerical example was used to investigate the behavioral features of the model and its sensitivity to the values of various parameters. These values were selected such as to be similar to those representating the chlorination of n-decane which is a typical gas-liquid reaction as discussed by Ding et al. (9). The values and the properties of the feed streams are presented in Figure 2 where $q_{gf} = q_{1f}$ and when the influence of a certain parameter was investigated, the values of all the other parameters were set equal to those of the base case labeled B.

When a single second-order homogeneous chemical reaction is carried out in a CSTR, the heat generation curve has a sigmoïdal shape such that no more than three steady state solutions exist. However, when such a reaction is carried out in a gas liquid CSTR the shape of the heat generation curve Q_{II} is such that under certain conditions five steady state solutions exist. This unique feature of a gas liquid reacting system is shown in Figure 2a which describes the effect of residence time distribution on the shape of the Q_{II} graph. It is seen that for this example a unique steady state exists for residence times of either 2 or 8 min., three steady states for $\tau = 4$ min., and five steady states for $\tau = 6$ min. This special shape of the heat generation curve is due to the conflicting influences of the temperature on the reaction factor and on the solubility of the gas.

Moreover the effect of changes in the rate constant k on the steady state temperature is shown in Figure 2b where the dashed lines represent the unstable steady state ($dQ_I/dT < dQ_{II}/dT$). The base case labeled B (k = 0.05 cc/mol.s) has five steady states for liquid residence time τ comprised between 5 and 7 minutes. Due to the special shape of the Q_{II} curve, an increase in the residence time ignites the low temperature steady state and shifts the reactor to a high temperature steady state. However, when the residence time is decreased, the extinction occurs by a shift of the high temperature steady state to the intermediate temperature branch. Upon a further decrease of the residence time, a second extinction occurs due to a shift from the intermediate to the low temperature branch. The graphs indicate that the intermediate temperature steady state solutions are most sensitive to variations in the rate constants. The low temperature solutions are rather insensitive to these changes. In general, a decrease of the reaction rate constant causes an increase in the ignition as well

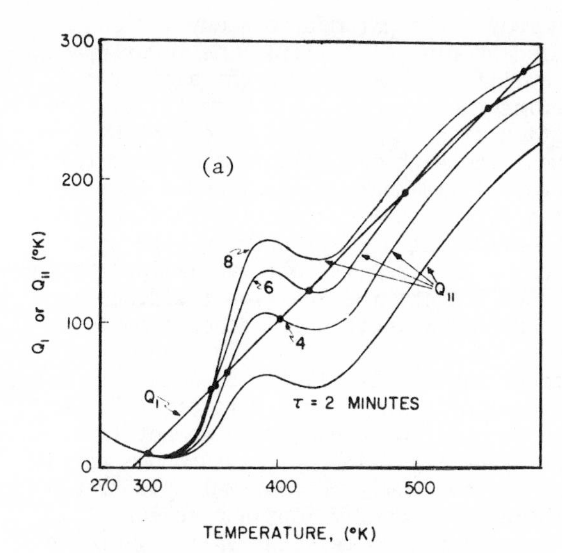

Q_I and Q_{II} as a function of temperature and liquid residence time.

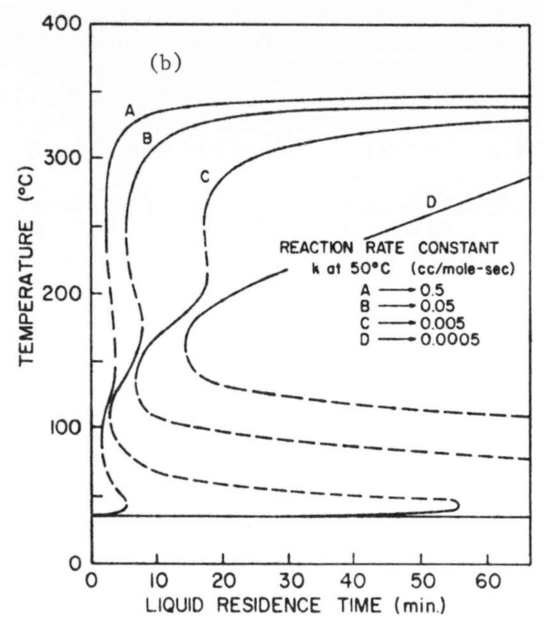

The influence of the reaction rate constant and the liquid residence time on the steady state temperature.

$E_1 = 30000$ cal/mole
$k_l = 0.04$ cm/s
$a_v = 2.0$ cm^{-1}
$C_{pl} = 70.0$ cal/mole-°K
$-\Delta H_s = 5000$ cal/mole

$k_1 = 0.05$ cc/mole-s at (50°C)
$x_A = y_A\, 10^{(1010/T - 4.16)}$
$D_A = 6 \times 10^{-5}$ cm^2/s
$C_{pg} = 6.5$ cal/mole-°K
$-\Delta H_R = 25000$ cal/mole
$\epsilon = 0.9$
$T_{gf} = T_{lf} = 25$°C
$y_{Af} = x_{Bf} = 1.0$
$\rho_{lf} = 0.725$ g/cc
$\rho_{gf} = 0.00298$ g/cc
$M_{wB} = 142$ g/g mole
$M_{wA} = 71$ g/g mole

Figure 2. The chlorination of n-*decane (after Ref.* 10)

as the extinction time. Moreover the parametric study indicates that for the example studied by Ding et al. (9), the lower branch was most sensitive to solubility of the gas and the activation energy of the system. The intermediate branch was most influenced by variations in the reaction rate and to a lesser extent by changes in the interfacial area. The high temperature steady states were sensitive mainly to the liquid heat capacity and heat of the reaction. All steady states were rather insensitive to changes in the liquid mass transfer coefficient and diffusivity. The upper temperature branch was usually more sensitive to variations in the parameters than was the lower branch. The ignition residence time was much more sensitive than the extinction residence time to variations in the parameters.

This study of the steady state multiplicity of adiabatic gas liquid reactors was then carried in the same laboratory (12) in developing a model for the case of two consecutive competitive reactions with second order kinetics, with the aims to demonstrate the possible existence and the maximum number of multiple steady states with different temperatures, conversions and selectivities under the same operating conditions, to test the validity of the model and the associated assumptions by comparing it with experimental results obtained as previously with the chlorination of n-decane and to determine the parametric sensitivity of the system which is essential for the rational design, operation and control of these gas liquid reactors. The model developed by Sharma et al. (12) uses again expressions which describe the rate of reaction in both the film and bulk liquid and account for the continuous shift from chemical to mass transfer control with increasing temperatures.

Numerical simulations reveales that changes in certain parameters may change both the nature of the multiplicity as well as the range of liquid residence time for which multiplicity exists. Such changes in reactivity of the n-decane can cause the appearance or the disappearance of a special multiplicity pattern (an isola) for which a shift from the low to the high temperature steady state cannot be attained by increasing the residence time. Thus the high temperature steady state can be attained only by preheating the reaction mixture. The presence of an isola thus may not be discovered in preliminary experiments and its existence may lead to rather unexpected pitfalls in the operation of the reactor. Note also that the higher the steady state temperature, the higher is the conversion but the lower is the selectivity. So an intermediate steady state may be interesting for selectivity. The key parameter which influence the multiplicity pattern and region are k, a, $(-\Delta H_R)$, C_p and the heat loss from the reactor. The model predicts the occurence of up to seven steady state solutions under certain operating conditions. These different steady states correspond to great changes of Ha and E^* and are due to a shift from chemical reaction to mass transfer control.

The fundamental work of Hoffman, Sharma and Luss has been

carried on in 1977 by Raghuram and Shah (11) that published criteria for uniqueness and possible multiplicity of the steady states which were not explicitly stated by the previous authors because of algebraic complexity. Finally the results of these three studies suggest that the information about the multiplicity pattern and the ignition and extinction points can serve as a remarkable tool for discriminating among rival models and assumptions and for estimation of kinetics parameters of these models which necessitates also the knowledge of physical chemical and interfacial parameters. These results are applicable to a large number of practical situations and should find extensive practical use. They need now complementary development for the situation at very high temperature close to or exceeding the boiling point of the liquid where evaporation into the gas phase takes place.

24. Thermal effects in gas absorption without and with chemical reaction

Non isothermal phenomena are present to some extent in all gas absorption operations. As seen above the importance of the non isothermal effects arises from their influence on the mass transfer rate and, in the case of chemical absorption, from their influence on reaction rates and selectivities. This means that to neglect the boundary condition coupling between solubility and interface temperature may lead to serious overestimates of the absorption rates and to model the non isothermal absorption process, it is necessary to consider the mass and energy balance equations simultaneously. Attention has been considerably focused both experimentally and theoretically on that important topic particularly during these 3 or 4 last years (14 - 23).

241. In absence of chemical reaction, Verma and Delancey (17) have studied the thermal effects in gas absorption with the simultaneous measurement of interfacial mass transfer rates and liquid temperature profiles over a range of physical conditions within which most gas absorption systems may be expected to fall (ammonia-water and propane-decane systems). The experiments were performed with a quiescent liquid phase (no momentum equation included) and the surface renewal theory was estimated to provide a vehicle for extending the results to the hydrodynamic conditions normally encountered in industrial absorbers. An analytical solution of the simultaneous energy and mass conservation equations in the volume average reference frame is proposed which applies to a semi-infinite liquid medium and includes the effects of heat of absorption, variable density, arbitrary condition at the interface and volume changes in the liquid phase in absence of solvent evaporation. It is able to predict interfacial mass flows and liquid temperature profiles. Comparisons with the experimental data from the ammonia-water systems (at 0.2, 0.467, 0.733 and 1 atm) spans surface temperature rise from 6.9 K to 18.2 K. The propane-decane

system at 1 atm exhibits a temperature rise of 1.3 K. In all cases the model adequatly predicts the temperature profiles.

242. For gas absorption and reaction which is so exothermic that also the assumption of constant interfacial temperature does not hold, Mann and Moyes (22) have recently published an extension of the film theory to predict analytically the enhancement factor E and surface temperature increase under slow and fast reaction conditions using a linear solubility/temperature relationship. Severe reductions in E values in conjunction with large surface temperature increases up to 100 K are predicted using data applicable to the direct sulfonation of dodecylbenzene with SO_3. This is done assuming that the film theory is applicable and temperature and concentration gradients exist across the film. As the physical reality is such that the thermal diffusivity α is much bigger than the diffusivity of the absorbing component (commonly α = 100 D) thus the film thickness x_H over which temperature gradients exist is very much larger than the film thickness $\delta = x_M$ over which concentration gradients are observed. This means that the temperature may be considered practically constant and equal to the interfacial temperature T^* across the mass transfer film. Thus if the effect of temperature on specific heat, liquid phase density and diffusivity as well as the Duffour and Soret effects taken into account by Verma and Delancey (17) are considered to be secondary in importance, the equations describing the diffusion and first order reaction and the heat transfer are

$$D_A \frac{d^2C}{dx^2} = k(T^*)\ C \quad \text{and} \quad \alpha \frac{d^2T}{dx^2} = -\frac{\Delta H_R}{C_p} k(T)\ C$$

subject to the boundary conditions

$$C = C^*(T^*) \text{ and } T = T^* \text{ at } x = 0$$

$$C = 0 \quad \text{and } T = T^* \text{ at } x = x_M$$

$$C = 0 \quad \text{and } T = T_b \text{ at } x = x_H$$

T^* is a lumped representation for the spatially dependent T(x) and the rate constant to use in the diffusion reaction balance is reasonably evaluated at the interfacial temperature T^*. This interface temperature value is governed by the heat release due solution and reaction that Mann and Moyes consider to be interfacial heat fluxes within the frame describing heat transfer. The result is that T(x) must be linear over the heat transfer film x_H while being approximatively constant over the mass transfer film x_M.

The solutions of the above equations using the employment of a linear relationship for the solubility C^* lead to explicit analytical forms for T^* and E under no or slow reaction conditions and under very fast reaction conditions. Finally the expression for the ratio of the enhancement factor at large Ha to the initial value for solely physical absorption is given by

$$\frac{E\ (\text{fast reaction})}{E\ (\text{physical absorption})} = 1 + \frac{1 - \mu_S\ \Delta H_R\ (k_L/h_L)}{\mu_S(\Delta H_R + \Delta H_S)(k_L/h_L)}$$

μ_S is the proportionality constant in the linear solubility relationship. It is seen that the advent of reaction can result in an ultimate increase or decrease from the initial (already depressed) value, depending upon whether $\mu_S \Delta H_R (k_L/h_L)$ is greater or less than unity : for large values of μ_S appropriate to high solubility systems and for large heat of reaction, the reaction can be expected to decrease the absorption potential and to hinder the absorption process. Finally the confines of the overall behavior under slow reaction conditions and beyond the fast reaction regime can be deduced analytically and explicitly and intermediate ranges of reaction rate can be solved for implicity by using the appropriate expression for $(dC/dx)_{x=x_M}$ in the boundary condition of the overall heat balance.

This theory is illustrated by Mann et Moyes with the presentation of the graphs surface temperature increase and E versus Ha for the high solubility system SO_3-trioxidedodecylbenzene. Large increases in interface temperature (up to 100K) and enhancement factors below unity (down to 0.3) are predicted for both linear approximations or theoretical ideal solubility. Moreover to compare theory and experiment, some experimental results obtained from the laminar jet technique with exposure time appropriate to bubble and packed bed reactors show that high interface temperatures can be reached and that reductions in absorption rate below that for physical absorption can occur. An important feature of the work published by Mann et Moyes (22) is that if applying the exact solubility relationship produces a closer correspondance between the experimental observations and the film theory predictions, however the linearized approach is seen to be appropriate for preliminary evaluation of the likely magnitude of thermal effects from the relevant basic physico-chemical data.

The theory and experimental findings emphasize the need for further experimental studies particularly concerning the effet of high surface temperature on by-product reactions (as explained by the authors) and for complementary theoretical work where interfacial turbulence (18, 19) or gas phase resistance (evaporation effect) are taken into account. This could be undertaken with the help of the formulation developed by Tamir et al. (20, 21).

243. Absorption with chemical reaction in the presence of heat generation, bulk flow and effects of the gaseous environment. A penetration model for gas-absorption into a laminar liquid stream in the presence of a pseudo first order chemical reaction and large heat generation was investigated theoretically by Tamir

et al. (20) accounting for the presence of the gaseous environment, non considered in the previous analysis, as well as the transverse bulk flow contribution to the total mass transfer. In this model, heat of solution is liberated at the interface while the heat of reaction is liberated within the absorbing stream. As a result of the vapor pressure exerted by the solvent, the gaseous atmosphere becomes a binary mixture where simultaneous evaporation of the solvent and absorption of the soluble gas takes place. The complete formulation is based upon the conservation of mass and energy in both phases with the appropriate matching conditions at the gas-liquid interface. Indeed the absorbed gas flux towards the interface is the sum of the fluxes due to diffusion and the bulk flow acting in the same direction and for the evaporating solvent the above flow contributions are in opposite directions, where the diffusion takes place from the interface towards the bulk. This leads to the interface matching conditions for mass conservation and for energy conservation. In the interface condition for energy conservation, there are four terms : the first term designates the condition of heat into the absorption streams while the second one takes care of heat transfer by conduction to the atmosphere (it is generally negligible). The third term accounts for evaporation of the solvent and the last term includes the effect of heat liberation due to dissolution. The originality of the formulation of Tamir et al. (20) was to consider the second and third terms. To complete the formulation, equilibrium is assumed to prevail at the interface. The simultaneous solution of the equations was carried out in a transform coordinate system in which the conservation equations are reduced to a form having a similarity solution at the inlet.

The experimental data of the study by Mann et Clegg (18) of the absorption with chemical reaction of chlorine in toluene which is associated with large heat effects is taken to illustrate the model presented above in assuming that the reaction is first order. It is found that the absorption flux m_1 of the gas corresponding to the inlet and with the bulk flow contribution are higher by 23 % relative to the values obtained when not considering the bulk flow. This difference remains almost constant for low values of the dimensionless reaction rate constant K_1 and reduced towards high values of reduced distance ξ and K_1. However the error introduced in the evaluation of the total absorption flux (which is usually the measured value in experimental studies) will continuously grow with increasing value of ξ. This is particularly observed for small value of ξ which are encountered in practice. It is shown that the bulk flow contribution at the interface to the total absorption flux m is appreciable at the entrance region (18 %) where the effect of the chemical reaction is negligible, and for relatively low values of the reaction rate constant for the overall range of the absorber length. The effect of the evaporation flux of toluene m_2 is in general much lower than m_1 and also behaves in a opposite way with ξ and K_1. The increase of the evaporation flux for increasing values of K_1 explains the reduction in the stream surface temperature as compared to the case where no heat losses to the surroundings as well as bulk flow effects are considered. Thus absorption fluxes are continuously decreasing and are even lower than those obtained under condition of absorption in absence of chemical reaction as also presented with the film theory by Mann and Moyes (22) concerning the gas solubility reduced due the initial increase in stream surface temperature but here the gas side resistance is also contributing to the reduction. However it should be concluded that in the presence of a gaseous atmosphere under forced convection the study published by Tamir and Taitel (21) let the impression that the gaseous influenced would be less marked but it should be interesting for a further work to evaluate it for the case of a highly soluble gas.

At the end of this chapter on gas-liquid reaction accompanied by a rise in temperature which may be great enough to affect the rate of gas absorption substantially, it may be observed that fundamental background formulations have been developped these few last years and this must be completed now by experimental studies. As in the case of the important work in isothermal conditions (more directed on kinetic scheme) that has also been developped recently and that we have not presented here, the suggested complementary work necessitates of course the simultaneous knowledge or determination of the physico-chemical parameters such as the solubility, mass and thermal diffusivity... and of the interfacial mass transfer parameters. Let us see now what has been done recently on that topics.

3. Solubility and diffusivity of gases in liquids

It is possible to find in the literature (recent or not) many experimental data on solubility and diffusivity that are scattered in a large number of journals. I would like in this part to discuss the references in which I think that is is now possible to find helpful and sufficiently general analysis and presentation with empirical or theoretical correlations for these fundamental physico-chemical parameters.

31. Solubility of gases in liquids

Experimentally determined solubilities have been reported in the literature for over 100 years but even now only a few comprehensive and critical compilations are published especially at temperature well removed from 25°C. Different approaches are made, depending upon whether the solutions are electrolytic or nonelectrolytic. First of all, the solubilities of gases are expressed in terms of many different units, depending on the particular application and it would be recommended that the standard form be the mole fraction solubility x^* or Henry's constant H at 1 atm partial pressure of gas at the temperature of measurement except for those cases in which the solvents do not have an easily characterizable molecular weights such as biological fluids for which the weight solubility is recommended. When the solubility is small, Henry's law provides a good approximation,

$$f_A = H_{A,B}^{(p_{BS})} \cdot x_A^* \qquad x_A^* << 1$$

where f_A is the fugacity and the Henry's constant ($H_{A,B}^{(p_{BS})}$ often designed H) of solute A in solvent B is defined as

$$H_{A,B}^{(p_{BS})} = \lim_{x_A^* \to o} (f_A/x_A^*)$$

Superscript p_{BS} indicates that the pressure of the system, as $x_A^* \to o$, is equal to the saturation pressure of the solvent at temperature T. At modest pressure, the fugacity is essentially equal to the partial pressure of the gaseous solute $p_A = y_A P$ and Henry's constant is the reciprocal of the solubility when the partial pressure of the solute is 1 atm. A crude estimate of solubility is obtained by extrapolating the vapor pressure of the gaseous solute on a linear plot log p_{SA} vs 1/T. The ideal solubility is thus given by $x_A^* = y_A\ (P/p_{SA})$. For non ideal solutions an activity coefficient γ_A is introduced and for mixture of non polars or slightly polar components a considerable degree of success is obtained in using the correlations of the regular solution theory (25) i.e., for solutions of gases in solvents involving small differences in molecular size (25, 28, 29). Hildebrand and Lamoreaux (27) have published an interesting empirical linear relationship and a graphical plot allowing for the calculation of the solubility of any nonreacting gas except H_2 in any nonpolar solvent except fluorochemicals. Alternate results are published by Fleury and Hayduck (26) that also apply to perfluorinated gases and solvents. However the correlations for nonpolar solvents are not satisfactory to predict solubilities in polar or associating solvents such as water or alcohols. Note that the solubilities of all gases in a single solvent approach a constant value as the solution temperature is increased towards the critical temperature of the solvent. So it is possible to estimate the temperature coefficient of solubility, by the use of a graphical plot, for any gas in any one solvent from data of other gases in that solvent (33, 34). More recently Cysewski and Prausnitz (31) have presented a semi empirical correlation that may be useful for estimating gas solubilities at modest pressure which is not restricted to nonpolar systems nor to temperature near 25°C. Their relationship for the reduced Henry's constant $H_{A,B}^{pBS}\ V_A^o/RT$ is proposed in terms of parameters, determined empirically from 76 binary systems, which characterize the interaction of a single solute molecule with a matrix of solvent molecules. It may be used in either of two ways over a wide range of temperature and for a variety of gases in typical nonpolar and polar solvents, including water : first, if a solubility is known at one temperature, the correlation predicts, often with a good accuracy, the solubility at another temperature ; second, if no solubility data are available, the correlation may be used to obtain a reasonable estimate (within a factor of 2). Simultaneously 289 literature data on the standard entropy of solution and 408 literature data on Henry's constant were correlated by De Ligny et al. (32) by regression analysis. The 20 gases molecules range in size from He to C_3H_8 and the 39 solvent molecules from CS_2 to polyethylene of molecular weight of 10^5 and the solvents range in polarity from hexane to methanol. Moreover useful correlations for the pressure dependence of gas solubility or solubility of gas in mixed solutions are presented in references (28, 29, 35). However, as explained by Cysewski and Prausnitz (31), despite many efforts, a truly satisfactory correlation of gas solubilities remains elusive yet now as theoretical correlations are inevitably limited to very simple systems and narrow range of temperature while empirical correlations are always limited by the particular data on which they are based and it must always be kept in mind that extrapolations to new systems and conditions are often subject to very large errors. However the recent tentative for generalization (31-33) must be underlined. Note also that recent practical data on the industrial important need of solubility of mixtures of acidic gases in amines may be obtained in ref. (48-50). Besides the solubility of electrolyte solutions is estimated by the well-known empirical equation $\log H/H_B = hI$ where H and H_B are the Henry's constant values in solution and in pure solvent and I is the ionic strength of the solution. The contributions for various species in the salting coefficient h have been presented several years ago (36, 37) and the literature of the last few years allow only for some sparse and complementary data for the species contributions such as $C_6H_5O^-$ and MnO_4^-.

32. Diffusivity in liquids. In contrast with the situation for gases, there are no yet satisfactory theoretical methods to predict diffusivities in liquid systems. Different approaches are needed, depending on whether the solutions are electrolytic or nonelectrolytic and most studies are devoted to the estimation of diffusion coefficients in very dilute solutions. However, some papers report substantial variations with increasing concentrations of the diffusing solute (29). The semi and empirical relations and the experimental methods available for estimating diffusivities in liquids have been reviewed recently in the references (29, 30, 38-41). For estimating the diffusivity D_A, a convenient relationship is still that well known of Wilke and Chang based on the Stokes-Einstein equation,

$$D_A = 7.4 \times 10^{-8} \, T \, (x \, M_B)^{0.5} / \mu_B V_{mA}$$

where M_B is the molecular weight of the solvent B and V_{mA} the molal volume of the solute at the normal boiling point. The association parameter "x" allows for differences in solvent behavior : x = 2.26 for water, 1.9 for methanol, 1.5 for ethanol, and 1.0 for benzene, ether, heptane, and other unassociated solvents. For amylalcohol, isobutyl alcohol, ethyleneglycol, and glycerol, Akgerman and Gainer (42) have approximated the associated paramehomolog defined by substituting a "$-CH_3$" group for the "-CH" group. However the uncertainty in assigning values to x has resulted in efforts to eliminate this factor either in introducing the ratio of the molar volumes of solvent and solute, or in replacing x by functions of the latent heats of vaporization at the normal boiling temperature. More recently Sovova (43, 51) proposed a new equipment and an empirical correlation in using existing correlations and expanded set of experimental data (365 in water and 126 in organic liquids). The liquids are divided into 3 groups : water, organic liquids with approximatively spherical molecules, liquids with linear molecules. In the range μ_B = 0.3 - 5 cp the accuracy of the above correlation is comparable with that of Akgerman and Gainer for diffusion in water and markedly better for diffusion in organic liquids except for small solute gas molecules such as H_2 or He. This correlation is simple and requires only a little information about the diffusing species (μ_B, V_{mA}). However for small solute molecules, for systems involving solvents having viscosities greater than 5 cp or for highly non-ideal systems such as aqueous alcohol mixture, use of the relationships proposed either by Akgerman and Gainer that require the physical properties involved and applies with associated and nonassociated systems or by Sridhar and Potter (44) that require the quantum parameter and critical molar volumes instead of the molar volume of the solute at the normal boiling point should be suggested. Anyway it should be noted that yet now the scatter of experimental data for H_2 is rather large owing to its low solubility and more accurate diffusion data are needed over a wide range of temperature in order to test the theories of liquid state.

Diffusion of a dissolved ionized electrolyte involves the diffusion of both cations and anions which, because of their smaller size, diffuse more rapidly than the undissociated molecules. On the assumption of complete dissociation, the diffusion coefficients of electrolytes can be predicted very accurately at infinite dilution using the equations by (a) Nernst-Haskell for solutions containing two species or (b) Vinograd and Mc Bain (45) for solutions containing more than two species or (c) Gordon (46) for higher concentration up to 2 N. In fact, for quick practical calculations, the formula of Wilke and Chang is often sufficient with the complementary assumption that the diffusivities in solutions vary roughly as $\mu_B^{-0.8}$ in the aqueous solutions usually taken to determine the interfacial parameters in gas-liquid contactors.

Finally it may be emphasized that many experiments must be carried out to exactly know gas solubility and diffusivity in liquids.

To this end it is interesting to note that Takeuchi et al. (47) have published results of diffusivity and solubility of CO_2 in non-electrolytic liquids simultaneously measured in a diaphragm cell. The results show good agreement with others in the literature.

Such difficulties of knowing the solubility or the diffusivity of gas with good accuracy are sometimes overcome by use of the laboratory models, such as a laminar-jet or a wetted-wall column, for determination of the interfacial parameters. Indeed in this case, only a mathematical combination of solubility and diffusivity such as $C_A^* \sqrt{D_A}$ is necessary, not the separate values of C_A^* and D_A. But for industrial design, the separate knowledge of these two parameters is most often necessary and the studies presently reviewed will be extremely helpful.

4. Measurement of interfacial areas and mass-transfer coefficients

Gas holdup, interfacial area, and mass transfer coefficients are also important variables determining the mass transfer rates in gas-liquid contacting devices. Because of the diversity of equipment used for contacting gas and liquid and, furthermore, the substantial change in the state and flow of phases in given type of equipment with variation of operating conditions, it is not surprising that quite a number of techniques is proposed for measuring interfacial parameters. They can be classified into two categories : local measurements with physical techniques such as light scattering or reflection, photography, or electric-conductivity methods, and global measurements with chemical techniques. Each method has its advantages and its drawbacks. Though the ground of these techniques is now well established (56, 57, 60) intensive researches have been published recently on that topics.

41. Physical methods. Physical measurements can be made of gas holdup α, bubble size, and specific surface area a' in gas liquid dispersions, as usually encountered in bubble columns, plate columns, mechanically agitated tanks, and spray towers. Any two of these interfacial parameters are sufficient to define all three, since they are interrelated ($a' = 6\alpha/d_{SM}$) where d_{SM} is the volume surface mean diameter or Sauter mean diameter. The gas hold-up is determined directly by measuring the height of the aerated liquid and that of the clear liquid without aeration. This method is rapid, but is not very accurate (15-20 % accuracy) especially when waves or foams are occuring on the top of the dispersion. An alternate and more accurate manometric technique consists in computing α from measurements of the clear liquid height in the dispersion at successive manometer tappings on the side of the froth container (52, 54). The electrical technique is based on measuring the surface elevation at certain selected points by means of an electrically conductive tip. The height is determined by the vertical position of the tip at which the sum of contact times equals one

half of the measurement period (55). The gamma-ray transmission technique depends on the use of the relationship $\ln I_0/I = \lambda \rho s$ where I_0/I is the intensity ratio between incident beam of radiation and transmitted beam λ is the mass absorption coefficient, ρ is the density to be measured (simply related to gas holdup) and s is the thickness of the absorbing medium. (56)

The Sauter mean particle size of dispersion is evaluated directly by a statistical analysis of high-speed flash photomicrographs when the dispersion is dynamically maintained. Photographs are taken through the wall of the transparent reactor or in the interior of the reactor with the aid of an intrascope. To avoid any wall effect or perturbation effect that may occur with this method, a sampling apparatus for the photographic technique may be used where bubbles are extracted from the tank containing the dispersion by means of a calibrated capillary tube or a tube connected to a small square-section column through which a continuous flow of liquid and bubbles rises. The photographic technique for measurement of bubble size is time consuming and most often gives local values on conditions near the wall ; it may miss a small number of large bubbles, and is best used in the case of small gas holdup, i.e., at low gas velocity.

The local interfacial contact area is determined directly by light-transmission and reflection techniques. In the light-transmission technique, a parallel beam of light is passed through the dispersion and a photocell is placed at some distance from it. Light scattered by the bubbles passes outside the photocell and is lost, while the unscattered part of the incident parallel beam is recorded by the photocell at the extremity of an internally blackened tube. Calderbank (57) showed that for scattering bubbles, which are large in comparison with the wavelength of light, the scattering cross-section is equal to its projected area. Furthermore the total interfacial area per unit volume of the dispersion a' equals four times the projected area per unit volume, giving the equation $\ln I_0/I = \tau = a' L/4 = \ln t/t_0$ where L is the optical path length. By connecting the photocell to a light-quantity meter and electric timer, it is possible to measure the times for a given quantity of light to be received by the photocell, when the light passes through the liquid (t_0), and through the dispersion (t). This technique holds for values of $a'L < 25$, when multiple scattering is negligible and for bubble diameters larger than 50μ. Quite recently Landau et al. (58) extended this method to conditions when the light source and photocell are placed outside the column and multiple scattering is taken into account. An interesting empirical correlation based on anisotropic scattering in all six mutually perpendicular directions is given,
$a'L/4 = (\ln I_0/I)/I\text{-}\Phi$ with $\Phi = 1\text{-}6.59\ (1\text{-}\exp\text{-}0.233\tau)/\tau$ which is valid for a'L up to 100. This represents a fourfold increase of the range of the applicability of the light attenuation technique which is not time consuming. The range of interfacial area is up to 8 cm^{-1} with fractions of light transmitted less than 0.02 and values of a' are approximating by 5 % those obtained simultaneously

by the photographic method.

42. Chemical methods

Chemical methods to determine gas-liquid interfacial areas and mass transfer coefficients have been intensively developed for the last ten years (60-69). The principles of these methods are deduced from the theory of isotherm absorption with chemical reaction : a gas A is absorbed into a liquid, where it undergoes a reaction with a dissolved reactant B : $A + zB \xrightarrow{k}$ Products. By choosing a reactant having a suitable solubility and concentration along with an adequate rate of reaction, either the mass-transfer coefficients or the interfacial area (or both) can be deduced from the overall rate of absorption depending on the limiting step being the diffusion or the reaction (or both). Generally a steady flow of each phase through the reactor is assumed.

To my opinion what has been really new during the two last years concerns the application of this technique to organic and viscous liquids by Sridharan and Sharma (64) and Ganguli and Van den Berg (65). The suggested reactions between CO_2 and selected amines in hydrocarbon solvents such as toluene, xylene..., in polar solvents such as cyclohexanol and in highly viscous solvents such as diethylene and polyethyleneglycol (64) and the hydrogenation of edible oil in the presence of an homogeneous Ziegler-Natta catalyst (65) deserve now some complementary kinetics study.

43. Comments on physical and chemical methods : their limits

The specific surface area of contact for mass transfer in a gas liquid dispersion (or in any type of gas-liquid reactor) is defined as the interfacial area of all the bubbles or drops (or phase element such as films or rivulets) within a volume element, divided by the volume of that element. It is necessary to distinguish between the overall specific contact area for the whole reactor with volume, and the local specific contact area for a small volume element in the reactor - a consequence of variations in local gas holdup and in the local Sauter mean diameter. So there is need for a direct determination of overall interfacial area, over the entire reactor, which is possible with use of the chemical technique.

At the outset, a technique that measures over all values cannot be used without restrictions that arise from the results observed with physical methods. For example the chemical method can hardly be used with fast coalescing systems, since the presence of a chemical compound may well reduce the coalescence rates. Also, as observed with physical methods, the wide variation of specific contact area at different locations in the reactor negates the meaning of an average value. In fact, physical and chemical techniques should be used simultaneously to identify more fully the phenomena that occur in gas-liquid reactors. While the chemical methods provide overall values of interfacial area that are immediately usable for design, we must also know the variations of the local interfacial parameters (α, d_{SM}) within the reactor in order to deal competently with scale-up. These complementary data, measured by physical methods, should be obtained from local simultaneous measurements of two of the three interfacial parameters (α, a', d_{SM}). In order to gather these complementary informations simultaneously the electroresistivity probe, proposed by Burgess and Calderbank (53, 59) for the measurement of bubble properties in bubble dispersions, is a very promising apparatus. A three-dimensional resistivity probe with five channels was designed in order to sense the bubble-approach angle, as well as to measure bubble size and velocity in sieve-tray froths. Gas holdup, gas-flow specific interfacial area, and even gas and liquid-side mass transfer efficiencies have been calculated directly from the local measured distributions of bubble size and velocity. This method has revealed an interesting result : the interfacial areas reported compare very favorably with those computed from experimental global measurements using liquid-phase controlled chemical gas absorption, but are lower than those measured using photography. Whereas photography through the container wall appears to truncate data above equivalent diameters of about 15 mm, the probe used by Burgess and Calderbank only deletes data below a diameter of 4 mm. So the difference between the interfacial values measured by the chemical and photographic methods may be due to a small number fraction of large bubbles dominating the interfacial area parameters of an assembly of small and large bubbles, as in a sieve tray, and to their apparent inadvertent omission by photography. This is a good example of the limitation of the physical methods, where here they provide one datum within a local volume (d_{SM}) and the other datum information α for the entire reaction volume.

Moreover, always in the case of the determination of interfacial parameters in gas-liquid dispersion in stirred tank, Midoux (61) has proposed an asymptotic modelisation for bubble dispersion in assuming that each bubble has a life time decreasing with the diameter. The model leads to conditions that allow for the evaluation of the actual interfacial parameters with a 10 % accuracy. These conditions give either the superior limits of the absorption efficiency E_A or the inferior limits of a criterion $\theta^\circ = d_{SM}\gamma^\circ / 6\overline{\varphi_o}\tau_G$ in which the inlet solute concentration γ_o, the initial average absorption $\overline{\varphi}_o$ and the gas space time are taken into account. For the determination of k_La in slow reaction regime, the conditions are $\theta^\circ > 3.33$ or $E_A < 0.27$. For the determination of the interfacial area a with the pseudo-mth order reaction, if the order m is 1 or 2, θ° mini are respectively 2.60 and 3.70 and E_A maxi are 0.30 and 0.21. Following the author such conditions explains why the interfacial areas are seldom measured in reactors of great size : the values of τ_G are too high leading to too small values of θ°.

A question may arise about the data obtained with the previous techniques which is how the parameters so evaluated can be extrapolated to other operating conditions ; for example, from chemical absorption to physical absorption, or vaporization. In other words is the value of k_L or a for the hydrodynamic conditions of chemical absorption the same as for the hydrodynamic conditions of physical absorption ? (66, 70, 71). In a packed column, some zones of liquid in the packing are almost motionless, becoming saturated by the absorbing gas during physical absorption and hence almost ineffective to mass transfer. When the absorbing capacity of the liquid is increased by a chemical reactant, these zones may still be effective, and measurement of the amount of gas absorbed may give the impression that k_L and a have greater values. Also in a mechanically agitated reactor, in the case of absorption with a fast chemical reaction, the mass-transfer coefficient is independent of the hydrodynamics and equal at every point in the vessel, and the interfacial areas in all parts of the agitated vessel contribute equally to mass-transfer. But in the case of physical desorption or absorption, the mass-transfer coefficient can have quite different values, e.g. around the agitator and far away from it.

Thus the assumption of the same value for interfacial area in physical and chemical absorption leads to uncertainty, especially if the mass transfer coefficient is deduced from k_La measured by physical absorption or desorption and from a in a chemical absorption. The effective interfacial area in the case of the fast reaction system where the absorbing capacity is increased by a chemical reactant is substantially larger than the effective interfacial area for physical absorption or desorption, as pointed out by Joosten and Danckwerts (70) that introduced a correction factor γ, the ratio between the increase of liquid absorption capacity $(1+C_{Bo}/zC_A^*)$ and the increase of mass transfer due to chemical reaction (E).

The technique of simultaneous absorption with fast pseudo-mth-order reaction and physical absorption or desorption concurrently, (66-68) is certainly a promising effort to understand the whole complex problem of transport in gas-liquid reactors, since it provides simultaneous measurement of k_La and a. But still it may leave some doubt as to a value of k_L, which can be changed by the occurence of chemical reaction (66). As discussed by Prasher (69), it will be even more promising to conduct such simultaneous experiments in a regime where both hydrodynamics and reaction have comparable effects. So it seems important to remark that adequate techniques now exist, and the coming years should provide the data needed to clarify the picture.

5. Some considerations on hydrodynamics and mass transfer and interfacial areas in gas-liquid reactors : application to trickle beds and well-stirred tank reactors

The choice of a suitable reactor for a gas-liquid reaction or absorption is very often a question of matching the reaction kinetics with the characteristics of the reactors to be used. The specific interfacial area a, liquid holdup β, and mass transfer coefficients k_La and k_Ga are thus very important characteristics of gas-liquid reactors. Table I gives the typical values of these parameters that are provided by some typical contactors for fluid with properties not very different of those of air and water (especially for liquid vicosity smaller than 5 cP and for nonfoaming liquid).

Only a few recent results for co-current downflow packed bed and well stirred tank will be presented in the present review for the illustration of the previous considerations and because of the complex and simultaneous mechanism involved. A similar and complete review for the case of bubble columns may be obtained in the texts published by Deckwer et al. (72) and by Kastanek et al. (73).

51. Trickle-bed reactors

Trickle-bed reactors or similar equipment are used in the petroleum, petrochemical and chemical industries as well as in the field of waste water treatment where the trickle-bed is an alternate to biological oxidation. Since three phases are present, analysis of reactor performances requires a careful study of the intrareactor, interphase, intraparticle mass transport and intrinsic kinetics. The topics as a whole is reviewed in the excellent papers (74-76) for hydrodynamics and kinetics for petroleum applications (especially hydrogenations) and (77) for mass transfer and kinetics for oxidation application.

Most recent works deal with modelling (75), solid-liquid contacting effectiveness (79 - 82) and gas-liquid mass transfer and hydrodynamics (83 - 90). Conflicting data about the degree of catalyst utilisation depending on phase flowrates under the localization of the reaction either in the gas phase in dry zone or in the liquid phase, mean that solid-liquid contacting effectiveness, with either the internal wetting (pore filling) or external catalyst surface wetting, plays a complex role, according to the reaction type, which must be considered in the design of trickle-bed reactors. Several scale-up criteria have been proposed to take into account the reactor performances with the liquid flowrate that are developed on the basis of the ideal pseudo-homogeneous model (75). This model takes into account only the liquid phase with (a) plug flow, (b) no mass transfer limitation, (c) first order isothermal, irreversible reaction with respect to the liquid reactant, (d) total wetting of the pellets, (e) no vaporization nor condensation and the relationship between the inlet C_{in} and outlet C_{out} concentration of the liquid reactant (or conversion X) is given by

$$\ln \frac{C_{in}}{C_{out}} = - \ln(1-X) = \frac{3600\ k_V\ (1-\varepsilon)\eta}{LHSV}$$

Table I : Mass transfer coefficient and effective interfacial area in gas-liquid reactor.

Type of reactor	β (% gas-liquid volume)	k_G $(gmol/cm^2.s.atm)x10^4$	k_L $(cm/s)x10^2$	a cm^2/cm^3 reactor	k_La $(s^{-1})x10^2$
Packed columns					
- counter-current	2-25	0.03-2	0.4-2	0.1-3.5	0.04-7
- co-current	2-95	0.1-3	0.4-6	0.1-17	0.04-102
Plate columns					
- Bubble cap	10-95	0.5-2	1-5	1-4	1-20
- Sieve plates	10-95	0.5-6	1-20	1-2	1-40
Bubble columns	60-98	0.5-2	1-4	0.5-6	0.5-24
Packed bubble columns	60-98	0.5-2	1-4	0.5-3	0.5-12
Tube reactors					
- horizontal and coiled	5-95	0.5-4	1-10	0.5-7	0.5-70
- vertical	5-95	0.5-8	2-5	1-10	2-100
Spray columns	2-20	0.5-2	0.7-1.5	0.1-1	0.07-1.5
Mechanically agitated bubble reactors	20-95	-	0.3-4	1-20	0.3-80
Submerged and plunging jet	94-99	-	0.15-0.5	0.2-1.2	0.03-0.6
- Hydrocyclone	70-93	-	10-30	0.2-0.5	2-15
- Ejector reactor	-	-	-	1-20	-
- Venturi	5-30	2-10	5-10	1.6-25	8-25

where η is the catalyst effectiveness factor for completely wetted particle calculated from Thiele Modulus $\Phi = (V_p/S_p)\sqrt{k_V/D_i}$, k_V is the intrinsic kinetic rate constant per unit catalyst volume, ε is the bed porosity and LHSV is the liquid hourly space velocity. Because of the hydrodynamic behaviour of the trickle beds, the apparent kinetic rate constant k_{app} and/or the liquid-solid contacting effectiveness η_{TB} (with the particles wetted partially or wetted with semi stagnant liquid pockets) are different of the ideal values. The scale-up criteria are thus based on the non-capillary liquid holdup (91), $k_{app} = k_V\beta_{nc}$, on the fraction of the external pellets area that is wetted (92), $k_{app} = k_V(a_W/a_V)$ or on the actual contacting effectiveness $\eta_{TB} = \eta.\eta_C$. In the last case η_C is defined as the degree of catalyst utilization (outside and inside the pellet) compared with that when the particles are completely and uniform in contact with the flowing liquid. Colombo et al. (81) expressed the solid-liquid contacting effectiveness by the ratio between the apparent diffusivity $(D_i)_{app}$ of a tracer in a porous particle in trickle bed reactor partially wetted and the same diffusivity D_i determined in the full reactor. They calculate η_{TB} from a Thiele modulus defined through $(D_i)_{app}$ as $\Phi_{TB} = \Phi\sqrt{D_i/(D_i)_{app}}$. Dudukovic (82) defined the effectiveness factor for a partially wetted catalyst for a reaction occuring only in the liquid filled pore region that may be calculated from a Thiele modulus defined with the diameter proportional to V_{eff}/S_{eff} where V_{eff} is the wetted pellet volume and S_{eff} is the external wetted area of the pellet $\Phi_{TB} = V_{eff}/S_{eff})\ \sqrt{k_V/D_i} = (\eta_i/\eta_{CE})\Phi$, i.e., η_{CE} and η_I represent the fraction of external area wetted and the fraction of internal volume wetted. A complete answer to the solid-liquid contacting effectiveness that necessitates complementary experimental supports will probably be given when the complex hydrodynamics of the liquid trickling over the packing is better known. Besides there are trickle-bed operations such as oxidation in aqueous solutions for which gas-liquid and liquid-solid mass transfer resistances are significant and reliable information for mass transfer coefficient and interfacial areas show that the values of such parameters are again strongly depending on the hydrodynamics (74, 75, 78). It has to be emphasized that the determination of the mass transfer parameters should be carried out under reaction conditions with porous catalyst thus including the simultaneous reaction and hydrodynamics influence. The hydrodynamics of trickle-bed reactors is mainly characterized by the liquid holdup and the pressure drop for the different gas-liquid flow patterns encountered. Indeed at low liquid and gas flowrates ($L < 5$ kg/m^2.s and $G < 0.01$ kg/m^2.s) a trickling flow exists where the flow of the liquid is little affected by the gas (small gas-liquid interaction regime). An increase of gas and/or liquid flowrates leads to pulsing and spray flow for nonfoaming liquids, and foaming, foaming-pulsing, pulsing and spray flow for foaming liquids (high gas-liquid interaction). By taking into account the properties of the fluids, a flow pattern diagram was proposed by Charpentier et al.

(86, 93) with lines (transition regimes rather than points of abrupt change) to separate the flow regimes.

Such a diagram is proposed for nonfoaming hydrocarbons, for hydrocarbon foaming in presence of a gas flowrate in the range 20-30°C, and for viscous organic liquids (in the ranges μ_L=0.31-70 cp ; σ_L=19-75 dyn/cm ; ρ_L=0.65-1.15 g/cm^3). Experimental results for foaming and nonfoaming aqueous solutions on nonwettable glass beads (89) and for foaming aqueous liquids on glass packings (88) indicate that use of the empirical Baker coordinates of this diagram does not cause the transition line from small gas-liquid interaction regime to high interaction to coincide especially for aqueous liquids. Thus it would be more desirable to develop a flow map, based on a sound theoretical foundation, which accounts properly for the influence of the physical and foaming properties of the fluid and the wetting characteristics of the packing. Guidance could be found in the interesting works by Taitel and Dukler (94, 95) for flows in empty tubes and by Talmor (90) for flows through catalyst beds where the mechanisms for transition are based on physical concepts, i.e., combination of Bernouilli effects, gravity forces, buoyant forces, transfer of energy to the liquid to create waves... or in terms of a force ratio relating inertia plus gravity forces to viscous plus interphase forces.

In the trickling flow of liquid, the noncapillary holdup β_{nc} is mainly a function of the superficial liquid flowrate L and properties as well as catalyst characteristics. From a detailed inspection of the literature data, it seems that β_{nc} is correlated as proportional to $L^a\mu_L^b$ where the exponent a is a function of the liquid texture (86, 93). Indeed the dynamic regime of the trickling liquid certainly changes from gravity-viscosity to gravity-inertia and then to gravity surface and the exponent a may have different values which are not systematically equal to 0.33 as very often suggested but depend on L, μ_L and the particle diameter. It is interesting to note that, if in the model for scaling-up based on the liquid holdup (91), β_{nc} is considered as proportional to L^a the corresponding relationship is replaced by $\ln C_{in}/C_{out} \propto k_v Z^a$ $(LHSV)^{1-a}$. The slope (1-a) of $\ln(C_{in}/C_{out})$ versus (1/LHSV) at otherwise constant reaction conditions was assumed to be 0.66 by Henry and Gilbert (91) which supposes a = 0.33. But in an experimental study on desulfurization, demetallization and denitrogenation of various gasoils by Paraskos et al. (96) the slope was found to range from 0.532 to 0.922 which corresponds to the experimental range from 0.58 to 0.86 obtained by Charpentier et al. (86, 87, 93) for β_{nc} experimental data concerning various hydrocarbons and organic liquids. In the high interaction regime, the liquid holdup and the pressure drop are determined by semi-empirical correlations using either momentum or energy balances and thus relating two-phase parameters proportional either to the resultant of the friction forces or to the frictional power to the values of the same parameters when only one phase is flowing with the same superficial flowrate as in the case of two-phase flow (85 - 88,

93). Note that this single parameters should be determined experimentally with the gas and the liquid phase to be used. So differents correlations are presented depending whether the liquid is viscous, nonfoaming or foaming (88, 93).

It is interesting to note that this foam ability cannot be estimated a priori with the physico-chemical properties such as surface tension (for example cyclohexane and kerosene have practically the same density, viscosity and surface tension, but in presence of a gas flowrate and for the same liquid flowrate, kerosene may foam leading to pressure drop ten times and even more higher than the nonfoaming cyclohexane). However it is well known that the foaminess of a liquid in the presence of a gas phase is caused by the decreased coalescence of gas bubbles trapped in the liquid. So the coalescence rate of pairs of identical bubbles generated and injected in the liquid can be quantitatively related to the foam ability (93). Indeed it was observed that for pure cyclohexane or for mixtures of cyclohexane and small weight percentages of desulfurized gas-oil there is 100 % coalescence (all the pairs of bubbles coalesce after their injection in the liquid) and such liquids are not foaming while between 6 and 14 % of added desulfurized gasoil, the coalescence rate is decreasing abruptly from 100 % to 0 % though the surface tension and the viscosity are varying very slightly (σ_L = 25.55 - 26 dyn/cm and μ_L = 0.95 - 1.10 cp). When the coalescence rate is zero the fluid may be characterized as foaming. If the pressure loss is measured simultaneously in the packed reactor at the same gas and liquid flowrate of the liquids, it continuously increase from the value obtained for the nonfoaming cyclohexane to the value obtained for the foaming desulfurized gasoil (93). So this quantitative empirical rate coalescence technique could be developed as a first step to characterize foaminess.

It should be also emphasized that data obtained with air and water and glass sphere systems should be used with a particular care if considered as representative of organic and foaming fluids and that the hydrodynamic results concern generally small column diameter (< 15 cm) with a good initial liquid distribution. For industrial reactors up to as much as 3 m in diameter and working at high temperature and pressure, it has not yet been proved that foaming occurs but it may be that some segregation in the flow of the phases lead to less important gas-liquid interaction and therefore to smaller pressure loss especially when the liquids are not foaming. So a considerable interest should be focused on the realisation of the initial distribution.

52. Well stirred tank reactor

Mechanically agitated bubble contactors are very effective with viscous liquids or slurries or at very low gas flowrates or at large liquid volumes. They are also noted for the ease with which the intensity of agitation can be varied and the heat can be removed. Their main disadvantage is that both gas and liquid are almost completely backmixed. A considerable amount of information is available in literature on these contactors and comprehensive reviews have been published and are summarized in ref. (97). This type of gas-liquid reactor shows also a good example where mass transfer and kinetics performances are strongly connected with the hydrodynamics and have focused many important researches these last years. It appears that values of H/T = 1, D/T in the range 0.4 to 0.5 and H_A/T in the range 0.33 to 0.5 with the gas superficial velocity u_G limited to less than 5 cm/s are likely to be most desirable for gas contacting with a simple impeller (T is the tank diameter, H the clear liquid height, D the agitator diameter situated at level H_A).

For a particular impeller type, the interfacial parameters (gas holdup α, interfacial area and mass and heat transfer coefficients) are strongly dependent on ionic strength (liquids inhibiting coalescence), ion valence number, viscosity, surface tension, by the presence of solid or immiscible liquid and once again by the foam ability of the aqueous or organic liquid (67, 98 - 102). Thus it is nearly impossible to predict a priori these parameters. However it is well known that the scale-up is practicable from experiments carried out with the actual gas-liquid system in a small agitator contactor (T = 10 to 20 cm). At higher tank diameter, to ensure the same specific interfacial area or liquid overall mass transfer coefficient, scale-up should be spent based upon constant total power input per unit volume of liquid $\varepsilon_T = \varepsilon_A + \varepsilon_D$, geometrically similar vessels and the same superficial gas velocity (ε_A and ε_D are the mechanical agitation and sparged-gas power contributions). The real difficulty is in calculating the mechanical agitation power ($P_a = \varepsilon_A V_L$). The mechanical agitation power requirement P_o of an ungassed newtonian liquid can be easily predicted for a number of impeller types from semitheoretical correlations of power number N_p and agitation Reynolds number. However the impeller power input P_a to the gas-liquid dispersion decreases compared to that of the gas free liquid. The reduction in power is dependent upon the agitator type, liquid phase physico-chemical properties, tank geometry and gas sparging rate Q_G. Hassan and Robinson (98) derived recently a semitheoretical equation from dimensional analyses to represent experimental data concerning water and aqueous solutions of either inorganic electrolytes or organics in two fully baffled stirred tanks (2.6 and 19 litres) with three types of turbine and paddle impellers over a range corresponding to a 100 fold variation in power input,

$$P_a/P_o = C_1 \; (N^2 D^3 \rho_L/\sigma_L)^m \; (Q/ND^3)^{-0.38} \; (1-\alpha)$$

The fitted exponent m was slightly dependent on impeller type (-0.19 to -0.25) and the constant C_1 was found to be dependent on impeller type, tank size and electrolytic nature of aqueous phase.

Loiseau et al. (100) have proposed to use empirical Michel and Miller (103) types of correlations to fit experimental data concerning water and aqueous, organic and foaming liquid systems inside two fully baffled small vessels (5.5 and 8.9 litres) agitated by a six flat-blade Rushton types of disk turbine.

$$P_a = C \left(P_o N D^3 / Q_G^{0.56} \right)^n = C\, M^n$$

with, for nonfoaming solutions and pure liquids C = 0.83 and n = 0.45 and for foaming solutions C = 0.69, n = 0.45 if $M<2.10^3$ and C = 1.88, n = 0.31 if $M \geq 2.10^3$. The accuracy is 20 %. Loiseau et al. (100) and Midoux (61) have compared the correlation for the nonfoaming liquid ($P_a = 0.83\ M^{0.45}$) with the literature data concerning different standard tank sizes and Rushton turbine impeller types (Figure 3a) and a relatively good fit was obtained (within 30 %) with the complementary conditions $0.05 < u_s < 9$ cm/s and $1 < M < 10^7$. This correlation also applies to experimental data presented by Edney et al. (101) for non-newtonian fluids or by Pollard et al. (106) for several superposed agitators. Midoux has proposed to take into consideration the number of blades n_b by the relation $C = 0.34\ n_b^{0.5}$. Values of C and n for other impeller types and different gas distribution are found in reference (61). It is interesting to note that such a correlation can be used to scale-up the experimental data of Foust et al. (104) for arrow head impeller (from 26 litres to 8.9 m^3) and Cooper et al. (105) for two flat-blades impeller (from 11 litres to 8.6 m^3) as shown in Figure 3b. This corresponds approximatively to a linear scaling up ratio of 10 (or a volumetric ratio of about 800-1000). But like in the case of the hydrodynamics of trickle bed reactors there still exist the problem of defining a priori the foam ability of the liquid in order to choose the corresponding scaling-up correlation because correlations for water an aqueous solutions of non electrolytes cannot be used to predict P_a (and α) in electrolyte solutions, and vice versa (98, 100).

6. Prediction of the effect of a chemical reaction in an absorber by laboratory scale apparatus : simulation

Among the different steps in designing industrial absorbers or reactors, we have seen that the determination of solubility and diffusivity of one or several solutes in a reacting solution with unknown kinetics can be a challenging problem. These difficulties have justified making relatively simple laboratory models with a well-defined interfacial area, and carrying out experiments to obtain data directly applicable to design. The aim is thus to predict the effect of chemical reaction in an industrial absorber from tests in a laboratory model with the same gas-liquid reactants, or to predict the reactor length for a specified duty, using data from the laboratory model, even though the means of agitating the liquid in the two types of equipment is quite different. This promising

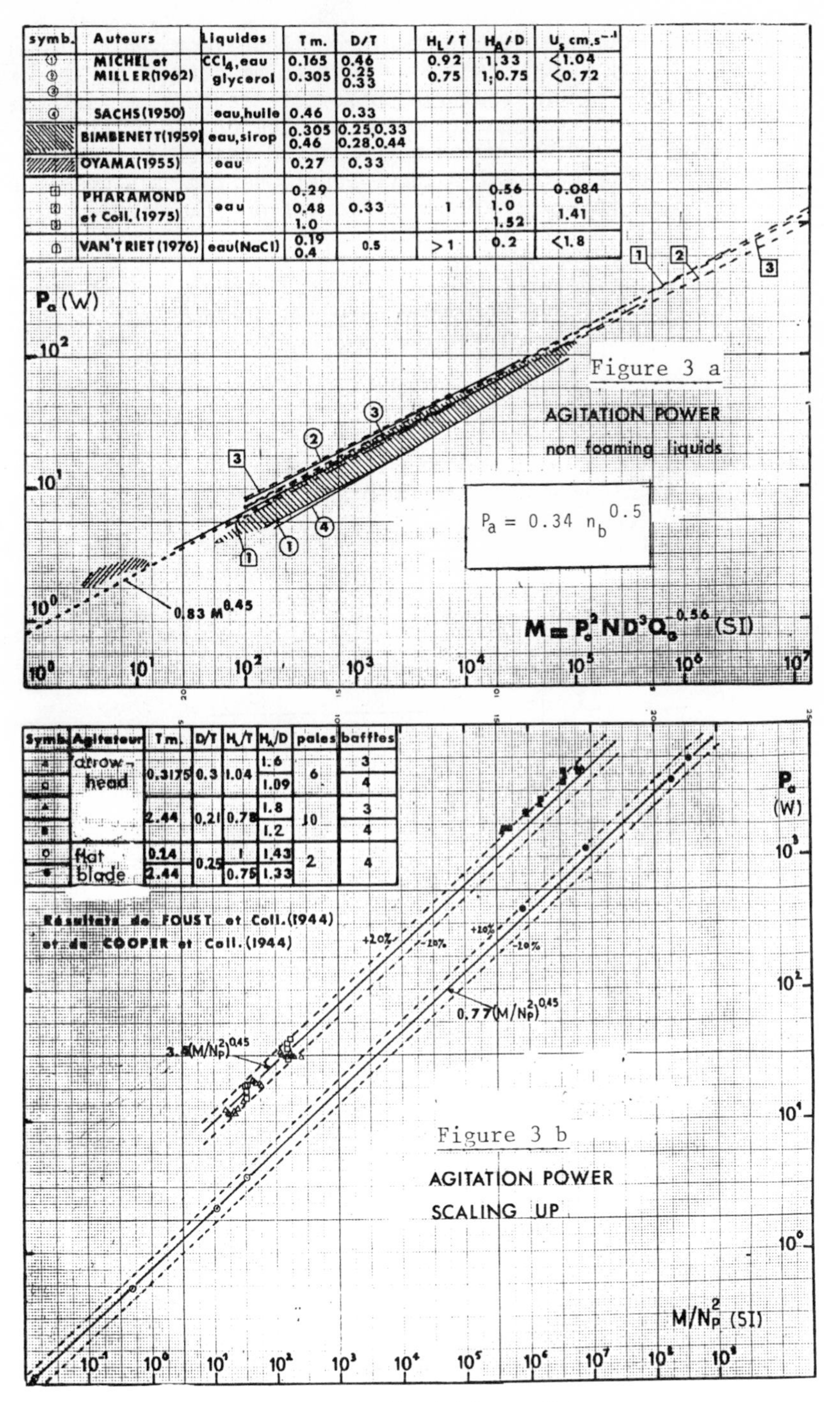

Figure 3. Top (a), bottom (b)

technique has been intensively developed these last years (107-118) and the methodology used deserves a particular attention. The criteria for simulating an industrial absorber are obtained in considering a small but representative volume element Ωdh of an industrial tubular absorber in which a gas-liquid reaction occurs ($A + zB \rightarrow$ Products). The material-balance equations for this case, when the gas contains only one soluble component and the liquid only one reactant, are

$$\varphi(k_L, k_G, C_A^*, C_{Ao}, C_{Bo}) a\Omega dh = -\frac{\Omega u_G}{V_M P} dp_h = u_L \Omega dC_{Ao} + r(C_{Ao}, C_{Bo})\beta\Omega dh$$

$$-u_L \Omega dC_{Bo} = z\, r(C_{Ao}, C_{Bo})\beta\Omega dh$$

Rearranging these two equations leads to double numerical integration over the length h of the absorber,

$$\int_{in}^{out} \frac{a\, dh}{u_L} = \int_{in}^{out} \frac{dC_{Ao}}{\varphi(k_L, k_G, C_A^*, C_{Ao}, C_{Bo})} - \frac{1}{z}\int_{in}^{out} \frac{dC_{Bo}}{\varphi(k_L, k_G, C_A^*, C_{Ao}, C_{Bo})}$$

$$= -\frac{1}{V_M P}\int_{in}^{out} \frac{u_G\, dp_h}{u_L \varphi(k_L, k_G, C_A^*, C_{Ao}, C_{Bo})}$$

$$\int_{in}^{out} \frac{\beta\, dh}{u_L} = \frac{1}{z}\int_{out}^{in} \frac{dC_{Bo}}{r(C_{Ao}, C_{Bo})}$$

Consider now a laboratory absorber in which the liquid and gas are agitated in a way that gives rise to mass-transfer coefficients k_G and k_L of the same magnitude as in the industrial absorber (when they are assumed constant over the height). Also assume that the ratio u_G/u_L, the concentrations C_A^*, C_{Ao}, C_{Bo}, the temperature, and the pressure in the laboratory absorber are the same as those of the volume element located at length, h of the industrial absorber. Then the specific rate of absorption $\varphi = \varphi(k_L, k_G, C_A^*, C_{Ao}, C_{Bo})$ is the same in both apparatuses, and the right side of previous equations are also the same, for the same concentration or partial pressure limits (i.e. entrance and exit bulk concentrations). It follows that the ratio ah/u_L is identical for both absorbers, hence

$$\frac{h}{h_m} = \frac{u_L/a}{(u_L/a)_m} \quad \text{and} \quad \frac{\beta h}{u_L} = \frac{\beta_m h_m}{(u_L)_m} = \tau = \tau_m$$

which defines a scaling ratio and means that the space time is the same in the volume element of the industrial absorber and in the laboratory model. Combination of the above equations leads to $(a/\beta) = (a_m/\beta_m) = A_m/\beta_m V_m = A_m/V_L$. This shows that, per unit volume of absorber or reactor, the ratio of interfacial area to liquid holdup is the same in the laboratory model and the industrial absorber (A_m/V_L).

Therefore the three "criteria" for simulation are identical values of k_L, k_G and a/β in the industrial and the laboratory absorber. "Simulation" means that, if the bulk compositions of gas and liquid in the laboratory absorber are the same as in a volume element of the industrial absorber, the absorption rate per unit interfacial area φ in this element will be the same as in the laboratory model φ_m, whatever the means of agitating the gas and the liquid in the two absorbers. Thus $\varphi_m = \underline{\varphi}$ can be determined experimentally as a function of C_{Ao}, C_{Bo}, or $\underline{p}$, and the above balance equations can be integrated numerically step by step between the limit compositions at the entrance and the exit of the industrial absorber to find the length h of the absorber. The third criterion (a/β) is required whenever the reaction between dissolved gas and a reactant in solution is slow. Indeed reactions will proceed in the bulk liquid, and the rate of absorption in industrial or laboratory absorber will depend upon the volume of bulk liquid available per unit area of interface. Comparison of Tables (I) and (II) leads to the choice of the laboratory equipment (62, 110) for a specified gas-liquid contactor. For example it may be seen that any wetted wall or stirred vessel can simulate a packed column (with respect to k_L and k_G) for reactions occuring in the liquid film.

A laboratory model which simulates a "point" in a tubular column is used when it is possible to calculate the corresponding compositions of the gas and the liquid streams at various points in the column (differential simulation). The specific absorption flux φ for the gas-liquid system concerned is thus systematically measured in the model for different solute partial pressures and liquid reactant concentrations corresponding to those that exist in different points in the column. Knowledge of these absorption rates is essential for predictive calculation of the column length h, as the consecutive values of φ_m from the model must be used to integrate the mass balance equation between the inlet and outlet conditions of the packed column,

$$h = \frac{u_G}{aV_M P}\int_{out}^{in} \frac{dp_h}{\varphi_m(k_L,k_G,C_A^*,C_{Bo})} = \frac{u_L}{za}\int_{out}^{in} \frac{dC_{Bo}}{\varphi_m(k_L,k_G,C_A^*,C_{Bo})}$$

TABLE II - Values of the simulation criteria in laboratory models

MODEL	ROTATING DRUM	MOVING BAND	LAMINAR JET	WETTED-WALL COLUMN			STRING OF DISCS-SPHERES	STIRRED VESSEL
				SPHERE	CYLINDER	CONE		
TIME OF CONTACT (s)	2.10^{-4} 10^{-1}	6.10^{-4} 6.10^{-2}	10^{-3} 10^{-1}	10^{-1} 1	10^{-1} 2	2.10^{-1} 1	10^{-1} 2	6.10^{-2} 10
INTERFACIAL AREA A_m (cm^2)	2 100	0.6 6	0.3 10	10 40	10 100	80	30 360	4 80
k_L ($cm.s^{-1}$)	0.016 0.356	0.021 0.210	0.016 0.160	0.005 0.016	0.0036 0.016	0.005 0.011	0.0036 0.016	0.0016 0.021
10^5 k_g ($mole.cm^{-2}.s^{-1}.atm^{-1}$)	—	—	10 40	1 — 9			1 25	1 15
A_m/V_m (cm^{-1})	100 1250	80 400	20 80	20 60	25 60	40 70	20 60	0.002 0.540

With this technique, Laurent (111) used a 10 cm i.d. double-stirred cell to simulate a 30 cm i.d. column packed with 20 mm glass Raschig rings in a height of 1.92 m. CO_2 from air was absorbed into sodium hydroxide and sodium carbonate solutions, both in the stirred cell and in the packed column, so as to compare the predictions from the model with the packed-column results for different values of u_L, u_G, and gas and reactant concentrations. The predicted and actual heights differed by less than 20 percent in all cases, indicating that this method is quite sound. Danckwerts and Alper (108) have even obtained better prediction (within 10 %) with the same gas-liquid system, using different packing heights for constant fluid flowrates.

Differential simulation is not applicable in cases where the absorption rate is influenced by reactions occuring in liquid bulk or between several reagents and gases. In this case the height of the model is related to that of the full sized reactor by a known scaling factor h/h_m and the model simulates all the essential features of the reactor when the inlet and outlet compositions are the same in both equipment without any assumption about the transfer mechanism or reaction kinetics (integral simulation). A classical example is the simulation of a "complete" packed column of known u_L, u_L/u_G, h, k_L, k_G and a/β (as regard both gas and liquid side phenomenae and bulk reactions) by a string of spheres with a cylindrical pool at the top of each sphere to increase the liquid holdup and to vary the values of the third simulating criterion

$10 < a/\beta < 25$ (109, 112). Once the gas and liquid volumetric flow-rates q_L and q_G are determined in the model so that k_L and k_G are the same as in the packed column and once the number of spheres and the dimension of the pool on the top of the spheres are calculated so that a/β is the same (which leads to the scaling ratio, h/h_m), the model now simulates all the essential features of the column and if the inlet compositions are the same so are the outlet compositions. The total absorption rate Φ_C of the packed column is thus predicted from the measurement of the total absorption rate in the model Φ_m. An excellent illustration of this technique has been presented by Alper and Danckwerts (109) where a 10 cm diameter column packed with 1.27 cm ceramic Raschig rings over 158 cm height is simulated by a string of 10 hollow spheres column, 49 cm height. The experiments have been carried out in cases where the bulk concentration of the reactants at a given level in the column could not be calculated using a material balance equation thus leading to theoretical predictions either too complicated, if indeed possible, or involving large errors. In each case the difference between predicted and measured total absorption rate was within 7 %.

Finally, it must be emphasized the considerable potential of the simulation technique. Though these small equipment have been intensively used to determine kinetics or interfacial parameters for gas-liquid reactions (113-118) it should be suggested that they may used now also for simulation of gas-liquid reactors different of packed columns and in the case of complex reactions with heat affects.

Literature cited

1 Astarita, G., "Mass transfer with chemical reaction", Elsevier, Amsterdam, 1967.

2 Danckwerts, P.V., "Gas-liquid reactions", Mac Graw Hill, London and New York, 1970.

3 Shah, Y.T., Sharma, M.M., Trans. Instn. Chem. Engrs., (1976) 54, 1.

4 Brian, P.L.T., Vivian, J.E., Matiakos, D.C., Chem. Eng. Sci. (1967) 22, 7.

5 Shah, Y.T., Pangarkar, V.G., Sharma, M.M., Chem. Eng. Sci., (1974) 29, 1601.

6 Shah, Y.T., Juvekar, V.A., Sharma, M.M., Chem. Eng. Sci., (1976) 31, 671.

7 Mitsutake, H., Sakai, M., AIChE J., (1977) 23, 599.

8 Hancock, M.D., Kenney, C.N., Proc. 2nd Int. Symp. on Chemical Reaction Engineering, Amsterdam, 1972.

9 Ding, J.S.Y., Sharma, S.S., Dan Luss, Ind. Eng. Chem. Fundam., (1974) 13, 76.

10 Hoffman, L.A., Sharma, S.S., Dan Luss, AIChE J., (1975) 21, 318.

11 Raghuram, S., Shah, Y.T., The Chem. Engng. Journal, (1977) 13, 81.

12 Sharma, S., Hoffman, L.A., Dan Luss, AIChE J., (1976) 22, 325.
13 Teramoto, M., Nagayasa, T., Matsui, T., Hashimoto, K., Chem. Eng. Japan, (1969) 2, 186.
14 Ponter, A.B., Vijayan, S., Craine, K., J. Chem. Eng. Japan, (1974) 3, 225.
15 Bourne, J.R., Von Stockar, U., Coggan, G.C., Ind. Eng. Chem. Process. Des. Dev. (1974), 13, 115 and 124.
16 Tripathi, G., Shukla, K.N., Pandey, R.N., The Canad. Jl. Chem. Engng., (1974) 52, 691.
17 Verma, S.L., DeLancey, G.B., AIChE J., (1975) 21, 96.
18 Mann, R., Clegg, G.T., Chem. Eng. Sci., (1975), 30, 97.
19 Sandall, O.C., The Canad. Jl. Chem. Engng., (1975) 53, 702.
20 Tamir, A., Danckwerts, P.V., Virkar, P.D., Chem. Eng. Sci., (1975), 30, 1243.
21 Tamir, A., Taitel, Y., Chem. Eng. Sci., (1975) 30, 1477.
22 Mann, R., Moyes, H., AIChE J., (1977) 23, 17.
23 Stockar, V., Wilke, U., C.R., Ind. Eng. Chem. Fundam., (1977) 16, 88.
24 Cable, M., Cardew, G.E., Chem. Eng. Sci., (1977) 32, 535.
25 Hildebrand, J.H., Prausnitz, J.M., Scott, R.L., "Regular and related solutions", Van Nostrand Reinhold (1970).
26 Fleury, D., Hayduck, W., Can. J. Chem. Engng., (1975) 53, 195.
27 Hildebrand, J.H., Lamoreaux, R.M., Ind. Eng. Chem. Fund., (1974) 13, 110.
28 Clever, H.L., Battino, R., "The solubility of gases in liquids", Edited by M.R.J. Dack, Wiley Interscience, (1977).
29 Reid, R.C., Prausnitz, J.M., Sherwood, T.K., "The properties of gases and liquids", 3rd Edition, Mc Graw Hill, (1977).
30 Hildebrand, J.H., "Viscosity and Diffusivity, a predictive treatment", Wiley Intersience, (1977).
31 Cysewski, G.R., Prausnitz, J.M., Ind. Eng. Chem. Fund., (1976) 15, 304.
32 De Ligny, C.L., Van der Veen, N.G., Van Houwelingen, J.C., Ind. Eng. Chem. Fund., (1976) 15, 336.
33 Goto, K., Ind. Eng. Chem. Fund., (1976), 15, 269.
34 Hayduck, W., Laudie, H., AIChE J., (1973) 19, 1233.
35 Togunoga, J., J. Chem. Engng. Japan, (1975) 8, 7 and 326.
36 Van Krevelen, D.W., Hoftÿzer, P.J., Rev. Trav. Chim. Pays-Bas, (1948) 67, 563.
37 Onda, K., Sada, E., Kobayashi, T., Kito, S., Ito, K., J. Chem. Engng. Japan, (1970) 3, 18 and 137.
38 Kruis, A., "Gleichgewicht der Absorption von Gasen in Flüssigkeiten", Landolt-Börstein, Springen-Verlag, Berlin, 1976.
39 Skelland, A.H.P., "Diffusional mass transfer", Wiley Interscience, New York, 1974.
40 Simons, J., Ponter, A.B., The Canad. J. Chem. Engng., (1975) 53, 541.
41 Ertl, H., Ghai, R.K., Dullien, F.A., AIChE J., (1974) 20, 1.
42 Akgerman, A., Gainer, J.L., Ind. Eng. Chem. Fundam., (1972) 11, 373.

43 Sovova, H., Coll. Czechoslov. Chem. Commun., (1976) 41, 3715.
44 Sridhar, T., Potter, O.E., AIChE J., (1977) 23, 590 and 946.
45 Vinograd, J.R., Mc Bain, J.W., J. Amer. Chem. Soc., (1941) 63, 2008.
46 Gordon, A.R., J. Chem. Phys., (1937) 21, 706.
47 Takeuchi, H., Fujine, M., Sato, T., Onda, K., J. Chem. Eng. Japan, (1975) 8, 252.
48 Isaacs, E.E., Otto, F.D., Mather, A.E., J. Chem. Eng. Data, (1977) 22, 71 and 317.
49 Isaacs, E.E., Otto, F.D., Lee, J.I., Mather, E., Canad. J. Chem. Engng., (1976) 54, 214 and (1977), 55, 210.
50 Sada, E., Kumezawa, H., Butt, M.A., J. Chem. Eng. Data, (1977) 22, 277.
51 Sovova, H., Prochazka, J., Chem. Eng. Sci., (1976) 31, 1091.
52 Reith, T., Renken, S., Israel, B.A., Chem. Eng. Sci., (1968), 23, 619.
53 Burgess, J.M., Calderbank, P.H., Chem. Eng. Sci., (1975), 30, 1107.
54 Landau, J., Boyle, J., Gomaa, H.G., Al Taweel, A., Canad. J. Chem. Engng., (1977) 55, 13.
55 Linek, V., Mayrhoferova, J., Chem. Eng. Sci., (1969) 24, 481.
56 Vermeulen, T., Williams, G.M., Langlois, G.E., Chem. Eng. Progr., (1955) 51, 85.
57 Calderbank, P.H., Trans. I. Chem. E., (1958) 37, 443.
58 Landau, J., Gomaa, H.G., Al Taweel, A.M., Trans. I. Chem. E., (1977) 55, 212.
59. Burgess, J.M., Calderbank, P.H., Chem. Eng. Sci., (1975) 30, 743.
60 Sharma, M.M., Danckwerts, P.V., Brit. Chem. Eng., (1970) 15, 522.
61 Midoux, N., Thesis University of Nancy (1978).
62 Charpentier, J.C., Laurent, A., AIChE J., (1974) 20, 1029.
63 Laurent, A., Charpentier, J.C., Journal Chim. Phys., (1974) 71, 613 and (1975) 72, 236.
64 Sridharan, K., Sharma, M.M., Chem. Eng. Sci., (1976) 31, 767.
65 Ganguli, K.L., Van den Berg, H., Chem. Eng. Sci., (1978) 33, 27.
66 Linek, V., Chem. Eng. Sci., (1972) 27, 627.
67 Robinson, C.W., Wilke, C.R., AIChE J., (1974) 20, 285.
68 Beenackers, A.A., Van Swaaij, W.P.M., Proc. Eur. Chem. Eng. Symp., Heidelberg (1976).
69 Prasher, B.D., AIChE J., (1975) 21, 407.
70 Joosten, G.E.H., Danckwerts, P.V., Chem. Eng. Sci., (1973) 28, 453.
71 Baldi, G., Sicardi, S., Chem. Eng. Sci., (1975) 30, 617 and 769, (1976) 31, 651.
72 Deckwer, W.D., Adler, I., Zaidi, A., I.S.C.R.E. 5, Houston, ACS Symposium Series (1978), 359.
73 Kastanek, F., Kratochvil, J., Pata, J., Rylek, M., Collect. Czech. Chem. Commun., (1977) 42, 2542-2665.

74 Satterfield, C.N., AIChE J., (1975) 21, 209.
75 Hofmann, H., Int. Chem. Eng., (1977) 17, 19.
76 Hofmann, H., to be published in advances in Catalysis, 1978.
77 Goto, S., Levec, J., Smith, J.M., Catal. Rev. Sci. Eng., (1977) 15, 187.
78 Charpentier, J.C., The Chem. Eng. J., (1976) 11, 161.
79 Germain, A.H., Lefebvre, A.G., L'Homme, G.A., ISCRE 2, ACS Monograph Ser., (1974) 133, 164.
80 Schwartz, J.G., Weger, E., Dudukovic, M.P., AIChE J., (1976) 22, 894.
81 Colombo, A.J., Baldi, G., Sicardi, S., Chem. Eng. Sci., (1976) 31, 1101.
82 Dudukovic, M.P., AIChE J., (1977) 23, 941.
83 Sylvester, N.D., Pitayagulsarn, P., I.E.C. Proc. Des. Dev., (1975) 14, 421.
84 Goto, S., Levec, J., Smith, J.M., I.E.C. Proc. Des. Dev., (1975) 14, 473.
85 Sato, Y., Hirose, T., Takahashi, F., Toda, M., Journ. Chem. Eng. Jap. (1973) 6, 147 and 315.
86 Charpentier, J.C., Favier, M., AIChE J. (1975) 21, 1213.
87 Midoux, N., Favier, M., Charpentier, J.C., Jour. Chem. Eng. Jap. (1976) 9, 350.
88 Specchia, V., Baldi, G., Chem. Eng. Sci., (1977) 32, 515.
89 Chou, T.S., Worley, F.L., Luss, D., I.E.C. Proc. Des. Dev., (1977) 16, 424.
90 Talmor, E., AIChE J., (1977) 23, 868.
91 Henry, G.H., Gilbert, J.B., I.E.C. Proc. Des. Dev., (1973) 12, 328.
92 Mears, D.E., ISCRE 2, ACS Monograph Ser., (1974) 133, 268.
93 Charpentier, J.C., Journal Powder and Bulk Solids Technol., (1978) 1, 68.
94 Taitel, Y., Dukler, A.E., AIChE J., (1976) 22, 47.
95 Taitel, Y., Int. J. Multiphase flow, (1977) 3, 597.
96 Paraskos, J.A., Frayer, J.A., Shah, Y.T., I.E.C. Proc. Des. Dev., (1975) 14, 315.
97 Charpentier, J.C., "Gas absorption and related separations" Adv. in Chem. Eng. (1978).
98 Hassan, I.T.M., Robinson, C.W., AIChE J. (1977) 23, 48.
99 Hassan, I.T.M., Robinson, C.W., Biotech. Bioengng., (1977) 19, 661.
100 Loiseau, B., Midoux, N., Charpentier, J.C., AIChE J., (1977) 23, 931.
101 Edney, H.G.S., Edwards, M.F., Trans. Inst. Chem. Engrs., (1976) 54, 160.
102 Van't Riet, K., Fortuin, J.M.H., Venderbos, D., Ibid, (1976) 54, 124.
103 Michel, B.J., Miller, S.A., AIChE J., (1962) 8, 262.
104 Foust, M.C., Mack, D.E., Rushton, J.H., Ind. Eng. Chem., (1944) 36, 517.

105 Cooper, C.M., Fernstrom, G.A., Miller, A.S., Ibid, (1944) 36, 504.
106 Pollard, R., Topiwala, H.H., Biotech. Bioengng., (1976) 18, 1517.
107 Laurent, A., Prost, C., Charpentier, J.C., Chemische Technik, (1974) 26, 471.
108 Danckwerts, P.V., Alper, E., Trans. Instn. Chem. Engrs., (1975) 5, 34.
109 Alper, E., Danckwerts, P.V., Chem. Eng. Sci., (1976), 31, 599.
110 Laurent, A., Charpentier, J.C., Journal Chim. Phys., (1977) 74, 1001.
111 Laurent, A., Thesis, University of Nancy, 1975.
112 Laurent, A., Charpentier, J.C., 3rd Conference on Applied Chemistry, Unit operations and Process, August 1977, Veszprem (Hungary), pp. 419-429.
113 Levenspiel, O., Godfrey, J.H., Chem. Eng. Sci., (1974) 29, 1723.
114 Weekman Jr., V.W., AIChE J., (1974) 20, 833.
115 Satterfield, C.N., Pelossof, A.A., Sherwood, K.T., AIChE J., (1969) 15, 226.
116 Takeuchi, H., Makoto, A., Kizawa, N., Ind. Eng. Chem. Proc. Des. Dev., (1977) 16, 303.
117 Van den Berg, H., Hoornstra, R., The Chem. Eng. Journ., (1977) 13, 191.
118 Teramoto, M., Ikeda, M., Teranishi, H., Int. Chem. Engng., (1977) 17, 265.

RECEIVED March 30, 1978

8

Biochemical Reaction Engineering

ARTHUR E. HUMPHREY

Department of Chemical and Biochemical Engineering, University of Pennsylvania, Philadelphia, PA 19104

The literature dealing with the kinetic behavior of biological reactor systems surely must be as extensive, if not more so, than that for chemical systems. Hence, any reasonable review of biological reactor systems must of necessity be rather cursory in character. Consequently, I would like to pick and choose my topics in this review, focusing on those items I have personally found most important in dealing with fermenting systems. In particular, I will deal with fermenting systems for the production of chemicals such as antibiotics, production of potential food and feed such as single cell protein, and biological conversion of wastes. This review will be divided into three parts:

1. basic biological kinetics
2. typical biological reactor configurations
3. specific examples

Biological Kinetics

The kinetics of biological systems may be expressed at four different system levels. These include

1. molecular or enzyme level
2. macromolecular or cellular component level
3. cellular level
4. population level

Each level of expression has a unique characteristic that leads to a rather specific kinetic treatment. For example, biological reactions at the molecular level invariably involve enzyme catalyzed reactions. These reactions, when they occur in solution, behave in a manner similar to homogeneous catalyzed chemical reactions. However, enzymes can be attached to inert solid supports or contained within a solid cell structure. In this case, the kinetics are similar to those for heterogeneous catalyzed chemical reactions.

Enzyme Kinetics. In their simplest form, enzyme catalyzed reactions, occurring in a well-mixed solution, are characterized

0-8412-0432-2/78/47-072-262$06.50/0

by the well-known Michaelis-Menten kinetic expression(1).This relationship depicts the substrate, S, combining reversibly with the enzyme, E, to form an enzyme-substrate complex, ES, that can irreversibly decompose to the product and the enzyme, i.e.

$$E+S \underset{k_{-1}}{\overset{k_{+1}}{\rightleftharpoons}} ES \xrightarrow{k_{+2}} E+P \qquad (1)$$

This leads to a kinetic expression for the velocity of the reaction, ν, of the following form:

$$\nu = \frac{\nu_{max} X}{K_M+S} = \frac{k_{+2}E_oS}{\frac{k_{-1}+k_{+2}}{k_{+1}}+S} \qquad (2)$$

where $\nu_{max}=k_{+2}E_o$ is the maximum observable reaction rate at high substrate concentration and, hence, is only limited by the initial enzyme concentration, E_o. K_M is the dissociation constant. When

$$k_{+2} << k_{-1}, \text{ then } K_M = K_S$$

and

$$ES \xrightarrow{k_2} E+P \quad \text{is limiting}$$

The saturation constant, $K_S=k_{-1}/k_{+1}$, is an indication of the affinity of the enzyme active site for the substrate.

Basically two kinds of catalytic poisoning or inhibition are considered.These include inhibition by competition for the active site by a non-reactive substrate and inhibition by a substance that modifies the enzyme activity but does not compete for the active site. This behavior is illustrated in equation (3).

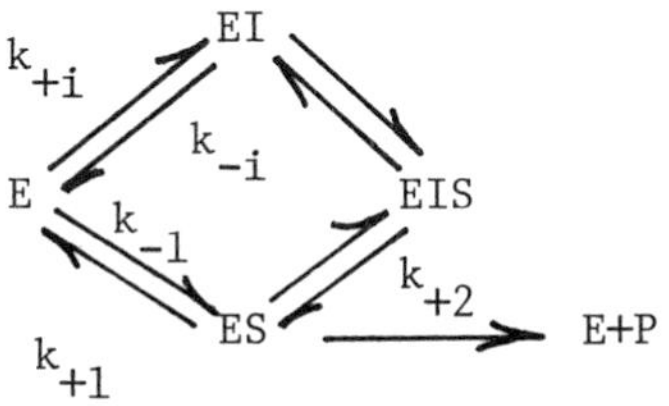

$$E \underset{k_{-i}}{\overset{k_{+i}}{\rightleftharpoons}} EI \rightleftharpoons EIS \rightleftharpoons ES; \quad E \underset{k_{-1}}{\overset{k_{+1}}{\rightleftharpoons}} ES \xrightarrow{k_{+2}} E+P \qquad (3)$$

This behavior can be expressed by

$$\nu = \frac{\nu_{max}^{app} S}{K_S+S} \qquad (4)$$

where

$$\nu_{max}^{app} = k_{+2}E_o \left(\frac{K_I}{K_I+I}\right) \tag{5}$$

and where K_I is defined by

$$K_I = \frac{k_{-i}}{k_{+i}} \tag{6}$$

Cell Growth Kinetics. Enzyme kinetic concepts have been utilized by Monod (2) and others (4-6) to express the kinetic behavior of cell growth on a single limiting substrate. Monod (3) postulated that the growth of cells by binary fission on a single limiting substrate probably had a single limiting reaction step and therefore behaved in a manner analogous to the Michaelis-Menten enzyme kinetics, i.e.

$$\frac{dX}{dt} = \mu_{max}X \frac{S}{K_S+S} \tag{7}$$

where X=the cell concentration, S=the growth limiting substrate concentration, t=time and μ_{max}=the maximum growth rate. Since the growth rate of cells, increasing by binary fission, is defined by

$$\mu = \frac{1}{X}\frac{dX}{dt} \tag{8}$$

equation (7) can be expressed as

$$\mu = \mu_{max}\left(\frac{S}{K_s+S}\right) \tag{9}$$

The behavior of equation (9) is depicted in Figure 1.

For a complete theory of cell growth and substrate utilization, it is necessary to know the relationship between the growth of cells and the utilization of substrate.

This relationship is expressed as a yield, Y, defined as

$$Y \equiv \frac{dX}{dS} \tag{10}$$

The simplest assumption is that Y is constant. This is essentially true only at high growth rates as will be shown later. From equations (9) and (10) one obtains the following relationship for substrate utilization:

$$\frac{dS}{dt} = \mu_{max}\frac{X}{Y}\left(\frac{S}{K_s+S}\right) \quad (11)$$

More recently, it has been shown by Marr (7), Pirt (8) and others (5,9,11)that the yield is not a constant. Rather endogeneous respiration utilizes the energy yielding substrates for maintenance functions; hence,the substrate utilization by cells can be better expressed by

$$\frac{dS}{dt} = \frac{1}{Y_G}\frac{dX}{dt} + mX \quad (12)$$

or

$$\frac{1}{X}\frac{dS}{dt} = \frac{1}{Y_G}\mu + m \quad (13)$$

where m is the maintenance requirement of substrate per unit of cell biomass per unit of time and Y_G is a true yield constant representing the substrate utilized only for growth. This relationship is depicted in Figure 2.

Recently it has been suggested that the rate of cell increase should be expressed as a net growth rate which involved both growth, μ, and death, δ, (12-14), i.e.

$$\frac{dx}{dt} = \mu X - \delta X \quad (14)$$

where

$$\mu = \mu_{max}\left(\frac{S}{K_s+S}\right) \quad (9)$$

and

$$\delta = \delta_{max}\left(1-\frac{S}{K'_s+S}\right) \quad (15)$$

In expressions (14) and (15) cell death is depicted as having a first-order kinetic behavior and as maximal, i.e. δ_{max}, when the growth limiting substrate is zero, i.e. S=0, and minimal when the substrate is in excess, i.e. $1-S/(K'_s+S)=0$. In this latter expression, K_s is a constant used to express the observed behavior (14).

Over the years, numerous models for depicting cell growth have evolved. Several will be discussed here. One such model is that for growth under multiple substrate limitation. It can be expressed as

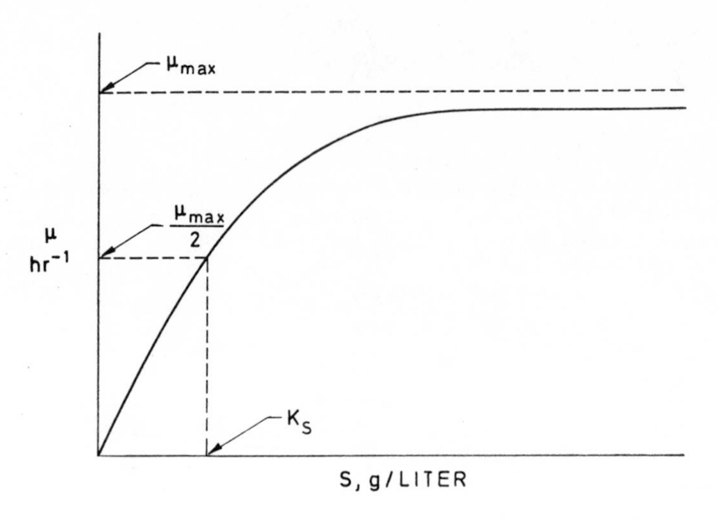

Figure 1. Behavior of Equation 9

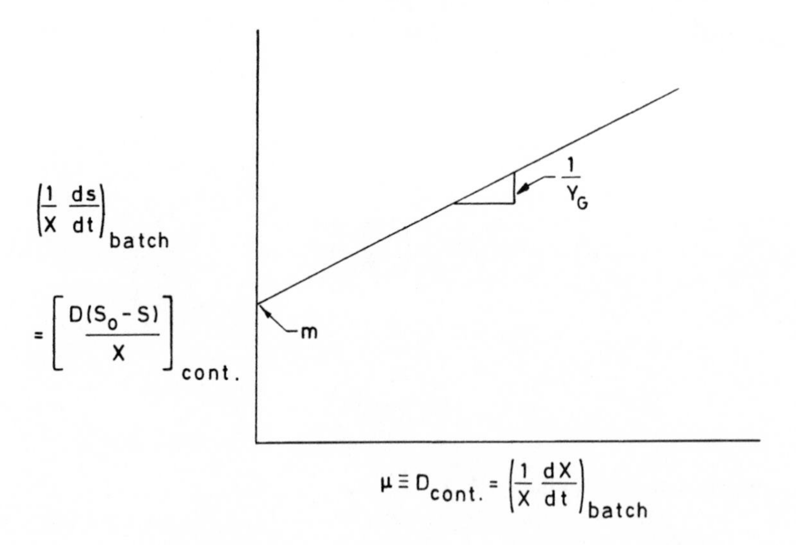

Figure 2. Relationship between substrate utilization and growth

$$\mu = \mu_{max} \left(\frac{S_1}{K_{s_1}+S_1}\right)\left(\frac{S_2}{K_{s_2}+S_2}\right) \cdots \quad (16)$$

Multiple substrate limitation frequently occurs in batch growth systems (15-17).

Another model for cell growth is that for situations of extremely low substrate concentrations, i.e. $S<K_s$. In this case equation (9) reduces to

$$\mu = KS \quad (17)$$

where K is constant and approximately equal to μ_{max}/K_s. This kinetic behavior is frequently observed in large single reactor waste treatment systems.

Another situation that commonly occurs in waste treatment is growth under conditions of "shock loading," i.e. conditions in which the substrate rises to a level in which it becomes inhibitory to growth. This situation is frequently modelled by waste treatment designers by the following kind of kinetic expression (18,19):

$$\mu = \frac{\mu_{max}}{1+\frac{K_s}{S}+\frac{S}{K_I}} \quad (18)$$

This expression is analogous to the enzyme kinetic expression for non-competitive inhibition when $K_I \gg K_s$. The three basic growth relationships are depicted in Figure 3. In passing, it is interesting to note that the observed μ_{max} under conditions of substrate inhibition is a fraction of the true μ_{max} if there were no inhibition, i.e.

$$\frac{\mu \text{ at } d\mu/dS=0}{\mu_{max}} = \frac{1}{1+2\sqrt{K_s/K_I}} \quad (19)$$

Another common kinetic expression for growth is one which takes into account a substrate diffusional limitation. It has been observed in various waste treatment systems that the K_s for "floc" or sludge growth was greater than that for single cell systems. Further, the growth rate of flocs or sludges appears to be a function of their size. This situation is depicted in Figure 4. This has lead Powell (20) to propose that K_s is an apparent K_s; i.e.

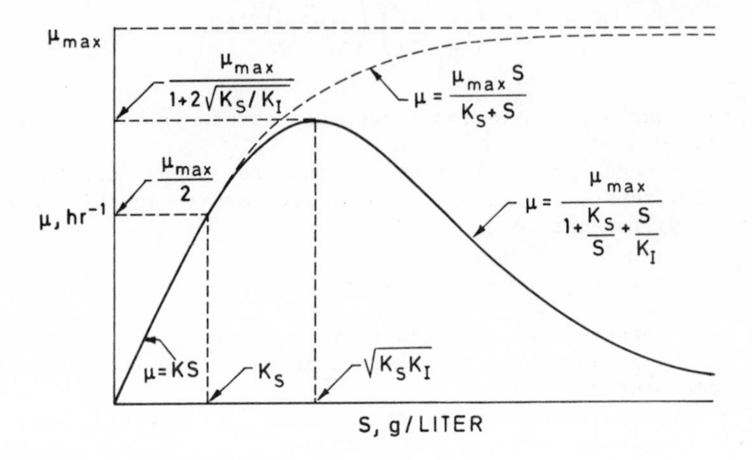

Figure 3. Kinetic expressions for growth

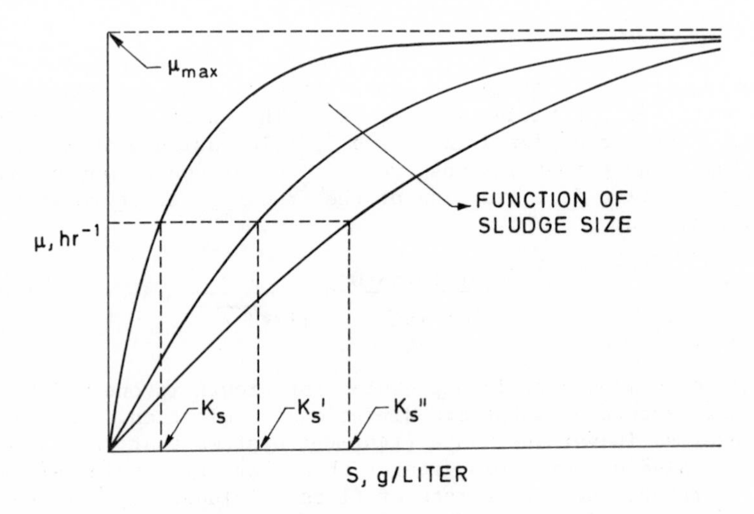

Figure 4. Effect of sludge size on growth rate

$$\mu = \frac{\mu_{max}S}{K_s^{app}+S} \tag{20}$$

He has mathematically shown that

$$K_D^{app} = K_s + K_D \tag{21}$$

where K_D=f(cell size, membrane permeability, diffusion coefficient, etc.). When K_D=0 there is no diffusion limitation to substrate uptake. The effect of K_D and K_s on μ is illustrated by the curves in Figure 5.

Temperature Effects. Before passing on to reactor configurations the effect of temperature and pH on growth rate will be briefly mentioned. The absolute reaction rate theory has been found applicable to both cell growth and death (9), i.e.

$$\frac{d\ \ell n\ k}{dT} = \frac{E_a}{RT^2} \tag{22}$$

or

$$d\ \ell n\ k = -(E_a/R)d(1/T) \tag{23}$$

where k is the rate constant for growth, i.e. μ_{max} or death, i.e. δ_{max}, T is the absolute temperature, E_a is the activation energy for the process, and R is the gas law constant. Generally speaking, E_a for growth, E_G, is the order of 8-12,000 cal/g-mole, ^oK and E_a for death, E_D, in the order of 50-100,000 cal/g-mole, ^oK. This behavior leads to an optimal temperature for growth for a given cell species. This is depicted in Figure 6.

pH Effects. The model for pH effect on growth has been based on the behavior of enzymes (21). Since enzymes are composed of amino acids, they exhibit "zwitterion" behavior, i.e. they have an acid, base, and neutral form. Cells have been depicted to have an active or neutral form and an inactive base or acid form, i.e.

$$X^+ \underset{K_1}{\rightleftharpoons} X \underset{K_2}{\rightleftharpoons} X^- \tag{24}$$

where K_1 and K_2 are the equilibrium constants of the above reaction.

If one defines the growing or active cell fractions as y, i.e.

$$y \equiv \text{active cell fraction} = \frac{X}{X_o} \tag{25}$$

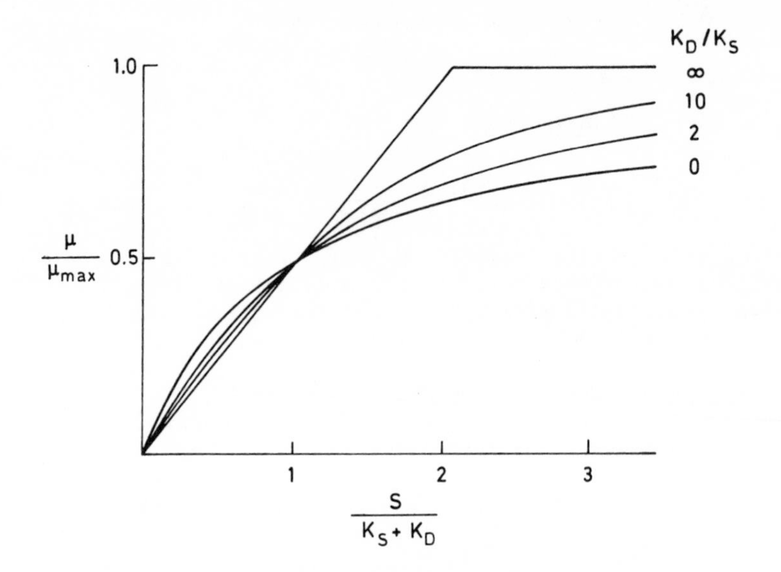

Figure 5. Effect of diffusion limitation on growth rate

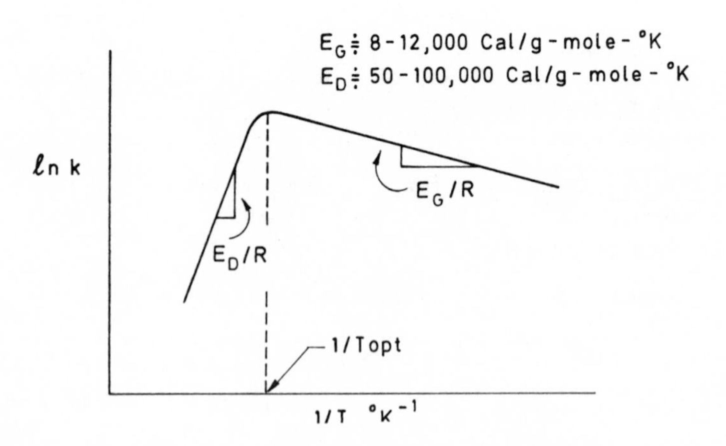

Figure 6. Effect of temperature on growth

where

$$X_o = X^+ + X + X^- \quad (26)$$

By analogy to enzyme behavior

$$y = \frac{1}{1 + \frac{[H^+]}{K_1} + \frac{K_2}{[H^+]}} \quad (27)$$

where $[H^+]$ is the hydrogen ion concentration.
Figure 7 depicts the effect of pH on the active cell fraction.
For most cell systems $pH_{opt}=6.5\pm1$ and $pK_2-pK_1=2\pm1$.

Biological Reactor Configurations

Numerous reactor configurations may be found in biological processes (6,13,18,19). These include:

- Batch
- Batch-fed
- Repeated draw-off
- Continuous
  - single stage
  - multiple stage
- Continuous with recycle
- Continuous with step feeding

For the most part, the antibiotic industry uses batch-type processes.The reason for this stems from the fact that most efficient antibiotic producing organisms are highly mutated and are readily replaced by fast growing, less efficient antibiotic producers in a continuous culture. In order to avoid substrate repression or inhibition some batch processes are continuously fed concentrated substrate on demand during the course of the batch cycle. This is referred to as a batch-fed fermentation. The production of Bakers yeast is an example of a batch-fed process. In some highly mycelial antibiotic fermentation 20 to 40 percent draw-off followed by fresh media make-up is practiced. In the trade, this is referred to as a repeated draw-off process. Strict continuous processes are only practiced in processes for the production of biomass for feed or food and the treatment of wastes. Continuous biomass producing systems are usually single stage reactors without recycle. Waste treatment systems always use multistage systems with recycle.They frequently use step feeding of several stages to improve system stability.

A typical single stage continuous biological reactor is depicted in Figure 8.

A biomass material balance around the reactor yields the following relationship. At steady state

$$\frac{dX}{dt} = 0(\text{steady state}) = \frac{F}{V}(0-X)-\mu X \tag{28}$$

from which

$$\mu = \frac{F}{V} \equiv D \tag{29}$$

This means that the dilution rate or the nominal residence time of the reactor sets the growth rate of the biomass in the reactor. A change in the dilution rate causes a change in the growth rate.

A similar balance around the reactor at steady state for the growth limiting substrate, assuming the the overall yield, Y, is constant gives

$$\frac{dS}{dt} = 0(\text{steady state}) = \frac{F}{V}(S_o-S)- \frac{\mu X}{Y} \tag{30}$$

from which

$$(S_o-S) = \frac{X}{Y} \tag{31}$$

Combining equation (29) and equation (9) yields

$$S = \frac{K_s D}{\mu_{max}-D} \tag{32}$$

and substitutions of equation (32) for S in equation (31) gives

$$X = Y(S_o - \frac{K_s D}{\mu_{max}-D}) \tag{33}$$

Strictly speaking, Y is not a constant, particularly at conditions of low growth rate because the maintenance requirement, m, of the biomass utilizes substrate without producing biomass(10). Equation (12) can be restated for a single stage bioreactor as

$$D(S_o-S) = \frac{\mu X}{Y_G} + mX \tag{34}$$

Table 1. illustrates the effect of maintenance and cell growth or reactor dilution rate on the overall cell yield.

TABLE I. EFFECT OF MAINTENANCE ON OVERALL CELL YIELD IN CONTINUOUS CULTURE

	CELL YIELD,Y,g/g		
D,hr^{-1}	m=0.01	m=0.02	m=0.05
0.01	0.333	0.250	0.143
0.02	0.400	0.333	0.222
0.05	0.455	0.416	0.333
0.10	0.476	0.455	0.400
0.20	0.488	0.476	0.444
0.50	0.495	0.490	0.476

NOTE: Y_G=0.5

Since most single stage continuous bioreactors are used to produce biomass, they are usually operated to optimize the biomass productivity. The unit volume biomass productivity of such a reactor is defined as DX. Utilizing equation (33) this unit volume productivity can be expressed as

$$DX = DY\left[S_o - \frac{K_s D}{\mu_{max} - D}\right] \quad (35)$$

By taking the first derivative of the productivity expression with respect to the dilution rate and setting it equal to zero (9), i.e.

$$\frac{d(DX)}{dD} = 0 = \frac{d\left[DY\left(S_o - \frac{K_s D}{\mu_{max} - D}\right)\right]}{dD} \quad (36)$$

the dilution rate of maximum productivity, D_m, can be found as

$$D_m = \mu_{max}\left[1 - \sqrt{\frac{K_s}{K_s + S_o}}\right] \quad (37)$$

and the maximum productivity, $D_m X$, as

$$D_m X = \mu_{max}\left[1 - \sqrt{\frac{K_s}{K_s + S_o}}\right] Y\left[S_o - \frac{K_s D}{\mu_{max} - D}\right] \quad (38)$$

The behavior of these relationships, i.e. equation (28) to (38), is depicted in Figure 9.

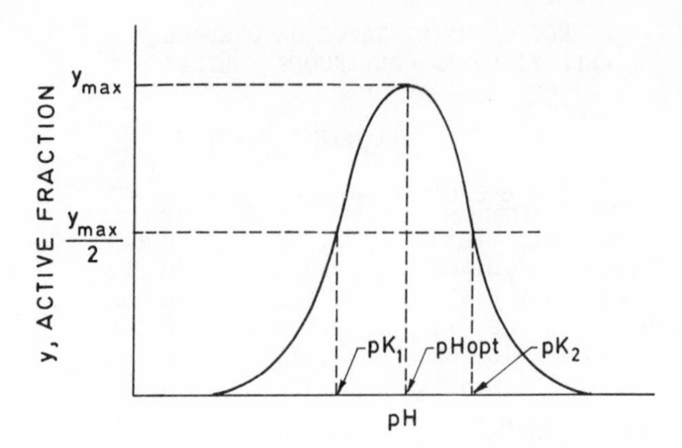

Figure 7. Effect of pH on growth

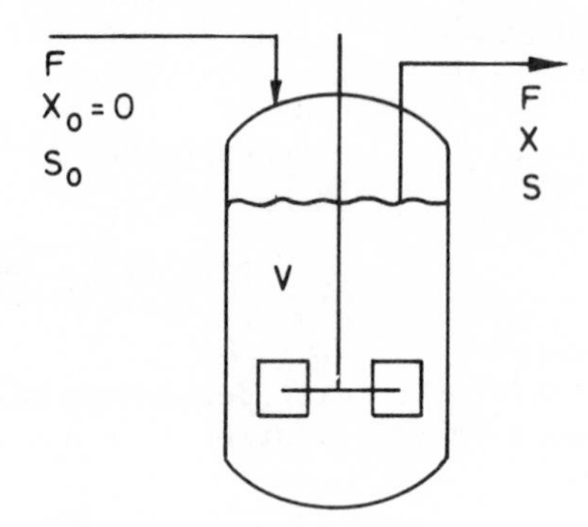

Figure 8. Typical single-cell stage continuous bioreactor. F = flow rate, L/hr; V = volume, L; D ≡ F/V = dilution rate, hr⁻¹; X = cell concentration, g/L; S = substrate concentration, g/L.

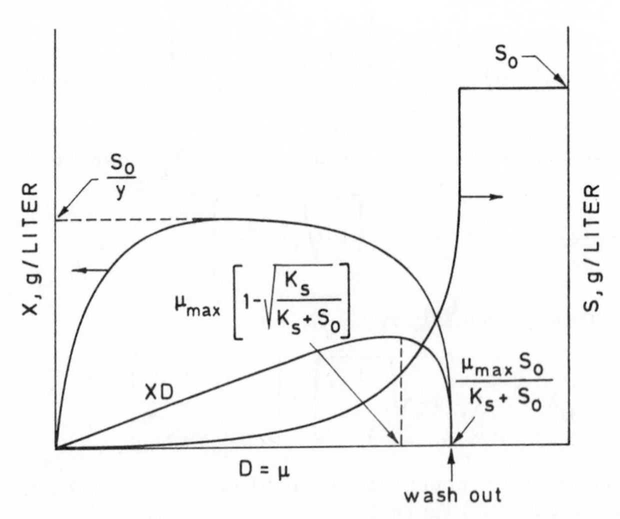

Figure 9. Relationships for a single-stage continuous bioreactor

Note that for this system "wash out" of the biomass occurs when D approaches the maximum growth rate, i.e. $D \to \mu_{max}$.

Biomass recycle is frequently used in bioreactors as a way of increasing the unit productivity (9). In order for the system to operate successfully, a biomass concentrator must be connected to the unit. (See Figure 10).

Usually a centrifuge or settling tank is used as the concentrator. The increased productivity achieved with a recycle bioreactor depends upon the recycle ratio, r, and the cell concentration factor, $C = X_r/X$, achieved in the concentrator. Equations expressing the recycle system behavior are derived from material balances around the reactor, i.e. for the cell biomass balance at steady state

$$\frac{dX}{dt} = 0 = \frac{F}{V}(0) + \frac{F}{V}rX_r - \frac{F}{V}(1+r)X + \mu X \tag{39}$$

from which

$$\mu = D\left(1+r-r\frac{X_r}{X}\right) = D(1+r-rC) \tag{40}$$

Since $(1+r-rC)<1$, it is possible to operate the system at dilution rates greater than the maximum growth rate. Utilizing the definition of μ from equation (9), the following expressions for the growth limiting substrate, S, system effluent biomass concentration, X_e, and biomass concentration in the exit stream from the bioreactor, X, are obtained:

$$S = \frac{K_s D(1+r-rC)}{\mu_{max} - D(1+r-rC)} \tag{41}$$

$$X_e = Y(S_o - S) \tag{42}$$

$$X = \frac{Y(S_o - S)}{(1+r-rC)} \tag{43}$$

These relationships are illustrated in Figure 11 for a particular set of conditions.

Another continuous reactor system encountered in the biological world is the multistage system with step feeding. This configuration is depicted in Figure 12.

In this system fresh feed is step fed, i.e. fed to all reactor stages. This configuration is used in waste treatment systems to provide greater stability and to minimize the effects of sub-

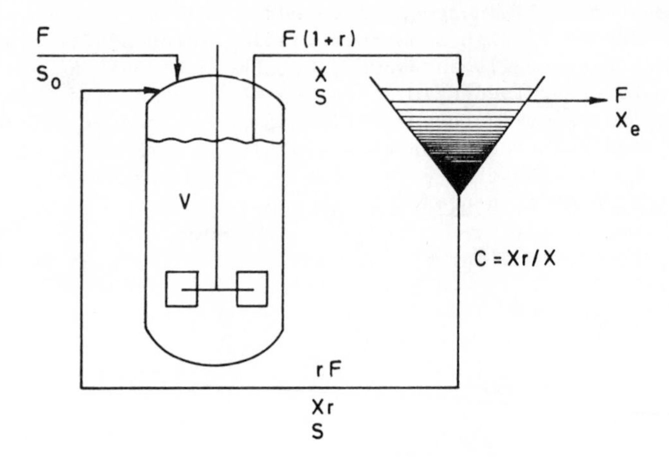

Figure 10. Single-stage continuous bioreactor with recycle. F = flow rate, L/hr; V = volume, L; X = cell concentration, g/L; S = substrate concentration, g/L; r = recycle ratio; C = cell concentration ratio.

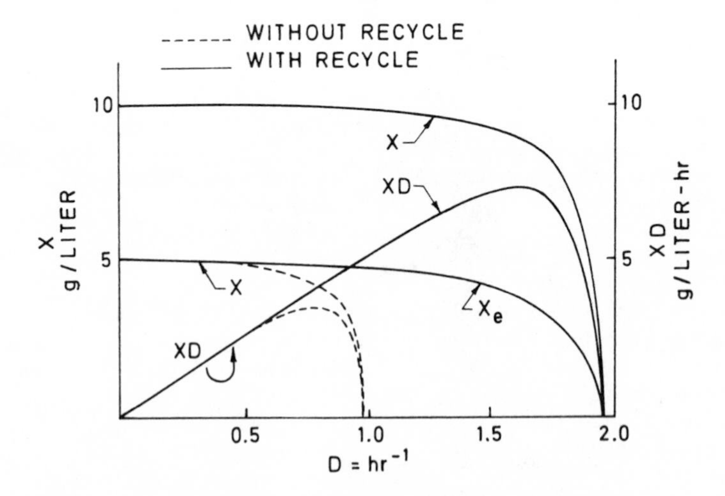

Figure 11. Relationships for a single-stage continuous bioreactor with recycle. (– – –) without recycle; (——) with recycle; μ_{max} = 1 hr^{-1}, Y = 0.5 g/g, K_s = 0.2 g/L, S_o = 10 g/L, r = 0.5, C = 2.

strate "shock loading" or inhibition through distribution of the concentrated feed along the system. A two stage step feeding system is also used as a research tool to look at the inhibitory effects of high substrate loading which can't be achieved in the single stage without wash out. The material balance relationships for step feeding systems are straightforward. What is usually troublesome is the selection of the proper kinetic expression for high substrate conditions.

Specific Examples of Bioreactor Systems

An important application of continuous bioreactor systems involves waste treatment, in particular the activated sludge process. This system is illustrated in Figure 13. It involves one or more stages with an associated cell clarifier and sludge recycle stream. If the clarifier is performing well, no sludge should be carried over in the effluent stream. As can be seen from equations (40) and (41), the two key operating parameters for efficient high capacity BOD removal are the recycle ratio and the cell concentration factor in the clarifier. Unfortunately, the cell concentration factor is a function of the manner in which the biomass is grown. In an early patent (22) it was declared that oxygen limitation was a factor in good cell settling properties and that settling problems could be overcome by aerating with O_2 enriched atmospheres. This claim has been shown to be only partially correct (23). The key to good sludge settling characteristics is the maintenance of the proper food to biomass ratio, F/M, in the system. This can be readily achieved by staging. It can also be achieved by other means (24).

For example, it has been shown that by combining anaerobic and aerobic systems you can have a high rate BOD removal system which selects for high phosphate content organisms. These organisms have good settling characteristics. Also, with such a system, phosphate removal from the waste water will occur (24). Such a system is illustrated in Figure 14. The objective of such a system is to operate it in a way that will give good sludge settling characteristics and minimum oxygen demand in the aerobic zone.

There are several interesting stability and dynamic characterization problems with the activated sludge systems. The first involves the diurnal variation of both the flow and the substrate concentration (BOD level) in the influent. The second problem involves the sludge wasting procedure. Normally, sludge is allowed to build up to give a 24-36 hour inventory in the clarifier and then wasted once every 24 hours. This means that the sludge age is continuously varying. This varying sludge age affects the kinetic properties of the system. A third problem involves the system design to insure stability. Two difficulties are encountered. One involves a system shock due to temporarily high BOD or substrate loading. This resultant high substrate concentration

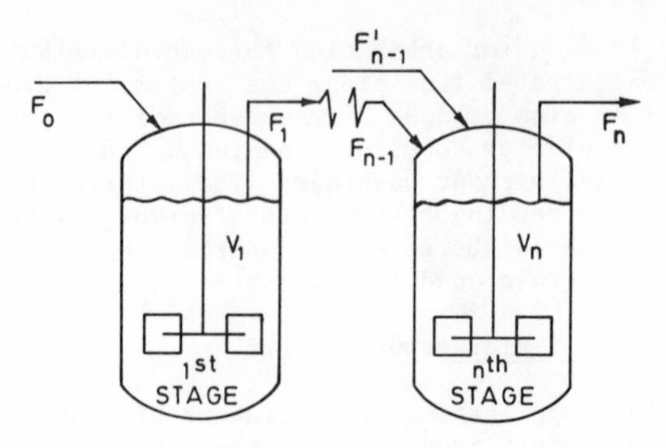

Figure 12. Multistage continuous bioreactor system with step feeding

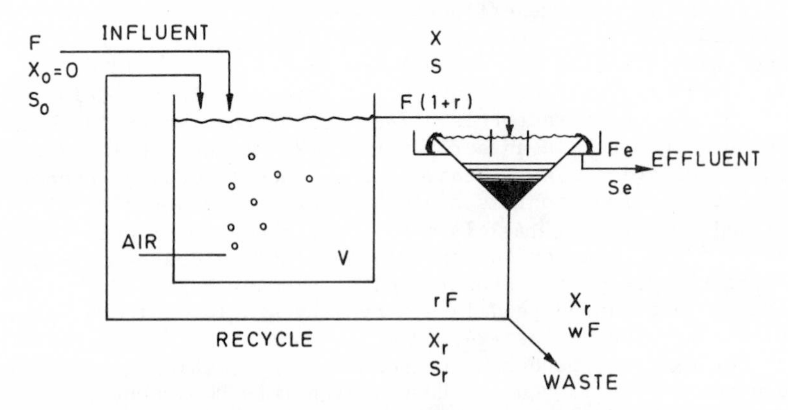

Figure 13. Single-stage aerobic waste treatment system. S_o = influent BOD, S_e = effluent BOD, X_r = recycle biomass concentration (VVS), rF = recycle flow rate, wF = wasting rate.

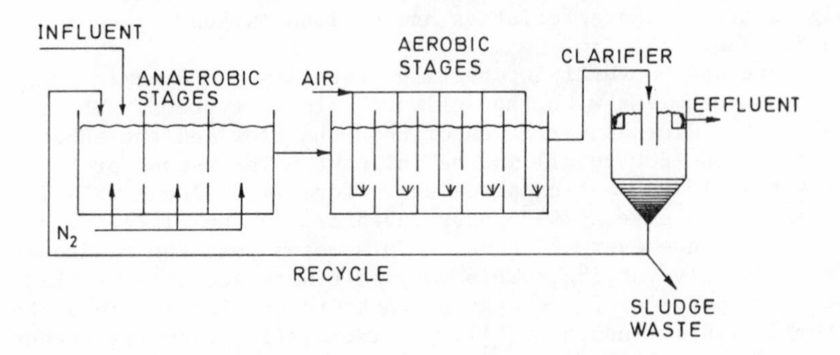

Figure 14. Combined anaerobic–aerobic waste treatment system

can be inhibitory to growth and can initiate washout from which the system may or may not recover. The other difficulty involves the problem of sludge washout due to high influent flows with little or no substrate. Some waste treatment systems are connected to storm sewers. This is particularly true in the large old Eastern cities. A hurricane or cloudburst may literally wash out the system. The question then arises about what measures to take to protect the system. Most frequently the waste treatment system, i.e. the bioreactor and clarifier are by-passed. Certainly, much more information is needed on the dynamic behavior of these systems.

The behavior of activated sludge systems has been best described in various papers of Andrews (18,25). A number of other persons have attempted to do dynamic analyses of waste treatment systems. For the most part, none of the analyses are completely satisfactory.

In recent times, the operating objectives of a waste treatment system have been extended beyond carbonaceous BOD removal. Today some localities require complete BOD removal including nitrification. Some very forward thinking communities are even going one step further to requiring denitrification and phosphate removal. Recent publications (26-28) and a key U.S. Patent (24) have suggested that all three, i.e. BOD, nitrate and phosphate removal are economically feasible using only a secondary treatment scheme. Until the present time, most waste treatment engineers have felt that the process of denitrification required two separate and distinct reactor sections. In the first section, nitrification would be performed by oxidizing the organic nitrogen to nitrates and nitrites. In the second section these nitrates and nitrites would be reduced to nitrogen.

Because the microflora that most optimally performed these two distinct operations was different, most felt that there had to be intermediate clarification. Also, because denitrification requires energy to derive the reaction, methanol or some other biologically oxidizeable carbonaceous material was added to the second section.

Several years ago, a very interesting paper appeared reporting on single pass denitrification (26) in which the anaerobic or denitrification section was placed at front of the system. This was made possible by a high internal recycle for the whole system with separate external sludge recycle. This had the effect of converting the first portion into a denitrification section and utilizing some of the influent BOD to drive the process. Further, it meant that less oxygen was required in the aerobic section since some of the BOD load had already been removed.

Just recently, a three unit system comprised of an anaerobic unit, followed by an anoxic and then an aerobic system with high internal recycle has been proposed (24) (See Figure 15). This system is a challenge to the biological reactor designer. It has four objective functions that must be achieved at the same time.

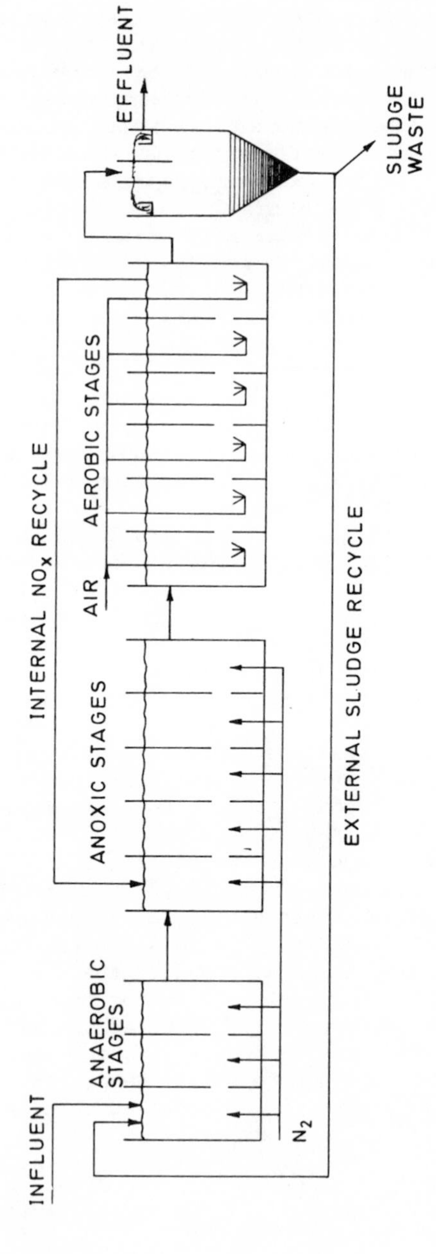

Figure 15. Future complex waste treatment systems for denitrification and phosphate removal

1. BOD removal
2. Nitrification
3. Denitrification
4. Phosphate removal

Further, the system has to operate to produce good settling sludge so that the clarifier provides a clear effluent. Also, the system must be managed to give reasonably stable operation. This can involve such techniques as step feed of the influent and by-passing of certain sections with near total recycle of other sections. I suspect that it will be several years before an adequate design and control model for this reactor system is evolved.

Bioreactor Stability. One troublesome and unsolved problem in bioreactor systems is their dynamic behavior. Simple analysis of the one stage continuous bioreactor yields the following cell and substrate balance equations (29,30):

For cells
$$\frac{dX}{dt} = D(0-X) + X\mu \qquad (44)$$

For substrate
$$\frac{dS}{dt} = D(S_o - S) - \frac{\mu X}{Y} \qquad (45)$$

Phase plane analysis of the system utilizing simple non-inhibitory kinetics, i.e.

$$\mu = \frac{\mu_{max} X}{K_s + S} \qquad (9)$$

and constant yield, i.e. no maintenance requirement,

$$Y = (S_o - S)/X = \frac{dS}{dX} \qquad (10)$$

gives the following phase plane relationships for the interrelationship between cell, X, and substrate, S, during an upset:

$$\frac{dS}{dX} = - \frac{D(S_o - S) - \frac{\mu X}{Y}}{DX - \mu X} \qquad (46)$$

The phase plane behavior is depicted in Figure 16 for two kinds of upsets - an instantaneous high substrate exposure (shock loading) and an instantaneous dilution (storm washout). The simple analysis has suggested that the system is stable and exhibits simple overshoot (29). The steady state conditions, at a fixed dilution rate, D, are given by

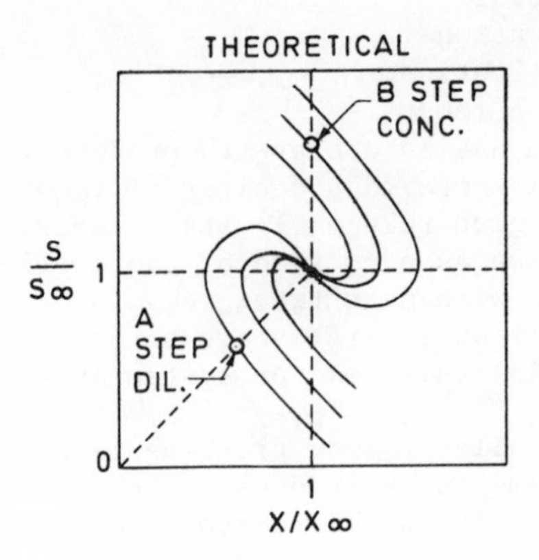

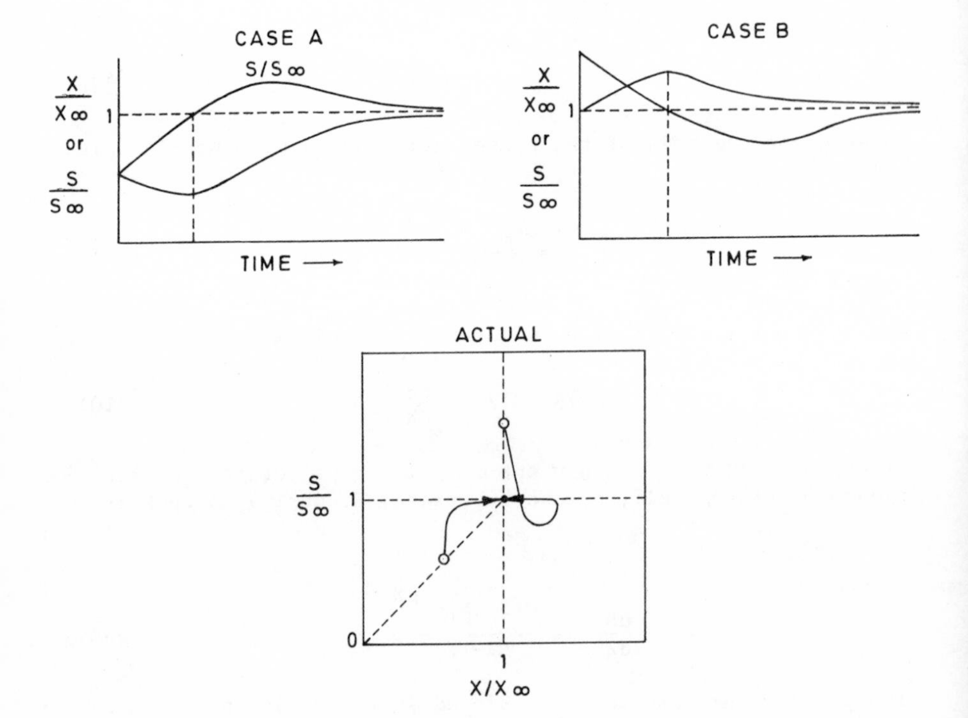

Figure 16. Transient behavior of single-stage bioreactor

$$X_\infty = Y\left(S_o - \frac{K_s S}{\mu_{max} - D}\right) \tag{47}$$

and

$$S_\infty = \frac{K_s S}{\mu_{max} - D} \tag{48}$$

When the kinetic behavior involves inhibition, such as that expressed by equation (18), it has been shown that the system can have two steady states, one stable and one unstable (30). Further, the system can oscillate in trying to reach steady state. The extent of the oscillations and the inherent instability have been shown to be a function of the size of the instability (31).

Dynamic experiments with real systems using a single stage continuous bioreactor under a single substrate limitation have exhibited behavior like that illustrated in the bottom diagram of Figure 16 (32). The real systems phase plane plot has two important characteristics. Firstly, it shows that the simple model is incorrect. Secondly, the system is capable of rapidly storing and excreting substrate in response to dynamic changes. This means that for transient behavior the kinetic model must have a storage function and some sort of structural function that will allow for lags and responses.

An attempt at providing biological kinetic models with structures has been reported in the papers by Ramkrishna, et al., (33, 34). To date, these have found little use, primarily because of the difficulty in monitoring and differentiating between the various bio-structures. Researchers in the waste treatment area have generally preferred to express the transient behavior in terms of BOD storage functions. This has occurred largely because it has been observed that in most multistage aerobic waste treatment systems, BOD uptake precedes biological oxidation as measured by the oxygen demand. Certainly, more research is needed on this point.

Mixed Populations. Considerable literature has evolved on mixed populations kinetics. Virtually every possible biological interaction from the classical prey-predator model of Lotka (35) to various forms of commensalism, synergism, etc.,have been modelled. Virtually all models end up exhibiting stable oscillations. Solutions are often expressed in triangular phase plane plots of the limiting substrate and the two species. Invariably the plots have a limit cycle as one of the stable steady state solutions. From this one gains the impression that oscillatory behavior is the norm rather than the exception in biological reactors.

Recently, this point was further analyzed in a paper by Aris

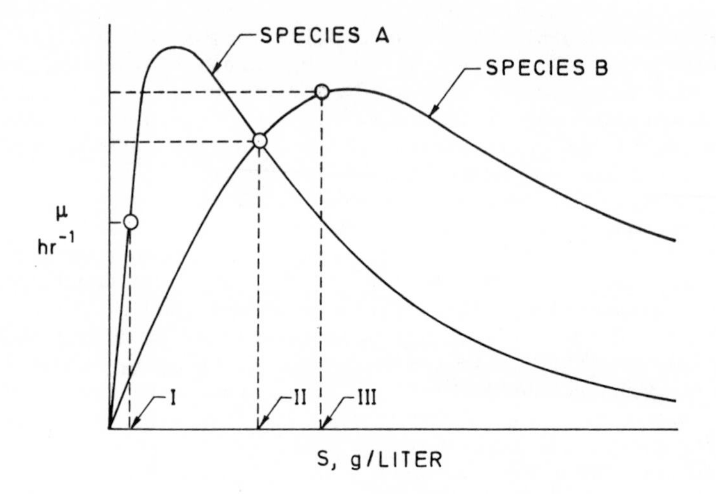

Figure 17. Mixed population bioreactor problems

and Humphrey (36). In mixed cultures exhibiting typical inhibitory kinetics, such as illustrated in Figure 17, numerous operating results are possible. Two pure steady states I and III can be achieved by careful start-up and management of the system. However, selection of an operating condition yielding the reaction rate or having the dilution rate of situation II, has virtually an infinite variety of states depending upon how the system is started up and how it is perturbed. At condition II any ratio of species concentration can be achieved and will give a steady state. One might think this situation is rare, certainly in normal industrial fermentations. However, other similar situations are possible. There are at least three distinct possibilities:

1. A single organism growing on a single growth limiting substrate which inhibits growth at high concentrations and which has two enzymes controlling the rate limiting metabolic step that can operate in concert. One enzyme is constitutive and the other is inducible at high substrate concentrations.

2. A single organism growing on two different energy producing substrates with separate metabolic pathways whose rates are additive and of which one enzyme system is repressed by the substrate of the other.

3. A single organism on a single growth limiting substrate having two separate and distinct pathways simultaneously metabolizing the substrate. One of the pathways is inhibited by the substrate at high concentrations. Other possibilities undoubtedly exist.

The author is of the opinion that oscillatory and variable behavior are not a rare occurrence in biological reacting systems. In fact, they may well be the basis by which organisms in the natural state can survive a variety of conditions.

Conclusions

Biological reacting systems present a variety of modelling challenges to the engineer. They are not only complex in their kinetic behavior but the systems are complex in terms of their process structure, i.e. they can be multistaged systems with step feeding, internal and external recycle, and internal catalyst (enzyme) regeneration. There certainly are real challenges for the future in bioreactor modelling and design.

Literature Cited

1. Roberts, D. V., "Enzyme Kinetics," pp451, Cambridge Press (London)(1977).
2. Monod, J., Ann. Inst. Pasteur (1950), 17, 309.
3. Monod, J. Ann. Rev. Microbiol. (1949), 3, 371.
4. Novick, A. and Szilard, L., Proc. Nat. Acad. Sci., Wash. (1950), 36, 708.
5. Herbert, D., Elsworth, R., and Telling, R. C., J. Gen. Micro-

biol. (1956), 14, 601.
6. Herbert, D., "Continuous Culture of Microorganisms," S.C.I. Monograph No.12 (London)(1960)pp21-53.
7. Marr, A. G., Nilson, E. H., and Clark, D.J., Ann. N. Y. Acad. Sci. (1963), 102, 536.
8. Pirt, S. J., Proc. Royal Soc., Series B (1965), 163, 224.
9. Aiba, S., Humphrey, A. E., and Millis, N., "Biochemical Engineering," pp434, Academic Press (New York)(1973).
10. Sykes, R. M., Journal WPCF (1975), 47, 591.
11. Zabriskie, D. E. and Humphrey, A. E., AIChE Journal (1977), 54 138.
12. Cooney, C. L. and Makiguchi, N., "Continuous Culture 6,"pp146-157, S.C.I. (London)(1967).
13. Pirt, S. J., "Principles of Microbe and Cell Cultivation," pp274, Blackwell Scientific Publications (London)(1975).
14. Yenikeyev, S. and Humphrey, A. E., "Modelling of Cell Growth on Cellulose,"(in press.)
15. Anderson, J. S. and Washington, D. R., Chem. Engr. Progr. Symp Ser. (1966), 62, 60.
16. Bungay, H. R., Chem. Engr. Progr. Symp. Ser. (1968), 64, 10.
17. Humphrey, A. E.,Chem. Reaction Engr. Adv. Chem. Ser. (1972), 109, 603.
18. Andrews, J. F., Biotechnol. Bioengr. Symp. (1971), 2, 5.
19. Humphrey, A. E., Chem. Engr. Progr. (1977),pp85-91.
20. Powell, E. O., "Microbial Physiology and Continuous Culture," pp34-55, Her Majesty Stationery Office (London)(1967).
21. Bailey, J. E. and Ollis, D. F., "Biochemical Engineering Fundamentals," pp567, McGraw-Hill (New York)(1977).
22. McWhirter, J. E., assigned to Union Carbide, U.S. Patent 3,547,814 (Dec. 15, 1970).
23. Casey, J. P., et al., assigned to Air Products, U.S. Patent 3,864,246 (Feb. 4, 1975).
24. Spector, M., assigned to Air Products, U.S. Patent 4,056,465 (Nov. 1, 1977).
25. Busby, J. B. and Andrews, J. F., Journal WPCF (1975), 47,1055.
26. Barnard, J. L.,Water Pollution Control (1973), 72, 705.
27. Barnard, J. L., Water, S. A., (1976), 2, 356.
28. McLaren, A. L. and Wood, R. J., Water, S. A. (1976), 2, 47.
29. Koga, S., Burg, C., and Humphrey, A. E., Appl. Microbiol.(1967) 15, 493.
30. Koga, S. and Humphrey, A. E., Biotechnol. Bioengr. (1967), 9, 375.
31. Humphrey, A. E. and Yang, R. D., Biotechnol. Bioengr. (1975) 17, 1976.
32. Burg, C., M.S. Thesis, Dept. Ch.E., University of Pennsylvania (1965),pp125.
33. Ramkrishna, D., Fredrickson, A. G., and Tsuchiya, H. M., Biotechnol. Bioengr. (1967), 9, 129.
34. Ramkrishna, D., Fredrickson, A. G., and Tsuchiya, H. M., J. Ferm. Technol. (Japan)(1966),44, 210.

35. Lotka, A. J., "Elements of Physical Biology," pp435, Williams and Wilkins (Baltimore)(1925).
36. Aris, R. and Humphrey, A. E., Biotechnol. Bioengr. (1977), 19, 1375.

RECEIVED March 30, 1978

9

Catalyst Deactivation

JOHN B. BUTT and RUSTOM M. BILLIMORIA

Department of Chemical Engineering and Ipatieff Catalytic Laboratory, Northwestern University, Evanston, IL 60201

In the years since the First International Symposium on Chemical Reaction Engineering, where a first review on catalyst deactivation was presented (1), there has been considerable activity in this area. In particular, there have been appreciable advances in the understanding of sintering processes, and of intraparticle and fixed bed reactor behavior under conditions of catalyst deactivation. Happily, much of this has consisted of experimental information as well as analysis.

The present effort makes no attempt to match in scope the previous review; we shall confine ourselves to work concerning chemical poisoning and coking as the primary mechanism of deactivation but retain the classification according to scale — individual kinetics and mechanism, intraparticle problems, and chemical reactor problems. Sintering has been admirably covered in a recent review (2), and the subject of automotive exhaust catalysis (which is almost wholly an exercise in catalyst mortality) will be treated in one forthcoming (3).

Mechanisms and Kinetics

Kinetic networks used to depict the processes of catalyst deactivation have typically been based on models which are simultaneous, parallel, or sequential. Both Carberry (4) and Khang and Levenspiel (5) have enlarged on these to include two additional cases, independent deactivation (characteristic of sintering) and simultaneous-consecutive deactivation (characteristic of coking via participation of both reactants and products).

As will be seen from the examples to be given here, deactivation kinetics have almost universally been correlated in terms of separable (6) rate factors. More detailed analysis, however, indicates that such assumptions are questionable for surfaces other than those ideal in the Langmuir sense (7). The argument can be developed along the lines employed to derive adsorption isotherms for nonideal surfaces starting with the concept of a subassembly of ideal surfaces distributed according to the heat of chemisorption. The differences in separable and nonseparable

0-8412-0432-2/78/47-072-288$08.75/0

kinetics arise because of differences in:

$$(r_T)_{NS} = \sum s_q n_q r_q = \int_0^{q_m} n_q s_q r_q dq \qquad (1)$$

for nonseparable kinetics, and:

$$(r_T)_S = \langle s \rangle \int_0^{q_m} n_q r_q dq \qquad (2)$$

$$\langle s \rangle = \frac{1}{q_m} \int_0^{q_m} s_q dq \qquad (3)$$

for the separable case. It is interesting that the analysis (7) reveals an extreme sensitivity of the validity of the separable approximation to pressure but only a modest effect of temperature. The reason that separable formulations seem to work is probably the same that Langmuir-Hinshelwood rate equations work: the qualitative results are of similar form for ideal and nonideal surfaces but physical interpretation of the parameters is incorrect.

As far as mechanism of deactivation — considering only chemical poisoning and coking — the situation remains much the same as outlined previously (1). Specific cases of chemical poisoning tend to be relatively well understood as far as detailing substances responsible for poisoning and the nature of the deactivated surfaces. Auger electron spectroscopy has become an important tool for the investigation of poisoned surfaces since the characteristic energies in that spectroscopy are sufficiently low to involve only the surface and immediately adjacent layers. Progress has also been made in understanding the poisoning of bifunctional catalysts; nice experimental examples are provided by Webb and Macnab (8) and Burnett and Hughes (9). The former investigated the hydroisomerization of butene on a Rh/SiO_2 catalyst and demonstrated the selective deactivation of the hydrogenation function by small amounts of mercury introduced into the system. This type of bifunctional reaction might be termed a "parallel" one:

$$\begin{array}{ll} n - C_4^{=} + H_2 \rightleftarrows n - C_4 & (Rh) \\ n - C_4^{=} \rightleftarrows i - C_4 & (SiO_2) \end{array} \qquad (I)$$

where the poisoning drastically affects one reaction but not the other. The reaction studied by Burnett and Hughes, on the other hand, was a "series" bifunctional reaction, disproportionation of butane over a mechanical mixture of Pt/Al_2O_3 and WO_3, where:

$$2C_4 \xrightarrow[-H_2]{Pt} 2C_4^{=} \xrightarrow{WO_3} C_2^{=} + C_6^{=} \xrightarrow[+H_2]{Pt} C_2 + C_6 \qquad (II)$$

They showed that coking or water poisoning of the Pt function shut the entire reaction down, characteristic of the sequential nature of the steps involved.

The description of the mechanism(s) of coke formation remain essentially the same as described before. With increasing experimental information on coking in the literature there appears now to be at least tacit agreement that it is not particularly rewarding to look for a mechanism of coke formation when there are probably as many mechanisms as there are reactions and catalysts. In this sense, simple reaction schemes may provide a starting point for analysis but may also fall far short of the mark in confrontation with experiment.

Individual Particles

An interesting problem, not much commented upon in the older literature, is the relationship between coke deposition and the physical properties of the catalyst. Pore blockage by coke deposition has been demonstrated in specific instances, but has been ignored in earlier analytical studies of coking (10,11).

Swabb and Gates (12) investigated methanol dehydration on H-mordenite (1 atm., 100-240°C) with the objective of investigating the influence of intracrystalline mass transport on initial activity and deactivation rates. Mass transport properties were varied by using three different samples of differing mean pore length; significant influences were found only above 200°C. There were no effects of pore dimension on the deactivation rates (due to coke formation) of fresh catalyst over the range investigated. Further studies of deactivation rates after one and two hours of utilization at 205°C also revealed no influence on pore structure. This would rule out intraparticle mass transport as controlling deactivation rates as well as the occurrence of any pore blockage resulting from coking in this reaction.

Experiments on larger size particles have also involved H-mordenite, but with cumene cracking as the reaction (13). Relations between coke content, activity, and intraparticle diffusivity were investigated on 1/16 in. Norton Zeolon extrudates for 230-250°C and space velocities from 0.2 to 0.65 wt/wt-hr. Effective diffusivities were determined (with SF_6 via chromatography) as a function of reaction time and coke content with the results shown in Figure 1. Diffusivity decreased twofold for reaction times of 2 hours or longer, but remained essentially constant after that. The effectiveness factor varied in the range 0.3-0.7 during these experiments; error in estimation of this factor using the effective diffusivity of the fresh catalyst was 20 to 30% at longer reaction times. SEM examination of coked particles revealed the formation of a glassy-like coating on the external surface, penetrating approximately 0.1 the radius into the interior, so it was concluded that pore blocking was responsible for the decrease in diffusivity.

Additional results on coke deposition and pore blocking are provided by Toei, et al. (14) and Richardson (15). Toei, et al. found no effect of coking on intraparticle diffusion for dehydrogenation of n-butane over chromia-alumina (80% γ-Al_2O_3, 5x5 mm pellets or 0.2-2.2 mm particles). Careful examination of the pore size distribution at different coke levels revealed only minor alterations of the micropore structure, and the system was modeled with the normal isothermal, quasi-steady state continuity equations with good agreement to the experimental data. Richardson found a linear decrease in effective diffusivity (Ar/He) with coke content for Nalco 471 cobalt molybdena (1/8 in. pellets) fouled in a pilot unit under hydrotreating conditions (700-800°F, 1000-1500 psig, 2 wt% N_2, 0.05% S, no metals). The Thiele modulus however, was essentially unchanged up to about 15 wt % coke on catalyst.

Several earlier workers have reported decreasing diffusivity on coke formation, but only three investigations (16,17,18) gave direct experimental measurement and none of these were for zeolites. Recent studies apparently have provided no more general picture concerning the effect of coking on transport properties than was available before; however, on weight of accumulated evidence it seems reasonable that for reactions of large molecules in catalysts of fine pore structure significant changes in diffusivity on coking can occur. Whether this in turn will change the effectiveness factor will depend on other chemical and physical parameters. We are unaware of any similar studies devoted to coking effects on thermal conductivity.

Recent theoretical studies of deactivation in individual particles have focused on bifunctional catalysts (19,20), on apparent overall kinetics of deactivation (5), deactivation in nonisothermal particles (21), and the effect of nonuniform distribution of the catalytic function within the particle on deactivation (22,23, 24).

Some factors associated with the deactivation of bifunctional catalysts have been explored computationally by Snyder and Matthews (19) and by Lee and Butt (20). These catalysts have the additional complication of the relative composition of the two functions, which may be employed as a means to control selectivity or deactivation rates (20). In (19) reaction networks of the form:

$$A_{(g)} \rightleftarrows B_{(g)} \begin{matrix} \xrightarrow{X} C \\ \xrightarrow[Y]{} R_{(g)} \end{matrix} \qquad \text{(III)}$$

were analyzed, with coke formation occurring on the X function and the desired reaction on the Y function. Both composite (mechanical mixtures of X and Y) and discrete formulations were investigated using a two phase, isothermal, quasi-steady state model, assuming a linear relation between coke content and activity. The major factors explored were the effects of the relative magnitudes of the rate constants for individual steps on the optional catalyst formulation. In (20) a similar problem was addressed, but with emphasis on the interaction of intraparticle diffusion, the rates of deactivation, and catalyst formulation. Bifunctional forms of the series (Type III) reaction model were investigated in which either the reactant and intermediate or the intermediate and product were responsible for coking of the two functions. The most important conclusion of this study was that in either case, given a set formulation, the inclusion of some degree of diffusional limitation could be used to enhance both catalyst efficiency (yield of desired product) and life -- a fact noted before for some types of coking mechanisms in monofunctional catalysis (10).

Khang and Levenspiel (5) investigated the relationship between the global activity of a particle, a, and the point activity, s. The values of a so determined were used to compute the overall order of the deactivation reaction according to:

$$-\frac{da}{dt} = k_d C_{A_s} a^d \tag{4}$$

where:

$$a = \frac{3}{R^3 C_{A_s}} \epsilon \int_0^R C_A s \, r^2 \, dr \tag{5}$$

for a spherical particle of radius R, porosity ϵ, and reactant concentration at the surface C_{A_s}. A number of parallel, series and poisoning schemes were studied; in the first two schemes reaction products were taken to be coke precursors. For parallel deactivation and no diffusional influence, net deactivation order d was close to unity; as the Thiele modulus increased from 1 to 100 the net order varied from 1 to 3. For series deactivation $d = 1$ for all values of the Thiele modulus providing the ratio of intermediate to reactant concentration at the surface was greater than one tenth the value of the modulus. Failing this, d approached 3. For poisoning, providing there was no or small diffusion limit on the poison, d was near unity for any situation with regard to diffusion limits on the main reaction. For other combinations of diffusional limits on main and poisoning reactions the net order was generally greater than unity, up to about 3.

The intraparticle deactivation of nonisothermal pellets has been analyzed by Ray (21) using a pore mouth poisoning, slab geometry model. Heat flux at the boundary of the active and inactive portions of the particle (Figure 2a) was computed via Bischoff's (25) asymptotic solution for large Thiele modulus. An effective diffusional modulus for this case can be defined as:

$$\Phi = K(\delta_s) \frac{\beta_s}{\beta_p} h_s \frac{\alpha}{\Lambda} \tag{6}$$

where:

$$K(\delta) = \frac{\sqrt{2}}{\delta} \left[e^{\delta} - (1 + \delta) \right]^{\frac{1}{2}} \quad ; \quad \delta = \beta \gamma \, , \quad \delta_s = \beta_s \gamma_s$$

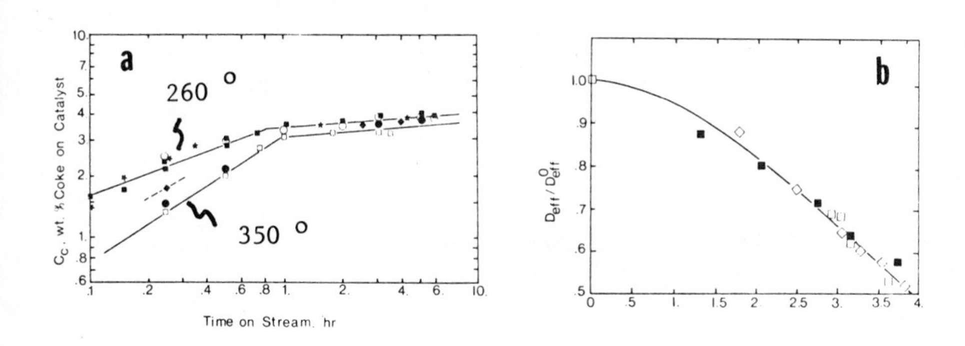

Chemical Engineering Science

Figure 1. (a) Voorhies correlation of coke formation, cumene cracking on H-mordenite, 260°–350°C, 0.33 g/hr-g (13); (b) effective diffusivity (SF_6) variation with coke content

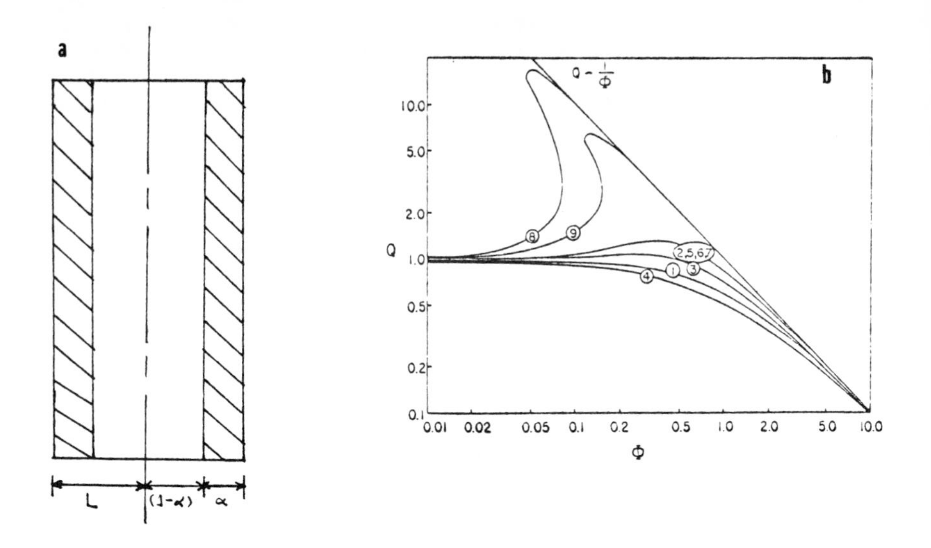

Chemical Engineering Science

Figure 2. (a) Geometry of partially deactivated pellet; (b) effective activity factor vs. *diffusional modulus for deactivation of nonisothermal pellet* (21)

$$\beta = \frac{\beta_s}{\beta_p}\left(\frac{1+\beta_p - y}{y}\right), \quad \beta_s = \frac{(-\Delta H)D\,C_s}{\lambda T_s}, \quad \beta_p = \frac{(-\Delta H)D_p C_s}{\lambda_p T_s}$$

$$\gamma = \frac{\gamma_s}{y}, \quad y = \frac{T_p}{T_s}, \quad \gamma_s = \frac{E}{RT_s}, \quad \Lambda = \frac{\lambda}{\lambda_p}$$

and

$$h_s = \left\{\frac{S_g A L^2 \exp[-E/RT_s]}{D}\right\}^{\frac{1}{2}}$$

Here S_g is the catalyst density and A the pre-exponential factor. Now the heat flux from the active part of the pellet, q_{II} at $(1-\alpha)L$ is:

$$q_{II} = K(\delta)\frac{\beta_s h_s}{\beta_p}(1+\beta_p - y) \tag{7}$$

and an effective activity factor can be defined as the ratio of this value to that for the undeactivated particle. This ratio, Q, is plotted against Φ in Figure 2b for various values of the parameters γ_s, β_s, β_p, and h_s. If $(\beta_s/\beta_p) < 1$, (Cases 2, 3, 5-9), the deactivated section has a lower mass transfer resistance (or higher heat transfer resistance) than the active portion, producing a beneficial effect on the diffusion effectiveness factor. For an exothermic reaction this can affect the loss in active surface such that Q can increase temporarily with deactivation. A sufficient condition for $Q > 1$ for some portion of the catalyst life was found to be:

$$\frac{\beta_s}{\beta_p} < \frac{e^{\delta} - (1+\delta_s)}{e^{\delta_s} - 1} \tag{8}$$

If $(\beta_s/\beta_p) > 1$ (Case 4), the opposite holds and deactivation is accelerated.

As indicated by Cases 8 and 9, some parameter sets result in multiple steady states. These are characterized by large δ_s and γ_s (60 in these examples) and $(\beta_s/\beta_p) < 1$; such results indicate that multiplicity can be induced in a deactivating particle even when the fresh catalyst shows no such possibility. However, $(\beta_s/\beta_p) < 1$ seems physically improbable. If deactivation by coking closes off a portion of the pore structure, then $(\beta_s/\beta_p) > 1$, while poisoning should have little effect on the ratio. This multiplicity would require a type of deactivation leading to an increase in porosity as decay proceeds.

In many industrial applications the performance of catalysts in which the activity distribution is nonuniform within the pellet is superior to those with a uniform distribution. Mars and Grogels (26) studied the performance of such a catalyst for acetylene hydrogenation and Friedrichsen (27) reports applications in oxidation of o-xylene to phthalic anhydride, to cite two examples. Recently Shadman-Yazdi and Petersen (22), Corbett and Luss (23) and Becker and Wei (24) have all investigated the deactivation of catalysts with spatial distribution of the catalytic function. These studies are similar in method and objective but differ considerably in the types of activity distribution considered; a summary of the models and deactivation mechanisms investigated is given in Table 1. In (22) the sensitivity of the deactivation with respect to the distribution parameter α was investigated. In line with prior results (10), $\alpha = 0$ gave rise to core poisoning at higher values of the Thiele modulus; $\alpha > 0$ led to uniform poisoning -- the result of compensating variations in k_B and C_B within the particle. An analytical solution presented in terms of the effectiveness factor for $\alpha > 0$ was:

$$\eta_A(\theta) = \sum_{m=1}^{\infty} \frac{(m/p)a_m}{h_A^2} \tag{9}$$

where:

$$p = 1/(\alpha + 2), \quad \delta = h_A(1 - w)^{\frac{1}{2}}, \quad a_m = \frac{(p\delta)^{2m-p}}{(m)!(m-p)!I_{-p}(2p\delta)}$$

$$\theta = C_A^o k_B^o t/q_o$$

and I_{-p} is a modified Bessel function of the first kind of order p. The parameter w appearing in δ was determined as a function of θ by numerical integration of:

$$\frac{dw}{d\theta} = (1 - w)\left[\frac{1}{1+\alpha} - \sum_{m=0}^{\infty} \frac{a_m}{(m/p) + 1 + \alpha}\right] \tag{10}$$

Figure 3 gives a plot of $\eta_A(\theta)$ vs. θ for several types of activity distribution. Clearly, increasing α enhances the long-term performance of the catalyst for this series reaction in a manner reminiscent of the response of the parallel scheme to incorporation of a certain amount of diffusional resistance (10).

In Figure 4 are given some results obtained numerically by Corbett and Luss (23) for impurity poisoning of the series reaction scheme for cases of small and large diffusional resistance. The effectiveness factor in these plots is computed with respect to the initial volume-averaged rate constant, which was the same for all the different distributions. The differential selectivity, s, appearing in the figure is defined as:

$$s = -\frac{(dv/dx)}{(du/dx)} \quad \text{at} \quad x = 1 \tag{11}$$

Table 1. Studies of the Deactivation of Catalysts with Distributed Activity Functions

	Author	Reaction	Mechanism of Deactivation	Geometry	Activity Distribution	Deactivation Function
1.	Shadman-Yazdi and Petersen (22)	$A \xrightarrow{k_A} B \xrightarrow{k_B} C$	Coking by C	Slab, Spherical (no decay)	$k_A = k_A^o\ a$ $k_B = k_B^o\ a$ $a = (x/L)^\alpha$	Linear, with $\frac{dq}{dt} = k_B^o \left(\frac{x}{L}\right)^\alpha s\ C_B$ $s = (1-q/q_o)$
2.	Corbett and Luss (23)	$A \xrightarrow{k_A} B \xrightarrow{k_B} C$	Impurity Poisoning, Coking by C	Spherical	$k_i = \bar{k}_i\ a$, $a = 4(x)^9$, $a = \frac{4x}{3}$ $a = 2.5 - 2x$	Poisoning: $\frac{ds}{dt} = -K\ s^n$ $s = \frac{k(x,t)}{k(x,0)}$ Coking: $s = (1-q/q_o)$
3.	Becker and Wei (24)	$A \xrightarrow{k_m} B$	Impurity Poisoning	Spherical	Discontinuous: inner, middle or outer	$k_m = k_m^o\ s$ $s = e^{-\alpha_p C_w}$ $\frac{dC_w}{dt} = k_p' C_p$

Notes:

a) In (22) q is amount of coke/cm^3 catalyst, q_o is maximum amount

b) In (23) $x = (r/R)$, $\bar{k}_i$ is volume averaged rate constant, q is amount of product adscribed/volume, q_o amount of product/volume causing complete deactivation

c) In (24) C_w is surface poison concentration, g mole/cm^2, C_p is concentration of poison

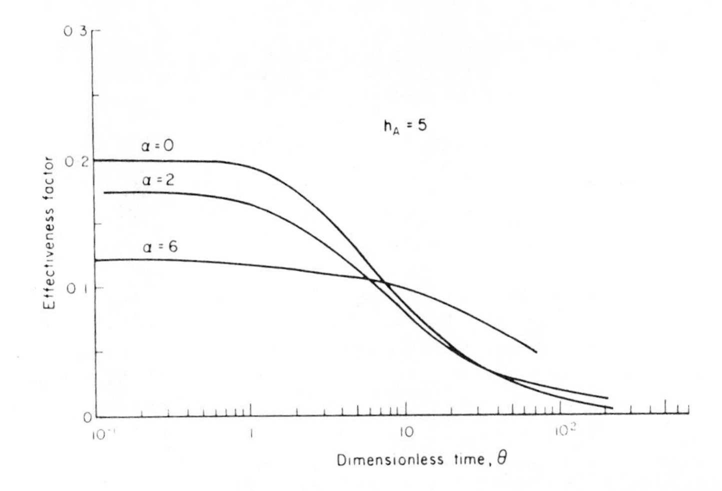

Chemical Engineering Science

Figure 3. Effect of activity distribution on long-term deactivation via coke formation in series reaction (22)

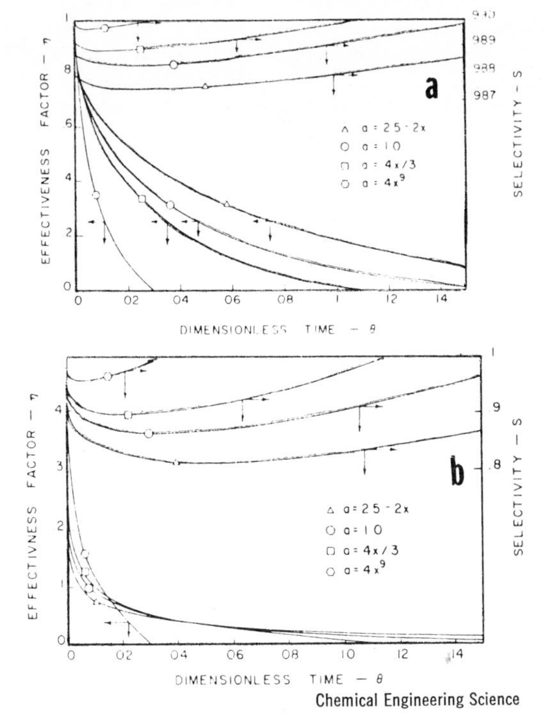

Chemical Engineering Science

Figure 4. Effectiveness and selectivity variation in impurity poisoning, series reaction, with different activity distributions (23). (a) Small influence of diffusion, $h_A = 1$, $h_B = 0.1$; (b) large influence of diffusion, $h_A = 10$, $h_B = 1$.

where: $u = \frac{C_A}{C_{A_s}}$, $v = \frac{D_B\, C_B}{C_{A_s} D_A}$, $x = (r/R)$

Dimensionless time, θ, is defined as $(\alpha\, D_P C_{P_s}\; t/NR^2)$ with α the number of active sites/volume, N the volume average active sites/volume, and C_{P_s} the poison concentration at the surface. When diffusional resistance is small it is seen that the best selectivity but the most rapid decline in effectiveness obtains when most of the catalytic function is near the pellet surface. Conversely, the smallest rate of decay is obtained by concentrating most of the catalytic function near the center (how to do this might present an interesting exercise in catalyst manufacture), so an optimal choice of catalyst would involve compromise between the selectivity and activity factors. When diffusional resistances become large both selectivity and effectiveness factor are improved by increasing the density of the catalytic function near the surface. Corbett and Luss also present results for catalyst fouling, as indicated in Table 1, which we shall not attempt to discuss here.

Becker and Wei (24), in the practical thought that catalysts with continuously varying distributions of activity might not be so easy to prepare, investigated several types of discontinuous activity distributions. Their qualitative results are similar to those detailed above, but an important assumption involved in the analysis is that poison is adsorbed on the support devoid of catalytic function as well as on the active surface. Thus, for example, when the poison is highly diffusion limited but the main reaction is not, a central active core surrounded by inert support (they call it an "egg yolk") would be the preferred configuration. Such would be predicted by the other models as well, but the detailed ranges of parameters where such preference exists would differ significantly.

Additional theoretical investigations of the intraparticle deactivation problem, which unfortunately we cannot treat in detail here, have been reported by Luss and co-workers (28,29) on the modification of selectivity upon poisoning, and by Hegedus (30) on the combined influence of interphase and intraparticle gradients on deactivation. It is of interest that deactivation in certain instances can actually have beneficial results on selectivity and in the long run the problem may be to achieve the best balance between diminished activity and enhanced selectivity; such results are reminiscent of those pertaining to deactivation of bifunctional catalysts (19,20).

Many experimentalists have been preoccupied over the past few years with the development of various types of gradientless reactors for the study of catalytic kinetics; it is interesting to see now the emergence of some reactor designs which are aimed at producing specific kinds of gradients which may be measured and used as the basis for model interpretation or parameter estimation in reaction engineering problems involving catalyst deactivation. These reactors share the common aspect of being single pellet reactors, but differ as to whether concentration gradients or temperature gradients are the measured quantity. The single pellet diffusion reactor developed by Petersen and co-workers (31,32, 33,34,35,36,37,38) is designed for concentration measurements, while somewhat different designs employed in this laboratory (39, 40,41), by Hughes and co-workers (42,43), Benham and Denny (44), and Trimm, et al. (45), provide temperature profile data. Schematic diagrams of the two types are given in Figure 5.

The general utility, properties and application of the single pellet diffusion reactor were reviewed by Hegedus and Petersen (36) in 1974, and it has been employed in extensive investigation of the interactions of diffusion, reaction and deactivation for the hydrogenolysis of cyclopropane on Pt/Al_2O_3 (34) and the dehydrogenation of methylcyclohexane, also on Pt/Al_2O_3 (38). The basis of the method resides in the fact that measurement of center plane concentrations (Figure 5a) vs. relative rate of reaction provide data sufficient to discriminate among various types of deactivation mechanisms and to allow parameter determination. In the original analysis (33) impurity poisoning, and parallel, series and triangular self-deactivation were considered. The appropriate equations were integrated for a set value of the Thiele modulus (2.5) for isothermal conditions; a summary of results for the various mechanisms is set forth in Figure 6a. Here the normalized center plane concentration is

$$\frac{\varphi_A\ (\tau,\ \eta=1) - \varphi_A\ (\tau=0,\ \eta=1)}{1 - \varphi_A\ (\tau=0,\ \eta=1)}$$

where τ is a scaled time variable, η is the effectiveness factor, and $\varphi_A = C_A\ (t,x)/C_A\ (t=0,\ x=0)$. The reaction rate ratio is that of the overall rate to that at zero time. Parallel and series self-deactivation were confined to relatively narrow bands designated by 2 and 3 in Figure 6a. Impurity poisoning could occur in regions 1 , 2 or 3 , while triangular self-deactivation is confined to zones 2 to 5 . Region 6 is inaccesible except for strongly diffusionally limited series self-deactivation. Decreasing the value of the Thiele modulus causes all zones to move toward the center diagonal. Clearly there can be coincidence among these regions for various values of the parameters and differing mechanisms; experimentally one makes the discrimination by evaluating results obtained at different temperatures or concentrations (where presumably only the Thiele modulus will change) for internal consistency as to parameter values such as reaction order, etc. The applications of these procedures (34,35) for cyclopropane hydrogenolysis resulted in postulation of a triangular self-poisoning model with the parallel mode predominating at

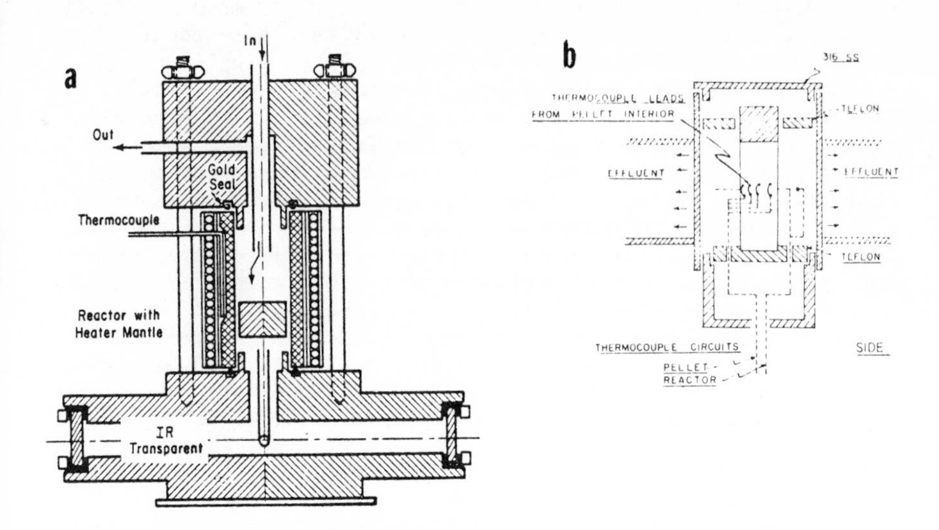

Figure 5. (a) Single pellet diffusion reactor—concentration measurements (36); (b) single pellet diffusion reactor—temperature measurements (39)

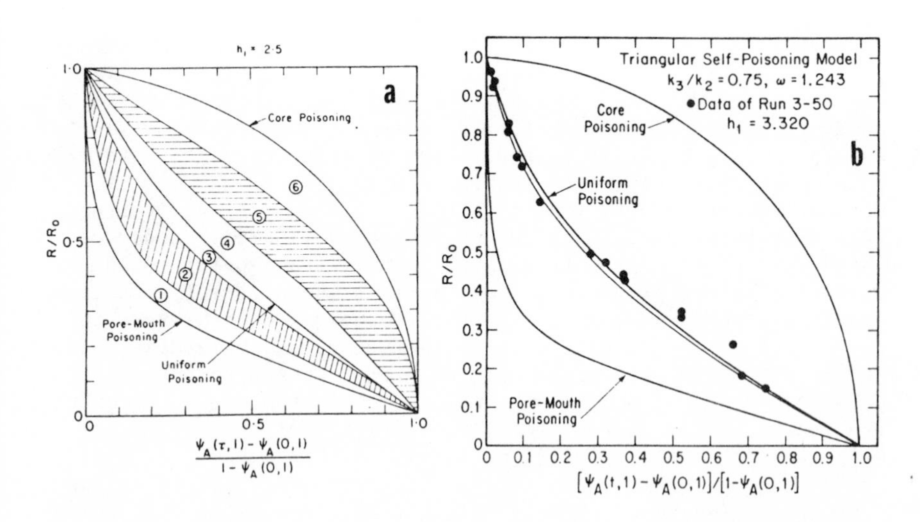

Chemical Engineering Science

Figure 6. (a) Qualitative discrimination among poisoning models (33); (b) results of interpretation of deactivation data for cyclopropane hydrogenolysis on supported Pt (34)

lower temperature and the series mode at higher temperature. One example of the final fit obtained is given in Figure 6b, where $\underline{k}_2$ is for the parallel step and $\underline{k}_3$ for the series step. It is seen that the deactivation here is relatively uniform.

More recent work by Wolf and Petersen (37,38) has investigated the influence of the order of the main reaction on single pellet diffusion reactor behavior, using the same model deactivation mechanisms studied before. Experimental application to a reaction with complicated kinetics (and complicated deactivation behavior) was reported for methylcyclohexane dehydrogenation.

Measurements of temperature profiles in a single pellet diffusion reactor were originally made for reactions in the absence of deactivation (39,42,44,45), but the technique is obviously easily extended to include this possibility, and experimental results have been reported for O_2 poisoning of Ni/SiO_2-Al_2O_3 in ethylene hydrogenation (43) and thiophene poisoning of Ni/kieselguhr in benzene hydrogenation (40,41). The latter work has included unsteady state as well as steady state experiments; steady state temperature profiles for both fresh and deactivated catalysts were obtained as the final state of two types of transient measurements, (i) start-up experiments measuring the development of temperature profiles on introduction of benzene into the reactor initially containing only flowing hydrogen, and (ii) perturbation experiments measuring the response of temperature profiles to step changes in concentration or flow rate starting from a given initial steady state. On deactivated catalysts these measurements were made with a poison-free feed after the catalyst had been pre-poisoned to a specified level of activity (as determined by the global rate of reaction). Typical data for the temperature response on startup for a typical series is shown in Figure 7 for the fresh catalyst and the same sample deactivated to 40% and 16% of original activity. Of particular note in these results is the relative insensitivity of the time scale of startup to extent of deactivation, and the obvious development for $\underline{s}$ = 0.16 of an outer, dead zone in the catalyst particle. A second interesting result was the documentation of wandering hot spots upon startup, a phenomenon treated theoretically by Hlavacek and Marek (46) and Lee and Luss (47) in prior work. Three types of transients were observed as shown in Figure 8a, the behavior characterized in general by the magnitude of the Dämkohler number (h^2/N_{Bi_m}). Such excursions were observed for both fresh and deactivated catalysts and so cannot be attributed to some particular aspect of the poisoning mechanism. Finally, it was noted in a number of cases for poisoned catalysts (41, 43) that intraparticle temperature gradients were greater than interphase gradients corresponding, contrary to conventional wisdom on that matter.

The interpretation of these experimental results differed somwhat in approach from that of Petersen and co-workers, since the mechanism of deactivation (parallel poisoning) is well

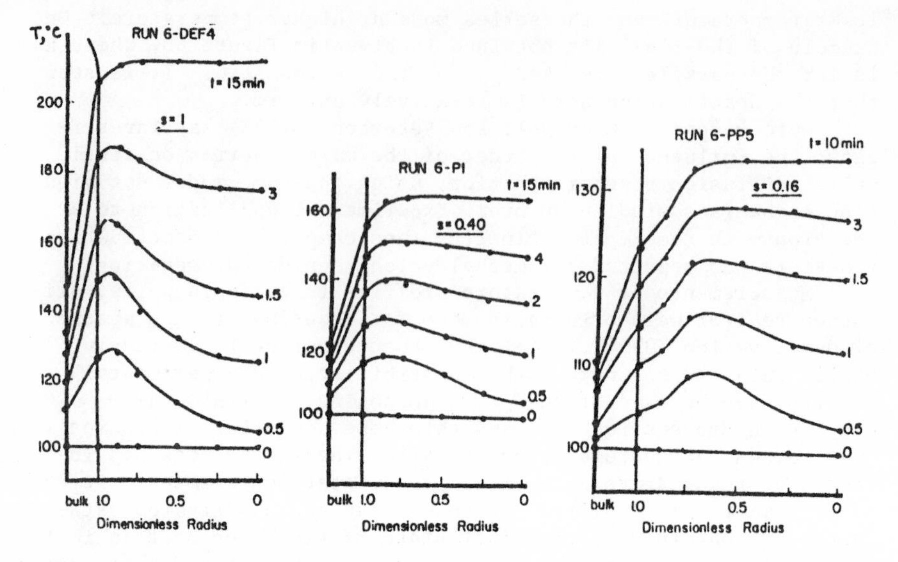

Figure 7. Temperature response on startup for fresh and poisoned Ni/Kieselguhr for benzene hydrogenation (41). C_6H_6 feed = 16 mol %.

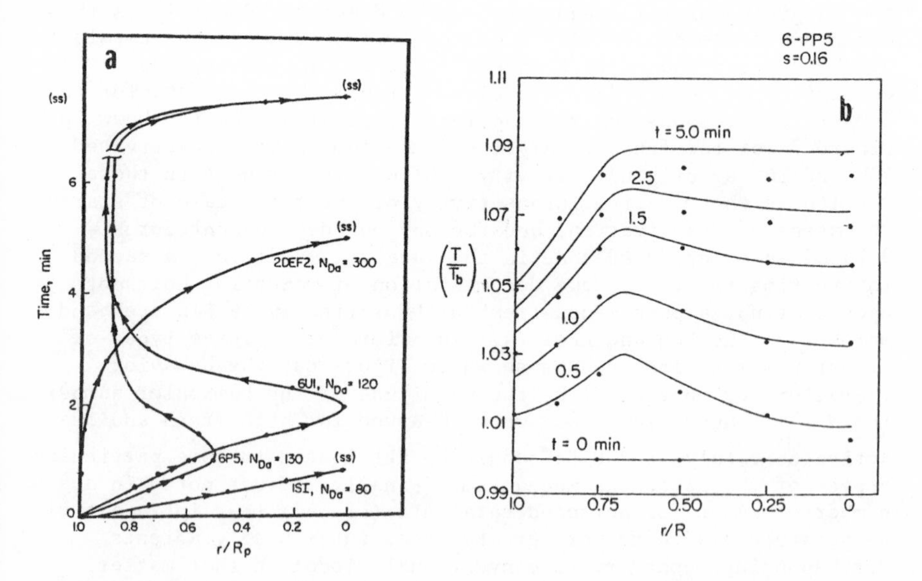

Figure 8. (a) Wandering hot spots in startup experiments. 1SI and 6U1 for s = 1, 6P5 for s = 0.4, 2DEF2 for s = 0.16 (41); (b) results of simulation of startup experiment for s = 0.16 (41).

identified and there is less uncertainty about appropriate models to use. In fact, all parameters of the system, kinetics, transport and poisoning, were measured in separate experimentation (39) and an attempt was made to simulate the steady state and unsteady state temperature profiles (both intraparticle and interphase) on an *a priori* basis. Typical results are shown in Figure 8b; the results seem moderately encouraging; however, it was still necessary to adjust the values of catalyst effective diffusivity by 50% to obtain the fit shown.

Reactor Modeling

An important development over the past few years is the large increase in those investigators who have been willing to determine the kinetics of deactivation in various reaction systems, mostly by coking, and then incorporate this information into appropriate models to interpret or predict integral reactor performance. Most of this has been done for fixed beds but important examples treat moving and fluidized beds as well.

Particularly thorough studies of two systems, isomerization of n-pentane on Pt-reforming catalysts and dehydrogenation of 1-butene on Cr_2O_3/Al_2O_3, have been provided by Froment and coworkers (48,49,50). Space precludes a complete discussion; here we summarize the n-pentane results. In (48) experiments were conducted in a fixed-bed at 2.5 atm, (H_2/n-pentane) from 1.0 to 1.6, and temperature from 388 to 427°C. The reactor was designed so that gas samples could be removed from different sections of the bed, however coke on catalyst could be measured only at the end of a run. Separate experiments were run at high H_2/n-pentane to determine the intrinsic kinetics in the absence of coking (51). In Figure 9 are given the experimental coke and n-pentane profiles measured at 388°C. The reaction scheme proposed was:

$$\begin{array}{ccc} & n - C_5 \rightleftarrows i - C_5 & \\ \swarrow & \searrow & \\ \text{coke} & \text{hydrogenated products} & \end{array} \qquad \text{(IV)}$$

in accord with the observed coking data. Under the assumption of constant density, isothermal operation, and no change in the number of moles, the following model was set up:

$$\text{n-pentane}: \frac{\partial y_A}{\partial t} + \frac{\partial y_A}{\partial z} = - \frac{\rho_B \Omega d_p}{F_T} r_A \qquad (12)$$

$$\text{hydrocracking}: \frac{\partial y_H}{\partial t} + \frac{\partial y_H}{\partial z} = \frac{\rho_B \Omega d_p}{F_T} r_H \qquad (13)$$

$$\text{coke}: \frac{\partial C}{\partial t} = \frac{\epsilon \Omega d_p P_T}{F_T} r_c \qquad (14)$$

with the kinetics of reaction and coking represented by:

$$r_I = \frac{k_I^o\,[y_A - y_B/K]}{y_w + K_B y_B} \quad \text{(from reference (51))} \tag{15}$$

$$r_A = r_I + 0.5\,r_H + S\,r_c \tag{16}$$

where r_H was fit to the equation:

$$r_H = k_H\,(y_A/y_w)^{n_1} \tag{17}$$

and $r_c = k_c\,(y_A/y_w)^{n_2}$; Dimensionless time $= \dfrac{t'F_T RT}{\epsilon\Omega d_p P_T}$ (18)

Separable deactivation kinetics were incorporated into the model via:

$$k_I = k_I^o \cdot s$$

$$k_H = k_H^o \cdot s \tag{19}$$

$$k_c = k_c^o \cdot s$$

with $s = e^{-\alpha C^{0.5}}$ (s identical for all rate functions).

In the notation above ϵ is void fraction, Ω reactor cross section, d_p particle diameter, P_T, total pressure, F_T total feed rate, and S a conversion factor for moles $n\text{-}C_5$/wt coke. From experimental data obtained under coking conditions the model parameters were determined with typical results given in Table 2. The absolute

Table 2. Parametric Values from Integral Reactor Modeling: n-Pentane Isomerization on Pt/Al_2O_3. (48)

Temp. (°C)	α	k_I^o	k_H^o	n_1	k_c^o	n_2
427	14.6	0.091	0.052	7.1	0.0022	1.9
413	17.5	0.087	0.067	4.8	0.0016	1.8
402	17.7	0.089	0.037	12.0	0.0020	2.2
388	15.0	0.026	0.025	11.0	0.00054	3.9

values shown are perhaps not as of much interest as their trends with temperature; in particular α and n_2 are nearly temperature independent while n_1 is not. The value of k_I^o evaluated from this fit under coking conditions was in agreement with prior work (51) and obeyed an Arrhenius correlation. Confrontation of the experimental and computed results with these parameters is given in Figures 9a and 10. The fact that α and n_2 are independent of temperature tends to lend *ex post facto* support to the use of

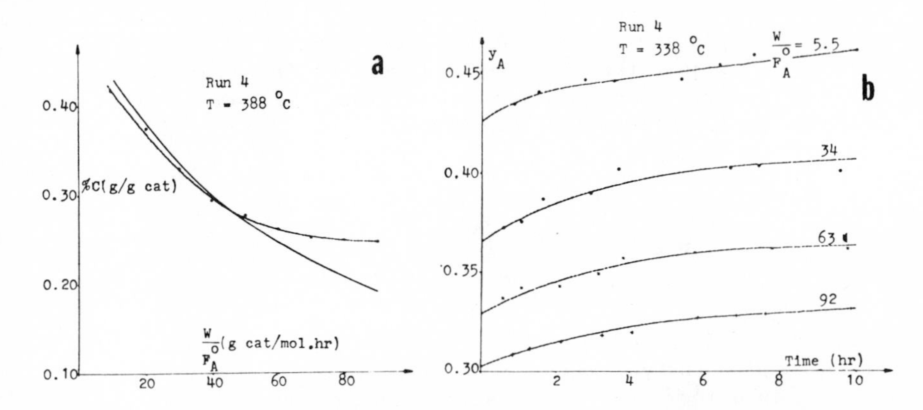

Proceedings of the 5th European Symposium on Chemical Reaction Engineering

Figure 9. (a) Wt. % coke on catalyst after 11 hr for different space velocities (48); (b) experimental n-pentane *mole fractions* vs. *time at different space velocities (48)*

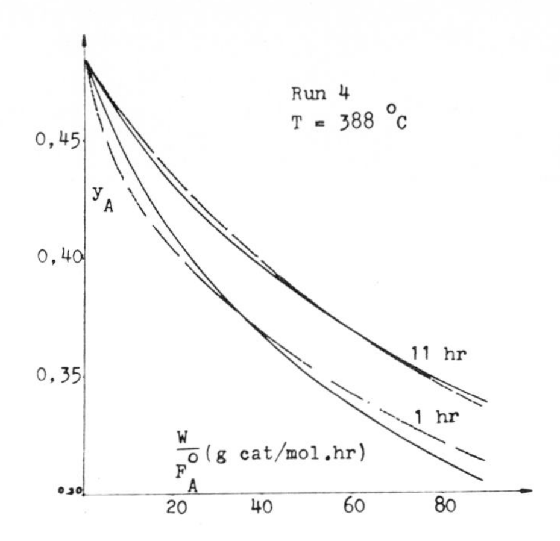

Proceedings of the 5th European Symposium on Chemical Reaction Engineering

Figure 10. Cross plot of n-pentane *mole fraction profiles (48); (_ _ _ _) experimental; (______) computed.*

separable kinetics in modeling of deactivation by coking.

In a second study of the n-pentane system, De Pauw and Froment (49) conducted a similar series of experiments but at much higher H_2/n-pentane ratios. Under these conditions they found coke profiles just the opposite of Figure 9a (i.e., increasing with W/F_A^o) and it was necessary to increase the complexity of the reaction model to interpret the data:

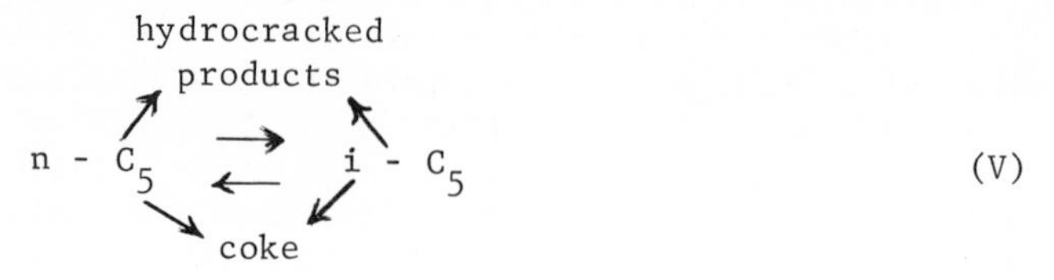

This schema is interesting since it admits the simultaneous deactivation of both hydrocracking and isomerization functions, as well as coke formation from reactants and products. A rather complicated expression was fit to separate thermobalance experiments on the rate of coke formation, and the effect of deactivation on kinetics again taken to be separable with s a negative exponential in coke content. Satisfactory fits to the data were obtained; different values for α for the isomerization and hydrocracking functions were obtained and in this case they were temperature dependent.

The results above for n-pentane are representative and indicative of a number of efforts at detailed modeling of reactors and reactions under conditions of coke formation. Yet, they are parameter-fit models which explain certain regions of experimentation very well and are undoubtedly useful for interpolation. What about extrapolation? For that purpose one would wish to have all parameters of a reasonable model determined in separate experimentation and confrontation with experiment accomplished on an *a priori* basis; if reasonable agreement is obtained model extrapolations may not be so uncomfortable. This has been accomplished in large measure for 1-butene dehydrogenation over Cr_2O_3/Al_2O_3 by Dumez and Froment (50). The model used is complex, accounting for diffusional as well as kinetic and coking effects, and a large amount of independent experimentation was required to determine the kinetic and deactivation parameters; excellent agreement was obtained between experimental and computational results. On the basis of this success, Dumez and Froment also report the simulation of an industrial scale reactor for butene hydrogenation.

Catalyst coking and integral reactor performance studies involving parametric fitting, similar to (48), have been carried out for a number of other reaction/catalyst systems and are summarized in Table 3. Although individual details vary among the various investigations, they share in common the mechanism of deactivation via coke formation. In summary one might conclude that the development and testing of reliable interpolation models for coking effects on reactor performance represents an important advance in the state of the art over the past eight years.

Detailed reactor modeling studies of dynamics as they are induced by or influenced by poisoning are reported by Blaum (52), and some experiments in which parallel poisoning is the mechanism of deactivation have been reported by Weng, et al. (53) and Price and Butt (54) for benzene hydrogenation on Ni/kieselguhr with thiophene poison. These studies were also attempted on an *a priori* basis similar to that of Dumez and Froment, with the many kinetic, reactor and poisoning parameters determined in separate experimentation (39,53). Both adiabatic and nonadiabatic operation were investigated; a characteristic of either type of operation for this highly exothermic reaction with rapid, irreversible poisoning is the formation of a moving reaction zone which passes through the catalyst bed with continuing time of operation. This is observed experimentally as a moving hot spot in nonadiabatic reaction or a moving thermal front in adiabatic reaction, as shown in Figure 11. Corresponding to the exit of the reaction zone from the bed is a catastrophic decrease in conversion. Modeling of the system can be carried out at several levels of effort, i.e., simple bed life, rate of motion of the thermal front, or detailed shapes and dynamics of the temperature profiles. The latter was attempted in both cases (53,54); it was found necessary to use a one-dimensional dispersion model for the energy balance, while a simple plug flow model was sufficient for the mass conservation relationships. The kinetics of main reaction and poisoning were correlated by:

$$-r_B = \frac{a\ \exp(-E_a/RT)C_B}{1 + b\ \exp(Q/RT)C_B} \cdot s \tag{20}$$

$$-r_d = -\frac{ds}{dt} = k_d^o\ \exp(-E_a/RT)C_T \cdot s \tag{21}$$

and

$$r_T = M_T\ r_d \tag{22}$$

where E_a, Q and E_d express temperature dependence of main reaction, adsorption and poisoning respectively, C_B and C_T are concentrations of benzene and thiophene, r_T the rate of uptake of thiophene per unit weight of catalyst, and M_T the total adsorption capacity of catalyst for thiophene. It is seen the poisoning kinetics are modeled as separable, and with this formulation it was found (53) that the simulation was extremely, even unreasonably, sensitive to the value of the parameter M_T. Similar results pertain to the simulation under adiabatic conditions (results for both cases are also given in Figure 11). Such parametric sensitivity with respect to the poisoning parameter indicates possible inadequacies in the separable model. Hopefully we will see some experimental studies of this in the near future.

Table 3. Catalyst Deactivation - Integral Reactor Studies

	Author	Reaction	Catalyst	Conditions	Reaction Scheme
1.	Noda, et al. (55)	i-pentane dehydrogenation	Cr_2O_3/Al_2O_3 (8/10, 16/32 and 40/60 mesh)	488-560°C, 1 atm 0.6-3.6 sec R.T. x_F = 0.14	(A) $iC_5 \xrightarrow{1}$ (B) $iC_5^{=} \xrightarrow{3}$ (C) isoprene; $iC_5 \xrightarrow{2}$ other products; $iC_5^{=} \xrightarrow{4}$ coke (E); isoprene $\xrightarrow{5}$ coke (E)
2.	Toel, et al. (56)	n-butane dehydrogenation	20% $Cr_2O_3/Al_2O_3(\gamma)$ 1% $Cr_2O_3/Al_2O_3(\gamma)$ 20% $Cr_2O_3/Al_2O_3(\alpha)$	550°C, 1 atm P_{nC_4} = 0.11-0.56 atm	$nC_4 \xrightarrow[(k_1)]{} nC_4^{=} \xrightarrow[(k_2)]{}$ coke
3.	Uchida, et al. (57) Otake, et al. (58)	n-butane dehydrogenation	Cr_2O_3/Al_2O_3 200 m^2/g	497-592°C, 1 atm P_{nC_4} = 0.32-0.71 atm P_{nC_4} = 0.10-0.52 atm	$nC_4 \longrightarrow nC_4^{=} \xrightarrow[(k_c)]{}$ coke
4.	Sadana and Doriaswamy (59)	theoretical investigation	--	--	a) $A \longrightarrow R$ b) $A \longrightarrow R \longrightarrow S$
5.	Prasad and Doriaswamy (60)	chlorination of tetrachloroethane	SiO_2, 30-40 mesh	200°C	tetra $\xrightarrow[(k_1)]{}$ penta $\xrightarrow[(k_2)]{}$ hexa
6.	Greco, et al. (61)	dehydration of 2-methyl-3 butene-2ol to isoprene	Al_2O_3	260-300°C	$A \longrightarrow R$ $A \longrightarrow$ coke
7.	Dumez and Froment (50)	1-butene dehydrogenation	20% Cr_2O_3/Al_2O_3 57 m^2/g	490-600°C[a] $P_{C_4^{=}}$ = 0.02-0.27 atm P_{H_2} = 0-0.10 atm P_{diene} = 0-0.10 atm	$C_4^{=} \longrightarrow$ diene; $C_4^{=} \xrightarrow{k_3}$ coke; diene $\xrightarrow{k_2}$ coke
8.	Weekman, et al. (62,63,64,65,66)[b]	catalytic cracking	SiO_2/Al_2O_3, amorph. & zeolite	~900°F	Feed $\xrightarrow{k_1}$ gasoline; Feed $\longrightarrow$ dry gas and coke; gasoline $\longrightarrow$ dry gas and coke
9.	Wojciechowski, et al. (67,68,69,70, 71,72,73,74)[c]	cumene cracking	La-Y zeolite 20/25-100/140 mesh	360-500°C (67)	
10.	Weng, et al. (53), Price and Butt (54)	benzene hydrogenation	Ni/kieselguhr 12/20 mesh	50-200°C x_B = 0.10-0.25 x_T = 300-3000 ppm	$C_6H_6 + 3H_2 \longrightarrow C_6H_{12}$ $T + S \longrightarrow T \cdot S$

NOTES

a) Conditions for hydrogenation kinetics, slightly different conditions employed for coking kinetics.
b) Results formulated for fixed, moving and fluid bed reactors; some details have been reviewed in (1).
c) Notation in reaction scheme: C = cumene; Y, Z = products; S = site; CS, etc. = adsorbed species.
k_2 : CS $\rightarrow$ YS, k_{-2} : YS $\rightarrow$ CS; K_1 : S $\rightleftarrows$ CS; K_3 : S $\rightleftarrows$ YS, K_4 : S $\rightleftarrows$ ZS.

	Form of Deactivation Kinetics	Form of Main Reaction Kinetics	Activity vs. Coke Content	Activation Energy for Coke Formation	Intraparticle Diffusion Effects
1.	$r_c = s(k_4 C_B^p + k_5 C_c^q)$ $p = q = 1$	$r_A = s(k_1 + k_2)C_A$ $r_B = sk_1 C_A - s(k_3 + k_4)C_B^p$	$s = 1 - \alpha C_c$	4 = 12.6 kcal/mole 5 = 10.6 kcal/mole	No, for < 16/32 mesh
2.	$r_c = \frac{sk_2 P_{nC_4^=}}{(1 + K_{C_4} P_{C_4})}$	$r_{C_4} = \frac{-sk_1 P_{C_4}}{(1 + K_{C_4} P_{C_4})}$	$s = 1 - \alpha C_c$	-	No
3.	$r_c = sk_c P_{C_4^=}$	$r_{C_4} = (sk_L + k_B)P_{C_4}$ k_L = Lewis acidity k_B = Bronsted acidity	$s = 1 - \alpha C_c$ (better than exp)	8.9 kcal/mole	Yes
4.	Contact time (linear, exp, algebraic)	$r_A = sk_A C_A^n$	-	-	No
5.	Contact time (linear, exp)	$r_T = -k_1 C_T s$ $r_p = (k_1 C_T - k_2 C_p)s$	-	-	No
6.	$r_c = k_c P_{MBE}$ $r_c = k_c$	$r_m = -s k_m$	unity unity	-	Yes (345°C) No (271°C)
7.	$r_c = \frac{(k_{CB}^o P_B^{n_1} + k_{CD}^o P_D^{n_2})s}{(1+K_{CH}\sqrt{P_H})^2}$	$r_H = \frac{-sk_H^o K_B (P_B - \frac{P_H P_D}{K})}{(1+K_B P_B + K_H P_H + K_D P_D)^2}$	$s = e^{-\alpha C_c}$ $(\alpha_H = \alpha_C)$	a) 32.8 kcal/mole, ($C_4^=$) b) 21.0 kcal/mole, (diene)	No (0.4-0.7 mm) Yes (4mm, industrial)
8.	Contact time (exp)	$r_F = -(k_1 + k_2)\ sy_F^2$ $r_G = k_1 sy_F^2 - k_2 sy_G$	$s = e^{-\alpha t_c}$ t_c = time	Data given as f(T)	Yes
9.	Time on stream (hyperbolic)	$r_c = \frac{-k_2 K_1 [s][C]}{(K_1[C] + K_3[Y] \cdots + k_{-2}K_3[Y][S][Z] + K_4[Z])}$	$s = (1 + Gt_c)^{-m}$ $(m = 1)$	360°-G = 0.0806 430°-G = 0.0259 500°-G = 0.0144	Yes (67) No (69)
10.	$r_s = -k_d C_T s$	$-r_B = \frac{a \exp(-E_a/RT) C_B s}{(1 + b \exp(Q/RT) C_B)}$	-	E_d = 1.1 kcal/mole	No

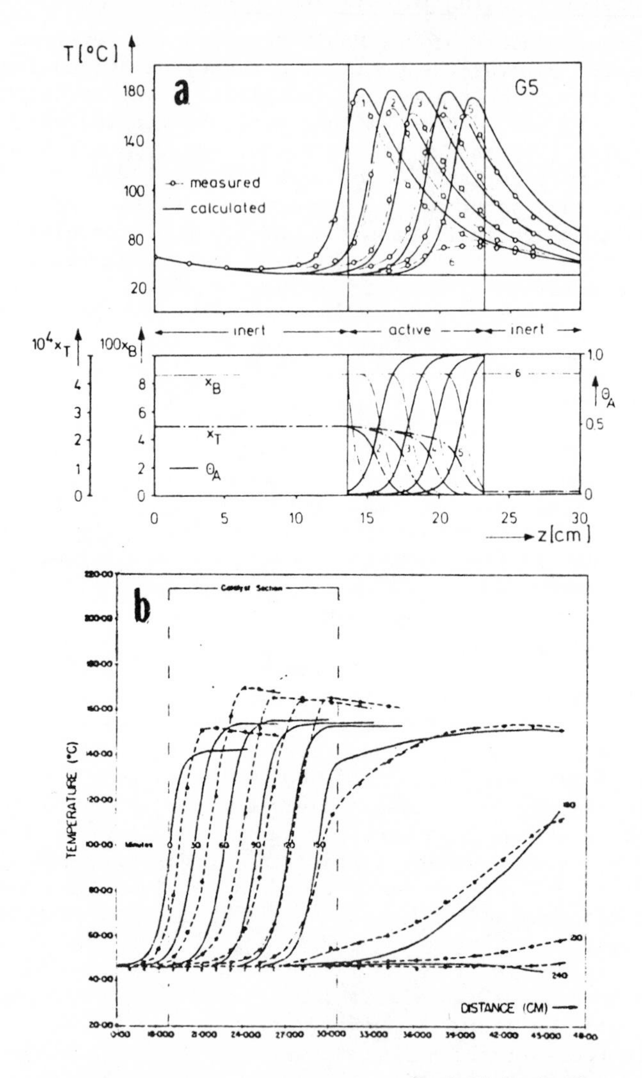

Chemical Engineering Science

Figure 11. Experimental and computed temperature profiles for a fixed bed reactor, parallel poisoning. (a) Hot spot migration, nonisothermal. Profiles at 60 min intervals (1 = 0 min). 4.3% C_6H_6, thiophene/C_6 = 5.65 $\times 10^{-3}$. $x_{B,T}$ = mole fractions, θ_A = fraction sites remaining (53); (b) active front migration, adiabatic. Profiles at 30 min intervals. 1.4% C_6H_6, 0.032% thiophene. Solid lines computed (54).

A Case History: Coking in Catalytic Cracking

Various members of the Mobil Research and Development Corp. have treated us over the years to a number of discussions on various aspects of their lumping, deactivation and reactor models for catalytic cracking. Some of this work was considered in the previous review (1), and we have done our own type of lumping in the form of entry 8 in Table 3 for work through 1971. The state of the art to that point may be summarized, perhaps too concisely, by the results given in Figures 12a and b. Good correlation of the activity of a large range of catlysts for feedstocks varying widely in composition was obtained simply on the basis of aromatic to naphthene weight ratio. Gross, et al. (75) subsequently extended this work in a thorough study of the cracking of three different feed stocks on zeolite catalysts in both fixed and fluidized beds. Their results confirmed that catalyst decay was independent of the reactor type, that gasoline selectivity was essentially the same in both reactors, and that the fixed bed was more efficient than the fluid bed. Such behavior was all predicted by the approaches developed previously, so the work of Gross, et al. stands as a rigorous test of those reactor and catalyst decay models.

The interesting point in Figure 12 is the failure of the correlation for the two feeds indicated as PC 32 and PA 37. These differed from all other feeds used in the correlation in that they were recycle stocks, and further experimentation revealed that the correlation failed as well for stocks containing poisons such as basic nitrogen compounds. This was investigated by Voltz, et al. (76) using the lumping, decay and reactor models of prior work. MCGO, MCGO plus quinoline, FCC fresh, and FCC with varying amounts of recycle were studied. MCGO plus quinoline (0.1% wt) only slightly reduced the catalyst decay constant at 900°F but reduced the rate constants for overall cracking and gasoline formation by about 50%; some decrease in gasoline selectivity was also noted. The addition of recycle stock to FCC fresh feed also had pronounced effects: at 50% recycle addition the rate constants for both overall cracking and gasoline formation decreased sharply while the gasoline cracking rate constant increased by a factor of 10. In both cases the alteration of feed stocks had profound effect upon the activation energy of the gasoline cracking reaction, increasing from a nominal value of 10 kcal/mole for MCGO to about 40 kcal/mole for MCGO plus 0.1% quinoline, and 23.5 kcal/mole for FCC plus 30% recycle stock.

Clearly the action of quinoline is to poison the acidic sites active for cracking; the deactivation model anticipated coking only and is clearly not capable of correlating results due to impurity poisoning. Recycle stocks differ from fresh feed primarily in the fact that their aromatic content is higher; also these aromatics in general do not have substituent groups, which in some way may account for their different reactivity in cracking.

Recently Jacobs, et al. (77) have set forth a lumping scheme which accounts in detail for the individual reactions of paraffins, naphthenes, aromatic rings, and aromatic substituent groups in both light and heavy fractions. This is shown in Figure 13. The major alteration, aside from the increased complexity, from the prior three lump scheme (Table 3) is with respect to aromatic rings with no substituent groups: these do not form gasoline and can crack ultimately only to the C lump. In addition, all reactions are treated as first order with an inhibition term for the adsorption of heavy aromatic rings. The deactivation function is still retained as separable but is more complex:

$$s = \frac{\alpha}{(\bar{P})^m (1 + \beta t_c)^\gamma} \tag{23}$$

where α, β and γ are deactivation parameters, $\bar{P}$ is oil partial pressure, and t_c catalyst residence time. The effect of nitrogen is also treated as an adsorption inhibition and the resultant function used as a scalar multiplier on the rate constant matrix. The overall scheme is demonstrated to provide excellent correlation of results with a seemingly endless range of feedstocks, catalysts, and reaction conditions.

Another substantial body of work dealing with cracking reactions has been reported over the past several years by Wojciechowski and coworkers. Results with cumene cracking over La-Y zeolite catalyst are summarized in entry 9 of Table 3. Catalyst deactivation in these studies is also treated as separable, but an hyperbolic function of time on stream is used for correlation of activity. The use of this type of correlation for a number of applications was summarized in a 1974 review (70). Since that time, in addition to recent work on cumene cracking (69) the model has been applied to an extensive series of studies on the cracking of gas oil distillates (71,72,73,74), as well as being employed in the correlation of Jacobs, et al. (77).

Another Case History: Coking of Nickel Catalysts

The mechanism of coke formation on various types of nickel catalysts during CO or CH_4 decomposition, or during steam reforming of hydrocarbons, has been a topic of intensive investigation during the past few years. Rostrup-Nielsen (78) investigated the decomposition reactions in the range 450-700°C; one would expect that coke originating from these decompositions could be controlled thermodynamically by operating with excess steam; however there has been uncertainty over the years concerning chemical equilibrium in these reactions (79,80,81). Hofer, et al. (82) had found via electron microscopy that the coke deposited on Ni was in the form of tubular, whisker-like threads; subsequent studies (83,84) have revealed two structures, Ni_3C (carbidic) and graphite

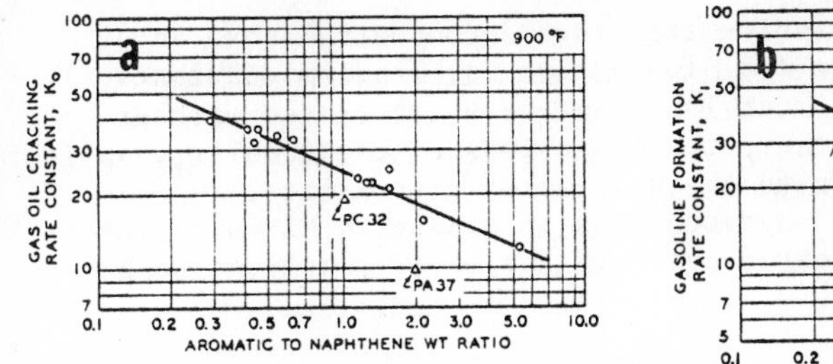

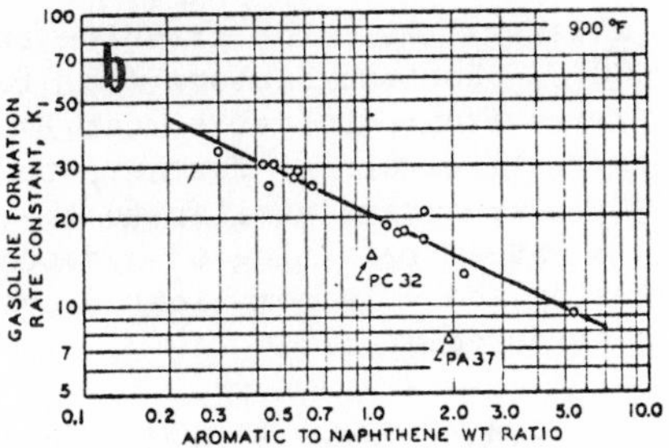

Industrial and Engineering Chemistry Process Design and Development Quarterly

Figure 12. Correlation of cracking kinetics with feed aromatic/naphthene (65). (a) Overall gas–oil cracking; (b) gasoline formation.

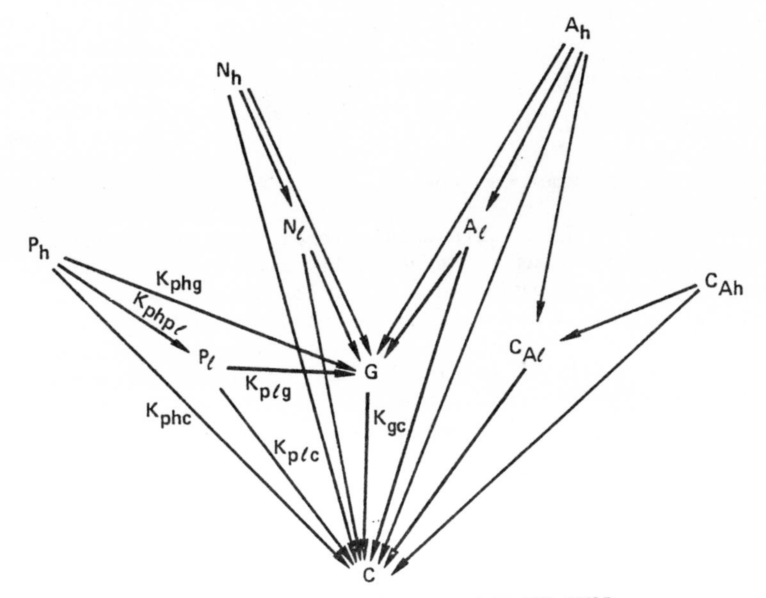

P_ℓ = Wt. % paraffinic molecules, (mass spec analysis), 430° - 650° F
N_ℓ = Wt. % naphthenic molecules, (mass spec analysis), 430° - 650° F
$C_{A\ell}$ = Wt. % carbon atoms among aromatic rings, (n-d-M method), 430° - 650° F
A_ℓ = Wt. % aromatic substituent groups (430° - 650° F)
P_h = Wt. % paraffinic molecules, (mass spec analysis), 650° F$^+$
N_h = Wt. % naphthenic molecules, (mass spec analysis), 650° F$^+$
C_{Ah} = Wt. % carbon atoms among aromatic rings, n-d-M method, 650° F$^+$
A_h = Wt. % aromatic substituent groups (650° F$^+$)
G = G lump (C_5 - 430° F)
C = C lump (C_1 to C_4 + COKE)
$C_{A\ell} + P_\ell + N_\ell + A_\ell$ = LFO (430° - 650° F)
$C_{Ah} + P_h + N_h + A_h$ = HFO (650° F$^+$)

American Institute of Chemical Engineers Journal

Figure 13. A ten lump model for the kinetics of catalytic cracking

(whiskers) with the graphite structure decomposing above about 400°C. Rostrup-Nielsen (78) found that equilibrium constants varied from catalyst to catalyst, but were correlated with Ni crystallite size; further, the dimensions of the tubular graphite structures were similar to the associated crystallite.

Steam reforming of hydrocarbons on supported Ni is an interesting and very complicated system. The major reactions to be considered are:

$$\begin{aligned} &C_nH_m + nH_2O \rightarrow nCO + (n + \tfrac{m}{2})\ H_2 \\ &CO + H_2 \rightarrow CO_2 + H_2 \\ &CO + 3H_2 \rightarrow CH_4 + H_2O \end{aligned} \qquad \text{(VI)}$$

with coke deposition possibly via:

$$\begin{aligned} &CH_4 \rightarrow C + 2H_2 \\ &2CO \rightarrow C + 2H_2 \\ &C_mH_m \rightarrow \text{polymer} \rightarrow \text{coke} \end{aligned} \qquad \text{(VII)}$$

Many workers have dealt with catalytic properties as they affect VI and VII. For coking it is generally agreed that both metal and acidic functions are involved; the incorporation of alkali or the use of less acidic supports retards coke formation (85,86,87). Kinetics of coke formation are reported to be nearly first order in hydrocarbon (for n-hexane) (88), and normally decrease with increasing steam/hydrocarbon ratio and increase with increasing unsaturation (86,89). A maximum in coke formation rate in the region 500-600°C has been observed (86,88), but there is some disagreement concerning the effect of hydrogen partial pressure on coking rates (86,90).

There are some interesting aspects to coke formation in the steam reforming reaction. First is the observation of an induction period for coke formation, shown in Figure 14a for n-heptane at 500°C (86), strongly dependent upon steam/C ratio. This has been noted for other hydrocarbons as well (91). Correlation of coke on catalyst in this instance can be provided by:

$$C_c = k_c(t - t_o) \qquad (24)$$

with t_o the induction time. Figure 14b shows the strong inhibition of coke formation with increase in steam/carbon (decrease in k_c), with an accompanying increase in induction time. An increase in hydrogen partial pressure increased coking rate (86). The correlation of Eq. (24) is obviously quite different from the familiar Voorhies form, and the thermal dependence much greater than that observed for Voorhies-type systems. Rostrup-Nielsen

(86) reports an activation energy of 40 kcal/mole compared to the 10 kcal/mole or less for coke formation on cracking catalysts (1).

A second factor is the relatively small dimunition in activity on coke formation in these catalysts, demonstrated by specific experiments in (86) for two catalysts containing 4.5 and 11.0 wt % coke on catalysts. The reason for this is not clear, but may reside in the tendency for carbon to segregate on nickel surfaces (92,93) or be due to the fact that a carbidic phase such as Ni_3C is the active catalytic surface under reaction conditions (84).

Coke formation reactions reported above in VII may be more complex than realized previously. In very recent work on steam reforming of n-hexane, Bett, et al. (94) conclude that carbon formation may result from interaction of unreacted hexane and the products of secondary cracking reactions, or from unstable intermediates in the cracking process. A selection of other thoughts concerning the mechanism of coke formation in reforming on Ni would include the works of Whalley, et al. (95), Presland and Walker (96), and Renshaw, et al. (97).

Carbon deposition on nickel catalysts for reactions other than steam reforming has also been studied extensively in recent years. The interaction of light hydrocarbons on Ni foils at 400-600°C has been investigated by Lobo and Trimm (98), and the pyrolysis of olefins on Ni foils at 400-750°C by Rostrup-Nielsen and Trimm (99). In the pyrolysis reactions the coking rates pass through a maximum, as with steam reforming, in the range 500-600°C. The kinetics of reaction are complex; below about 500°C the apparent activation energy is 32 $\pm$ 2 kcal/mole for all olefins and the reaction is zero order. Above 600°C the activation energy is ca. -44 kcal/mole and the reaction order is unity for both olefin and hydrogen. Similar results have been reported by Derbyshire and Trimm (100). A number of reasons including regasification (85,101), coke deactivation (101), poisoning by hydride formation, and competitive adsorption-surface diffusion phenomena (99,100) have been set forth to explain these kinetics.

Finally, there is a rather unique type of coking of supported Ni under certain conditions which leads to catastrophic destruction of the physical structure (i.e., reduction to dust). Kiovsky (91) has reported this phenomenon in some detail for low surface area (~ 10 m^2/g) supported Ni (10% wt), used for production of annealing gas via O_2 + CH_4. In one test, a commercial Norton NC-100 formulation in the form of 1 in diameter rings was completely destroyed within eight hours under partial oxidation conditions with inlet temperature of 1010°C, 3/1 air/CH_4 mole ratio and 400 hr^{-1} space velocity. Operation at higher SV (800 hr^{-1}) resulted in no destruction. Physical degradation was accompanied by coke formation, but an induction period was observed in all cases before destruction occurred; the catalyst could be completely regenerated by air oxidation if time on stream were less than the induction period. Some typical data are shown in Figure 15a for experiments at 730 hr^{-1}, 490°C, with a toluene-saturated

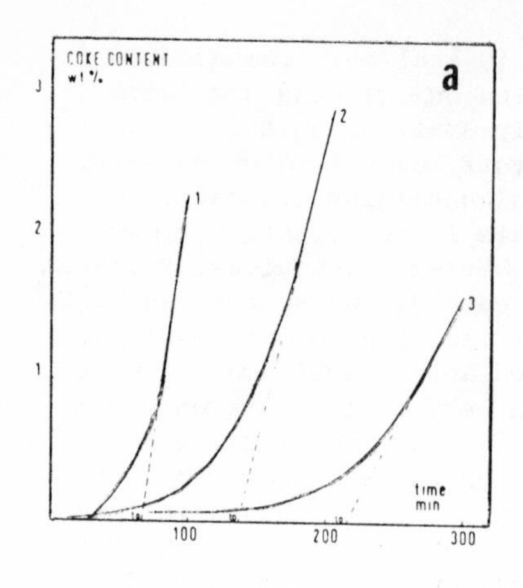

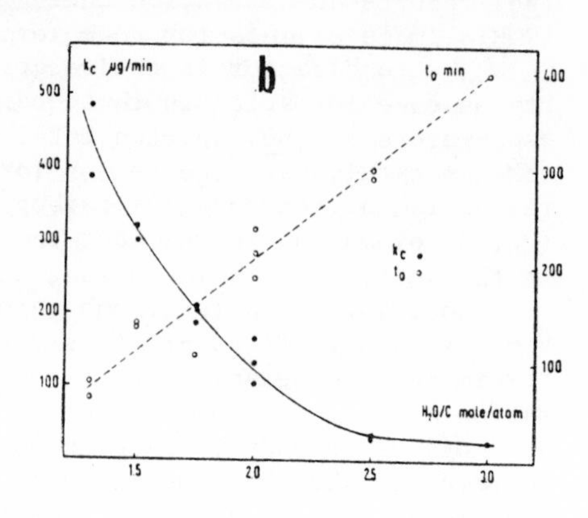

Journal of Catalysis

Figure 14. (a) Mean coke content of catalyst vs. *time on stream,* n-*heptane reforming at 500°C. 1.) $H_2O/C = 1.3$, 2.) $H_2O/C = 1.5$, 3.) $H_2O/C = 2.0$; (b) coke correlation constants, Equation 24, and steam/carbon ratio. Catalyst: 23.8% Ni on MgO with 6% wt Al, 0.07% Na, 2 m^2/g Ni, 20 m^2/g BET (86).*

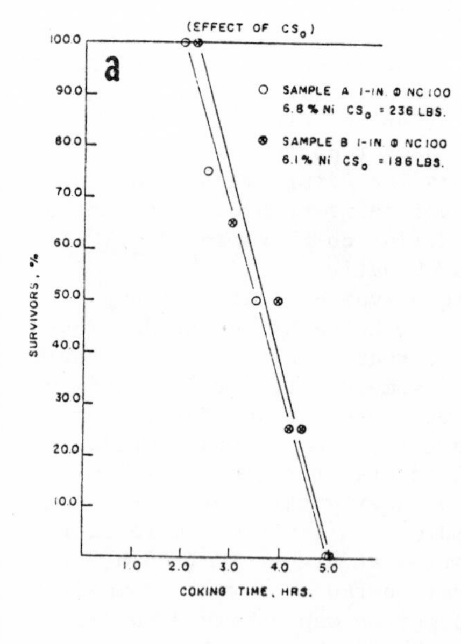

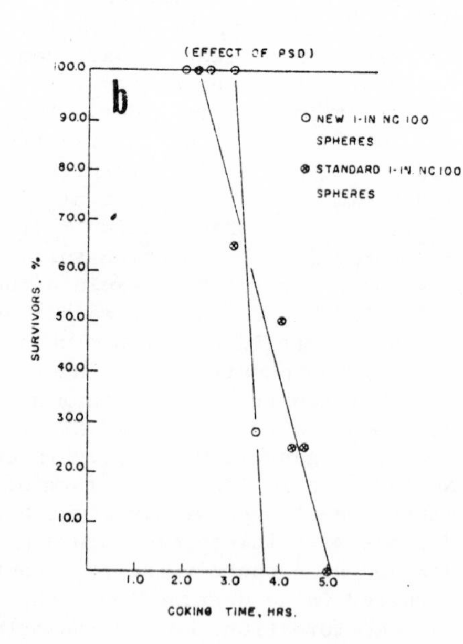

American Chemical Society

Figure 15. Destruction of supported Ni by coke formation during partial oxidation of hydrocarbons. (a) Survivors vs. *time at two air/HC ratios; (b) effect of average pore diameter upon survival rate (91)*

methane at two different air/hydrocarbon ratios. Survivors are defined as the number of catalyst pieces in the original charge retaining at least 80% of their original mass; typical coke levels are only on the order of 3-5 wt % while destruction is occurring and it is clear that the kinetics of the process are relatively rapid. Initial crush strength was not well correlated with survival rate, but the induction period was increased and rate of destruction decreased by reducing catalyst external surface area per volume and reducing median pore diameter. The latter effect is shown in Figure 15b for two catalysts; "new", with pore diameter 5μ, and "standard", 17μ. The ability of coke formation to destroy the physical structure of the catalyst is surprising, but hardly equivocal.

Acknowledgments

This work was supported via the generous help of the Central Research Department, Dow Chemical, U.S.A. Figures 1, 2, 3, 4, 6, 11 by permission of Pergamon Press, Ltd., Figures 12, 15 by permission of the American Chemical Society, Figures 9, 10 by permission of Elsevier Publishing Co., Figure 13 by permission of the American Institute of Chemical Engineers, and Figure 14 by permission of Academic Press, Inc.

Appendix

Two occupational hazards associated with the writing of reviews such as this are the inadvertent oversight of particularly germane literature and the publication of relevant work in the period between completion and publication of the review. We have examples of both here.

In the text we commented upon the possible inadequacies of the separable form of representation for deactivation kinetics. In fact, Bakshi and Gavalas (1A) have demonstrated this experimentally for methanol and ethanol dehydration on SiO_2/Al_2O at 150-225°C, poisoned by n-butylamine. The kinetics of reaction were correlated by:

$$(-r_T) = \frac{k\,K_A C_A^{\frac{1}{2}}}{1 + K_A C_A^{\frac{1}{2}} + K_W P_W} \tag{1A}$$

for both the fresh and deactivated catalyst, but the adsorption constants $\underline{K}_A$ and $\underline{K}_W$ varied with extent of deactivation which is not as the separable formulation would have it. Such changes were interpreted as a manifestation of the nonuniformity of surface sites, as we have suggested in Eq. (1). Kam, et al. (2A) have also used a rate expression of the form of Eq. (1A) in a theoretical analysis of isothermal, intraparticle fouling involving combined series and parallel fouling. The results of the combined mechanism are contrasted with those pertaining to the individual steps alone.

An example of structure sensitivity in deactivation is found in the work of Ostermaier, et al. (3A) for 2-15 nm Pt/Al_2O_3 and Pt black in ammonia oxidation at 368-473°K. The effects of crystallite size and temperature in deactivation were investigated; it was found that the extent of deactivation increased with decreasing temperature, and there was a difference in the Arrhenius behavior between sintered and unsintered materials. Deactivation was more severe with smaller crystallites, but the surface could be completely reactivated by H_2 at 673°K. It was suggested that PtO was the deactivated surface, and an excellent correlation of activity was provided by:

$$\frac{dN}{dt} = - K_D N^2 \tag{2A}$$

with
$$N = \frac{N_o}{1 + N_o K_D\, t} \tag{3A}$$

where N is density of sites, K_D a rate constant for deactivation, and N_o an initial site density.

In recent work Hegedus and Summers (4A) have investigated the poisoning of noble metals supported on Al_2O_3 by lead and phosphorous in the oxidation of carbon monoxide and hydrocarbons. The parameters studied were pore structure, support area and impregnation depth. Correlations are given for poison penetration and effectiveness as a function of time of operation. As in the case of coking via the parallel mechanism, the overall catalyst life can be optimized by manipulation of the macropore structure. Designing for a given diffusional limitation in the fresh catalyst retards the rate of the main reaction but also retards penetration of poison into the catalyst matrix, hence an optimum may be sought for maximum time-averaged effectiveness.

Finally, Mikus, et al. (5A) have reported studies of fixed bed reactor transients for CO oxidation on Pt/Al_2O_3 (0.1% Pt) with CS_2 poison. Their experimental results are in qualitative agreement with the computations of Blaum (52) and Ervin and Luss (6A), but no simulations are reported in the paper. The temperature profiles reported have a cruious shape for what is reported to be an adiabatic reactor.

Literature Cited

1. Butt, J.B., Adv. Chem., (1972), 109, 259.
2. Wanke, S.E. and Flynn, P.C., Catal. Rev.-Sci. and Eng., (1975), 12, 93.
3. Shelef, M., Otto, K. and Otto, N.C., Adv. Catal., (1978), 27.
4. Carberry, J.J., "Chemical and Catalytic Reaction Engineering", McGraw-Hill, New York, 1976.
5. Khang, S.J. and Levenspiel, O., Ind. Eng. Chem. Fundls., (1973), 12, 185.
6. Szepé, S. and Levenspiel, O., Proc. European Fed., 4th Chem. Reaction Eng., Brussels, 1968; Pergamon Press, 1971.

7. Butt, J.B., Wachter, C.K. and Billimoria, R.M., Chem. Eng. Sci., in press.
8. Webb, G. and Macnab, J.I., J. Catal., (1972), 26, 226.
9. Burnett, R.L. and Hughes, T.R., J. Catal., (1973), 31, 55.
10. Masamune, S. and Smith, J.M., AIChE Jl., (1966), 12, 384.
11. Murakami, Y., Kobayshi, T., Hattori, T. and Masuda, M., Ind. Eng. Chem. Proc. Design Devel., (1968), 7, 72.
12. Swabb, E.A. and Gates, B.C., Ind. Eng. Chem. Fundls., (1972), 11, 540.
13. Butt, J.B., Delgado-Diaz, S. and Muno, W.E., J. Catal., (1975), 37, 155; (1976), 41, 190.
14. Toei, R., Nakanishi, K. and Okazaki, M., J. Chem. Eng. Japan, (1975), 8, 338.
15. Richardson, J.T., Ind. Eng. Chem. Proc. Design Devel., (1972), 11, 12.
16. Ozawa, Y. and Bischoff, K.B., Ind. Eng. Chem. Proc. Design Devel., (1968), 7, 67.
17. Levinter, M.E., Panchekov, G.M. and Tanatarov, M.A., Int. Chem. Engr., (1967), 7, 23.
18. Suga, K., Morita, Y., Kunugita, E. and Otake, T., Int. Chem. Engr., (1967), 7, 742.
19. Snyder, A.C. and Matthews, J.C., Chem. Eng. Sci., (1973), 28, 291.
20. Lee, J.W. and Butt, J.B., Chem. Eng. Jl., (1973), 6, 111.
21. Ray, W.H., Chem. Eng. Sci., (1972), 27, 489.
22. Shadman-Yazdi, F. and Petersen, E.E., Chem. Eng. Sci., (1972), 27, 227.
23. Corbett, W.E., Jr. and Luss, D., Chem. Eng. Sci., (1974), 29, 1473.
24. Becker, E.R. and Wei, J., J. Catal., (1977), 46, 372.
25. Bischoff, K.B., Chem. Eng. Sci., (1967), 22, 525.
26. Mars, P. and Grogels, M.J., Chem. Eng. Sci. Supplement, 3rd Europ. Symp. Chem. Reaction Eng., Pergamon Press (1964).
27. Friedsichsen, W., Chem. Ing. Tech., (1969), 41, 967.
28. Karanth, N.G. and Luss, D., Chem. Eng. Sci., (1975), 30, 695.
29. Pareja, T.J. and Luss, D., Chem. Eng. Sci., (1975), 30, 1219.
30. Hegedus, L.L., Ind. Eng. Chem. Fundls., (1974), 13, 190.
31. Balder, J.R. and Petersen, E.E., Chem. Eng. Sci., (1968), 23, 1287.
32. Hegedus, L.L. and Petersen, E.E., Ind. Eng. Chem. Fundls., (1972), 11, 579.
33. Hegedus, L.L. and Petersen, E.E., Chem. Eng. Sci., (1973), 28, 69.
34. Hegedus, L.L. and Petersen, E.E., Chem. Eng. Sci., (1973), 28, 345.
35. Hegedus, L.L. and Petersen, E.E., J. Catal., (1973), 28, 150.
36. Hegedus, L.L. and Petersen, E.E., Catal. Rev.-Sci. and Eng., (1974), 11, 245.
37. Wolf, E. and Petersen, E.E., Chem. Eng. Sci., (1974), 29, 1500.

38. Wolf, E. and Petersen, E.E., J. Catal., (1977), 46, 190.
39. Kehoe, J.P.G. and Butt, J.B., AIChE Jl., (1972), 18, 347.
40. Butt, J.B., Downing, D.M. and Lee, J.W., Ind. Eng. Chem. Fundls., (1977), 16, 270.
41. Lee, J.W., Downing, D.M. and Butt, J.B., AIChE Jl. (in press).
42. Hughes, R. and Koh, H.P., Chem. Eng. Jl., (1970), 1, 186.
43. Hughes, R. and Koh, H.P., AIChE Jl., (1974), 20, 395.
44. Benham, C.B. and Denny, V.E., Chem. Eng. Sci., (1972), 27, 2163.
45. Trimm, D.L., Corrie, J. and Holton, R.D., Chem. Eng. Sci., (1974), 29, 2009.
46. Hlavacek, V. and Marek, M., Proc. European Fed., 4th Chem. Reaction Eng., Brussels, 1968; Pergamon Press, 1971.
47. Lee, J.C.M. and Luss, D., AIChE Jl., (1970), 16, 620.
48. Lambrecht, G.C., Nussey, C. and Froment, G.F., Proc. 5th European Symp. Chem. Reaction Eng., B-2-19, Elsevier, Amsterdam, 1972.
49. De Pauw, R.P. and Froment, G.F., Chem. Eng. Sci., (1975), 30, 789.
50. Dumez, F.J. and Froment, G.F., Ind. Eng. Chem. Proc. Design Devel., (1976), 15, 291.
51. Hosten, L.H. and Froment, G.F., Ind. Eng. Chem. Proc. Design Devel., (1971), 10, 280.
52. Blaum, E., Chem. Eng. Sci., (1974), 29, 2263.
53. Weng, H-S, Eigenberger, G. and Butt, J.B., Chem. Eng. Sci., (1975), 30, 1341.
54. Price, T.H. and Butt, J.B., Chem. Eng. Sci., (1977), 32, 393.
55. Noda, H., Tone, S. and Otake, T., J. Chem. Eng. Japan (1974), 7, 110.
56. Toei, R., Nakanishi, K., Yamada, K. and Okazaki, M., J. Chem. Eng., (1975), 8, 131.
57. Uchida, S., Osuda, S. and Shindo, M., Can. J. Chem. Eng., (1975), 53, 666.
58. Otake, T., Kunugita, E. and Kugo, K., Kogyo Kagaku Zasshi, (1965), 68, 58.
59. Sadana, A. and Doraiswamy, L.K., J. Catal., (1971), 23, 147.
60. Prasad, K.B.S. and Doraiswamy, L.K., J. Catal., (1974), 32, 384.
61. Greco, G., Jr., Alfani, F. and Gioia, F., J. Catal., (1973), 30, 155.
62. Weekman, V.W., Jr., Ind. Eng. Chem. Proc. Design Devel., (1968), 7, 90.
63. Weekman, V.W., Jr., ibid (1969), 8, 388.
64. Weekman, V.W., Jr. and Nace, D.M., AIChE Jl., (1970), 16, 397.
65. Nace, D.M., Voltz, S.E. and Weekman, V.W., Jr., Ind. Eng. Chem. Proc. Design Devel., (1971), 10, 530.
66. Voltz, S.E., Nace, D.M. and Weekman, V.W., Jr., ibid, (1971), 10, 538.
67. Wojciechowski, B.W., Can. J. Chem. Eng., (1968), 46, 48.

68. Campbell, D.R. and Wojciechowski, B.W., J. Catal., (1971), 20, 217.
69. Best, D.A. and Wojciechowski, B.W., J. Catal., (1973), 31, 74; (1977), 47, 11; (1977), 47, 343.
70. Wojciechowski, B.W., Catal. Rev.-Sci. and Eng., (1974), 9, 79.
71. John, T.M., Pachovsky, R.A. and Wojciechowski, B.W., Adv. Chem., (1974), 133, 422.
72. John, T.M. and Wojciechowski, B.W., J. Catal., (1975), 37, 240; (1975), 37, 348.
73. Pachovsky, R.A. and Wojciechowski, B.W., J. Catal., (1975), 37, 120.
74. Pachovsky, R.A. and Wojciechowski, B.W., Can. J. Chem. Eng., (1975), 53, 308; (1975), 53, 659.
75. Gross, B., Nace, D.M. and Voltz, S.E., Ind. Eng. Chem. Proc. Design Devel., (1974), 13, 199.
76. Voltz, S.E., Nace, D.M., Jacob, S.M. and Weekman, V.W., Jr., Ind. Eng. Chem. Proc. Design Devel., (1972), 11, 261.
77. Jacob, S.M., Gross, B., Voltz, S.E. and Weekman, V.W., Jr., AIChE Jl., (1976), 22, 701.
78. Rostrup-Nielsen, J.R., J. Catal., (1972), 27, 343.
79. Dent, F.J. and Cobb, J.W., J. Chem. Soc. (London), (1929), 2, 1903.
80. Dent, F.J., Moignard, L.A., Eastwood, A.H., Blackburn, W.H. and Hebden, D., (1946), Trans. Inst. Gas Eng., 602.
81. Leidheiser, H., Jr. and Gwathmey, A.J., J. Am. Chem. Soc., (1948), 70, 1206.
82. Hofer, L.J.E., Sterling, E. and McCartney, J.T., J. Phy. Chem., (1955), 59, 1153.
83. Coad, J.P. and Riviere, J.C., Surf. Sci., (1971), 25, 609.
84. McCarty, J.G., Wendreck, P.R. and Wise, H., Division of Petroleum Chemistry Preprints, Am. Chem. Soc. (1977), 22, 1315.
85. Andrews, S.P.S., Ind. Eng. Chem. Prod. Res., (1969), 8, 321.
86. Rostrup-Nielsen, J.R., J. Catal., (1974), 33, 184. See also J.R.R-N., "Steam Reforming Catalysts," Danish Technical Press, Copenhagen, 1976.
87. Saito, M., Tokuno, M. and Morita, Y., Kogyo Kogaku Zasshi, (1971), 74, 673.
88. Saito, M., Tokuno, M. and Morita, Y., Kogyo Kogazu Zasshi, (1971), 74, 693.
89. Moseley, F., Stephens, R.W., Steward, K.D. and Wood, J., J. Catal., (1972), 24, 18.
90. Bhatia, K.S.M. & Dixon, E.M., Trans. Farad Soc., (1967), 63, 2217.
91. Kiovsky, J.R. Division of Petroleum Chemistry Preprints, (1977), 22, 1300.
92. Smith, R.D., Trans. Met. Soc., AIME, (1966), 1224.
93. Blakely, J.M., Kim, J.S. and Poltec, H.C., J. Appl. Phy., (1970), 41, 2693.
94. Bett, J.A.S., Christner, L.G., Hamilton, R.M. and Olson, A.J., Div. of Petroleum Chemistry Preprints, (1977), 22, 1290.

95. Whalley, L., David, B.J. and Moss, R.L., Trans. Farad. Soc., (1970), 66, 3143.
96. Presland, A.E.B. and Walker, P.L., Jr., Carbon, (1969), 7, 1.
97. Renshaw, G.D., Roscoe, C. and Walker, P.L., Jr., J. Catal., (1971), 22, 374.
98. Lobo, L.S. and Trimm, D.L., J. Catal., (1973), 29, 75.
99. Rostrup-Nielsen, J.R. and Trimm, D.L., J. Catal., (1977), 48, 185.
100. Derbyshire, F.J. and Trimm, D.L., Carbon (1975), 13, 189.
101. Baker, R.Y.K., Barker, M.A., Harris, P.S., Feates, F.S. and Waite, R.J., J. Catal., (1972), 26, 51.

Literature Cited - Appendix

1A. Bakshi, K.R. and Gavalas, G.R., AIChE Jl., (1975), 21, 494.
2A. Kam, E.K., Ramachandran, P.A. and Hughes, R., J. Catal., (1975), 38, 283.
3A. Ostermaier, J.J., Katzer, J.R. and Manogue, W.H., J. Catal., (1976), 41, 277.
4A. Hegedus, L.L. and Summers, J.C., J. Catal., (1977), 48, 345.
5A. Mikus, O., Pour, V. and Hlavacek, V., J. Catal., (1977), 48, 98.
6A. Ervin, M.A. and Luss, D., AIChE Jl., (1970), 16, 979.

RECEIVED February 17, 1978

INDEX

D

E

F

G

K

L

M

N

O

R

S